Lecture Notes in Physics

Springer
Berlin
Heidelberg
New York
Hong Kong
London
Milan
Paris
Tokyo

Physics and Astronomy ONLINE LIBRARY

http://www.springer.de/phys/

The Editorial Policy for Edited Volumes

The series *Lecture Notes in Physics* (LNP), founded in 1969, reports new developments in physics research and teaching - quickly, informally but with a high degree of quality. Manuscripts to be considered for publication are topical volumes consisting of a limited number of contributions, carefully edited and closely related to each other. Each contribution should contain at least partly original and previously unpublished material, be written in a clear, pedagogical style and aimed at a broader readership, especially graduate students and nonspecialist researchers wishing to familiarize themselves with the topic concerned. For this reason, traditional proceedings cannot be considered for this series though volumes to appear in this series are often based on material presented at conferences, workshops and schools.

Acceptance

A project can only be accepted tentatively for publication, by both the editorial board and the publisher, following thorough examination of the material submitted. The book proposal sent to the publisher should consist at least of a preliminary table of contents outlining the structure of the book together with abstracts of all contributions to be included. Final acceptance is issued by the series editor in charge, in consultation with the publisher, only after receiving the complete manuscript. Final acceptance, possibly requiring minor corrections, usually follows the tentative acceptance unless the final manuscript differs significantly from expectations (project outline). In particular, the series editors are entitled to reject individual contributions if they do not meet the high quality standards of this series. The final manuscript must be ready to print, and should include both an informative introduction and a sufficiently detailed subject index.

Contractual Aspects

Publication in LNP is free of charge. There is no formal contract, no royalties are paid, and no bulk orders are required, although special discounts are offered in this case. The volume editors receive jointly 30 free copies for their personal use and are entitled, as are the contributing authors, to purchase Springer books at a reduced rate. The publisher secures the copyright for each volume. As a rule, no reprints of individual contributions can be supplied.

Manuscript Submission

The manuscript in its final and approved version must be submitted in ready to print form. The corresponding electronic source files are also required for the production process, in particular the online version. Technical assistance in compiling the final manuscript can be provided by the publisher's production editor(s), especially with regard to the publisher's own LaTeX macro package which has been specially designed for this series.

LNP Homepage (http://www.springerlink.com/series/lnp/)

On the LNP homepage you will find:

– The LNP online archive. It contains the full texts (PDF) of all volumes published since 2000. Abstracts, table of contents and prefaces are accessible free of charge to everyone. Information about the availability of printed volumes can be obtained.

– The subscription information. The online archive is free of charge to all subscribers of the printed volumes.

– The editorial contacts, with respect to both scientific and technical matters.

– The author's / editor's instructions.

T. Brandes S. Kettemann (Eds.)

Anderson Localization and Its Ramifications

Disorder, Phase Coherence and Electron Correlations

Springer

Editors

Tobias Brandes
Department of Physics
UMIST, PO Box 88
Manchester M60 1QD, UK

Stefan Kettemann
1. Institut für Theoretische Physik
Universität Hamburg
Jungiusstr. 9
20355 Hamburg, Germany

Cataloging-in-Publication Data applied for

A catalog record for this book is available from the Library of Congress.

Bibliographic information published by Die Deutsche Bibliothek
Die Deutsche Bibliothek lists this publication in the Deutsche Nationalbibliografie; detailed bibliographic data is available in the Internet at http://dnb.ddb.de

ISSN 0075-8450
ISBN 3-540-40785-5 Springer-Verlag Berlin Heidelberg New York

Springer-Verlag Berlin Heidelberg New York
a member of BertelsmannSpringer Science+Business Media GmbH

springer.de

Typesetting: Camera-ready by the authors/editor
Cover design: *design & production*, Heidelberg

Printed on acid-free paper
54/3141/du - 5 4 3 2 1 0

Preface

The manifestation of quantum mechanical effects in the transport properties of electronic systems has become one of the central issues of condensed matter physics in the 21st century. In fact, we are now entering an era where quantum coherence is used to manipulate electronic transport properties of nanoscale devices.

The phenomenon of localization of the electronic wave function in a random medium can be regarded as the key representative of quantum coherence in a condensed matter system. Named after a seminal paper by Anderson in 1958, an 'Anderson transition' occurs in the transition from a metal (conductor) to an insulator when the concentration and strength of impurities (disorder) in a three-dimensional material is varied.

The discovery of Aharonov–Bohm oscillations of the magnetoresistance of thin metallic cylinders provided direct evidence for quantum interference in 'real', disordered systems. One may say that this discovery, together with the discovery of the quantum Hall effects at the beginning of the 1980s, lead to the new branch of physics that has become known as 'mesoscopic physics' since. As one of the most remarkable phenomena in condensed matter physics of the 20th century, the localization problem is an indispensable part of the theory of the quantum Hall effects and rivals superconductivity in its significance as a manifestation of quantum coherence at a macroscopic scale.

Consequently, many theoretical concepts such as diagrammatics with Green's functions, scaling theory, and field theoretical methods related to localization due to quantum coherence have appeared in the last thirty years. Random matrix theories have been used to classify symmetries and universality classes of the Anderson transition. A powerful and precise numerical scheme to calculate quantities like the conductance or scaling exponents was developed and further extended in the 1980s. The investigation of the statistical properties of eigenvalues and the semi-classical limit of localization have established a link to the extremely active field of quantum chaos, or the localization of classical waves in random media. Furthermore, transport theories in mesoscopic, sub-micro scale systems have been formulated to incorporate the effect of coherence and interactions.

On the other hand, it is the very role of interactions which has remained the most controversial over the last few years both in experiment and theory. Today, the combination of disorder and interactions has indeed turned out to be one of the most challenging questions in this field. A partial understanding of, e.g., the effects of time-dependent fields, or electron–electron or electron–phonon interactions in weakly localized systems, has been achieved. New experimental and theoretical results, however, have shown that the situation is far more complicated than it has been anticipated in the 'early days' of localization, where one-particle wave functions were the main object of discussions.

First attempts have been made to incorporate interactions into the localization problem in a serious manner. Exact few-particle calculations are a first step to understand the interplay between Coulomb interactions and disorder. Much progress has been made in the analysis of one-dimensional, exactly solvable models, although the real breakthrough towards the understanding of the higher-dimensional case has still to be awaited for.

The present volume is intended to highlight some of the recent progress in the field of localization, with a particular emphasis on the effect of interactions on quantum coherence. The chapters are written in textbook style and should serve as a reliable and thorough introduction for advanced students or researchers already working in this field of mesoscopic physics. The authors are among the leading experts in this field. They gathered in September 2002 at the 283th Wilhelm and Else Heraeus Seminar 'Localization, Quantum Coherence, and Interactions' (Warburg-Haus, Hamburg). The workshop was held to celebrate the 60th birthday of Bernhard Kramer and his contributions to the area of physics presented in this volume.

Manchester and Hamburg,
July 2003

Tobias Brandes
Stefan Kettemann

Bernhard Kramer and His Contributions to Physics

Angus MacKinnon

The Blackett Laboratory, Imperial College London, London SW7 2BW, UK

I first met Bernhard Kramer in the spring of 1977. I can't have made a wholly bad impression as in July I travelled to Dortmund to take up a Post Doc position. I was quickly introduced to the problem on which I was expected to work: Compton scattering in amorphous Silicon, and a week later Bernhard left to spend a month in Edinburgh, leaving me to get on with it. During the next 4 years we progressed to other aspects of amorphous semiconductors and eventually to the Anderson Metal-Insulator transition. In late 1978 Bernhard was offered the position of head of the theory group in the Physikalisch-Technische Bundesanstalt (PTB) and we and 2 PhD students moved to Braunschweig.

Bernhard Kramer is a native of north Hessen and was a student in the University of Marburg where he completed his PhD. His work has covered a wide range of theoretical solid state physics, although disorder and computer simulation have rarely been absent for long.

His early work, while still in Marburg, was on the properties of glasses, especially amorphous semiconductors. A particular interest was the effect of disorder on the electronic properties of Silicon and Germanium and of Selenium and Tellurium.

When he moved to Dortmund there was a group of experimentalists under Winfried Schülke who were interested in using Compton scattering as a probe of the properties of semiconductors. In particular there had been a suggestion that the Compton profile might be sensitive to long range order and might display significant differences between (e.g.) crystalline and amorphous Silicon.

In January 1978 we moved to the Physikalisch-Technische Bundesanstalt (PTB) in Braunschweig. The culture of the PTB was very different from a University. It was dominated by a Civil Service mentality. Things had always been done in a particular way, and woe betide anyone who attempted to buck the system. The PTB is responsible for standards, similar to NIST in the US and NPL in the UK. The theory group didn't fit that pattern and often felt like a cuckoo in the nest. Nevertheless Bernhard managed to make a significant contribution to raising the scientific profile of the lab.; initially by publishing papers and attending conferences which appealed to a much broader audience than the usual PTB fare, later by inviting prestigious speakers to give seminars and by organising international conferences and workshops. I well remember the colloquia by Friedrich Hund and Neville Mott. Any student of science is impressed by meeting a mythical figure whose name appears in the textbooks.

Still, the moment the financial belt had to be tightened the group felt under threat. Salvation arrived, however, in the form of a young man who had done some interesting new measurements on the Hall effect in MOSFETs. Some of us were sceptical about the results presented by Klaus von Klitzing. He was, after all, making the seemingly absurd claim that you could measure a combination of fundamental constants in a dirty system with an accuracy of 1 part in 10^6. Bernhard, however, recognised the importance of the result and we spent many hours, and drank far too much coffee, arguing about the theoretical explanation. It proved the salvation of the group, however. It established the missing link between the Gruppe 8.1 and standards. Suddenly we were working on a potential standard Ohm.

The quantum Hall effect also provided an opportunity for an annual visit to the Alps. A workshop took place for many years in the village of Schleching, on the German-Austrian border. My main memory is of sitting in a room with a huge alphorn, with 3 mouthpieces, hanging from the ceiling, threatening to fall on us at any moment. Bernhard, however, discovered a gliding school at a small airfield in the next village and for many years he returned to the area to ride the alpine thermals.

At about the same time we started to take an interest in the Anderson metal-insulator transition or *Anderson junction* as it was memorably mistranslated in one annual report of the PTB[1] Initially we used an old idea from Dennis Weaire, the equation of motion method, but later moved on to Green's function recursion and the Transfer Matrix. It soon became apparent that we would have to find a way of analyzing the results in terms of a scaling theory so that we could make a direct comparison with the recently published paper of Abrahams, Anderson, Licciardello and Ramakrishnan (almost universally dubbed the "Gang of Four") which had contained the surprising result that all states were localised in 2 dimensions as well as in 1D. After various failed attempts we were able to show that the ratio of the localisation length to the system width $\Lambda = \lambda_M/M$ could be plotted such that data for various values of disorder fell on a common curve and that the result was consistent with the predictions of the Gang of Four. It also became apparent why previous attempts at numerical simulation of the problem had failed to pick up the 2D localisation effect: in every case the point at which the localisation length became larger than the system size had been misinterpreted as a transition; in fact this particular square wheel has been reinvented several times since.

I left the PTB in October 1981 but continued to collaborate with Bernhard for several years afterwards.

One important side effect of this work was that Bernhard was able to organise the 1st of a series of international conferences at the PTB. LITPIM (Localization, Interaction and Transport Phenomena in Impure Metals) took place in August 1984 and attracted well over 100 participants. In fact the 6th in the series, Localization 1999, was also organised by Bernhard; this time in Hamburg. On the latter occasion Bernhard's aptitude for public relations had the local press

[1] The German word *Übergang* may be translated as both *transition* and *junction*.

reporting that the participants were about to build a quantum computer that would revolutionise the lives of all Hamburgers. We are still waiting.

The Quantum Hall Effect became the main interest of the group 8.1 in the PTB and that work continues under Ludwig Schweitzer. A few years later, together with Bodo Huckestein, Bernhard introduced a new model for studying the QHE. It quickly became apparent that this model was capable of describing the critical behaviour very accurately, easily confirming that the localisation length diverges at a single energy in each Landau level and providing an estimate of the critical exponent accurate to about 1%. This was particularly surprising as, at that time, estimates of the corresponding exponent for the 3D Anderson model varied by about 50%. Nevertheless the results were consistent with experimental results (e.g.) from Friedl Kuchar and Ronald Meisels.

Work on the the metal-insulator transition continued with a series of papers on systems in magnetic fields with Tomi Ohtsuki and Yoshi Ono and, after moving to Hamburg in 1992, with Isa Zarakheshev, who carried out large scale calculations of energy level statistics.

The current state of this work is reflected in various chapters of the current volume, so I won't attempt to review it here. Suffice it to say that the initial concept of combining the transfer matrix method with finite size scaling remains the most reliable way of studying critical properties but the accuracy with which the critical exponent can be calculated has improved by close to 2 orders of magnitude in the meantime.

During the 80's many of those who had been interested in disordered systems began to think about the emerging field of mesoscopics, which had much in common with localisation. In Bernhard's case there was work on quasi-1D systems, quantum wires, with Jan Mašek and his colleagues in Prague. Various combinations of tunnel barriers and quantum dots, frequency dependent transport and interaction effects were studied together with Dietmar Weinmann, Wolfgang Häusler and Tobias Brandes. The effects of interactions on the states in quantum dots and on their coupling to leads were shown to be highly non-trivial. The combination of Coulomb blockade effects and interaction with phonons provides fertile ground for continued investigation.

When he first moved to Hamburg Bernhard told me of his dream of having a boat on the Alster[2] and sailing it during his lunch break and in the evenings. That never happened. However, he did manage to combine business with pleasure by striking up a collaboration with Maura Sassetti from the University of Genoa which has provided ample opportunity to sail on the Mediterranean.

The Luttinger liquid provides a very useful model for interacting particles in 1-dimension. From this starting point and in collaboration with the Genoese group Bernhard has been able to reveal a wealth of rich structure in the physics of quasi-1D systems.

This brief survey of Bernhard Kramer's work fails to do justice, however, to his contribution to Physics. A brief glimpse at his web site, http://www.physnet.uni-hamburg.de/hp/kramer/, reveals his editorship of *Ad-*

[2] A lake in central Hamburg, just a short walk from the Physics department.

Fig. 1. Bernhard relaxing ?????

vances in Physics and co-editorship of 3 other journals, one of which *Physik in unserer Zeit – Physics in our time* is a semi-popular journal, roughly equivalent to *Physics Today* in the US or *Physics World* in the UK. This latter reveals another side of Bernhard, not apparent from his bare publications list. He is a significant contributor to the public relations side of science. Various colleagues in Hamburg have told me about his physics show in a Hamburg shopping mall and numerous other events aimed at teenagers considering a career in science.

On his web site you will also find a list of books, mostly conference proceedings. Whereas most professional scientists have organised the occasional conference or workshop, Bernhard is a serial conference organiser. These extend over several international conferences, various workshops and summer/winter schools aimed at young researchers, to the aforementioned events for teenagers. Somehow while doing all this he has found time to do significant research and to to act as head of the physics department[3] in Hamburg during a time when the lack of demand from students seemed to place its very existence under threat.

The workshop on which this volume is based attracted participants from 13 different countries, including several from Japan. These were mostly people who have collaborated with Bernhard over the years, including numerous former PhD students and Post Docs. Many of these, like myself, count Bernhard's influence and support as a major contribution to the success of their own careers.

[3] Fachbereichssprecher

Contents

Part I

Scaling at the Anderson Transition

Numerical Investigations of Scaling at the Anderson Transition

Rudolf A. Römer[1] and Michael Schreiber[2]

[1] Department of Physics, University of Warwick, Coventry CV4 7AL, UK
[2] IUB, School of Engineering and Science, P.O.Box 750 561, 28725 Bremen, Germany

1 Introduction

At low temperature T, a significant difference between the behavior of crystals on the one hand and disordered solids on the other is seen: sufficiently strong disorder can give rise to a transition of the transport properties from conducting behavior with finite resistance R to insulating behavior with $R = \infty$ as $T \to 0$ as was pointed out by Anderson in 1958 [4]. This phenomenon is called the disorder-driven metal-insulator transition (MIT) [3,1,4] and it is characteristic to non-crystalline solids. The mechanism underlying this MIT was attributed by Anderson not to be due to a finite gap in the energy spectrum which is responsible for an MIT in band gap or Mott insulators [5]. Rather, he argued that the disorder will lead to interference of the electronic wave function $\psi(x)$ with itself such that it is no longer extended over the whole solid but is instead confined to a small part of the solid. This *localization* effect excludes the possibility of diffusion at $T = 0$ and thus the system is an insulator.

A highly successful theoretical approach to this disorder-induced MIT was put forward in 1979 by Abrahams *et al.* [2]. This "scaling hypothesis of localization" details the existence of an MIT for non-interacting electrons in three-dimensional (3D) disordered systems at zero magnetic field B and in the absence of spin-orbit coupling. The starting point for the approach is the realization that the sample-size (L^d) dependence of the (extensive) conductance

$$G = \sigma L^{d-2} = g\frac{e^2}{\hbar} \tag{1}$$

should be investigated [1] with σ denoting the conductivity, d the spatial dimension and g the dimensionless conductance. On the other hand, for strong disorder, the wave functions will be exponentially localized with localization length λ and thus the conductance in a finite system will be

$$g \sim \exp(-L/\lambda). \tag{2}$$

Defining the logarithmic derivative

$$\beta(g) = \frac{d \ln g}{d \ln L}, \tag{3}$$

we see from (1) that $\beta < 0$ for $d = 1$ and 2 and thus an increase in L will drive the system towards the insulator, there are no extended states and no MIT.

However, the β curve for $d = 3$ is positive for large g and negative for small g and an increase of L will drive the system either to metallic or to insulating behavior.

The scenario proposed by the scaling hypothesis is that of a continuous second-order quantum phase transition [4]. Then in the vicinity of the critical energy E_c the DC conductivity σ and the localization length λ should behave as

$$\sigma \propto (E - E_c)^s \quad \text{for } E \geq E_c, \tag{4}$$

$$\lambda \propto (E_c - E)^{-\nu} \quad \text{for } E \leq E_c. \tag{5}$$

with $s = \nu(d-2)$ due to further scaling relations [3]. Introducing a similarly defined dynamical exponent z for the temperature scaling as $\sigma(T) \propto T^{1/z}$, we can write the full finite-temperature scaling form as

$$\sigma(\mu, T) \propto [(\mu - E_c)/E_c]^s \mathcal{F}\left[\frac{T}{[(\mu - E_c)/E_c]^{z\nu}}\right], \tag{6}$$

with μ the chemical potential, $\mathcal{F}$ the scaling function and $z = d$ [7]. The special energy E_c is called the mobility edge [4] and separates localized states with $|E| > E_c$ from extended states with $|E| < E_c$. States directly at the transition with $E = E_c$ are called critical and will be examined later in much detail. In this way the disorder-driven MIT has been reformulated in terms of the theory of critical phenomena [3].

2 Experimental Evidence in Favor of Scaling

Much work has subsequently supported these scaling arguments at $B = 0$ experimentally, analytically and numerically [1]. The MIT can be observed by measuring the conductivity on the metallic side and the dielectric susceptibility on the insulating side of the transition [8]. For doped Si:P, many experiments have been performed following the original work of Paalanen and Thomas [9]. The one-parameter scaling hypothesis has been beautifully validated in these experiments by, e.g., constructing scaling curves for the conductivity [10]. The recent experiments in Si:P [10–13] are concerned with the exact estimation of the critical exponent ν as in (4). The current estimate is $\nu \approx 1$. The interest in the exact value of ν arises since compensated semiconductors apparently have $\nu \approx 1$ as do amorphous metals [8,14,15]. On the other hand, for uncompensated semiconductors one had previously found $\nu \approx 0.5$ [9,8]. The recent estimations of $\nu \approx 1$ for *uncompensated* Si:P [10], based on a careful scaling analysis according to (6) and a consideration of various temperature regimes, may suggest at last a resolution of this "exponent puzzle" [16].

Other experiments in 3D have been performed, e.g., on Si:B [17,18]. Scaling according to (6) yields $\nu = 1.6$. The large value of ν – as compared to the Si:P data – was attributed to the presence of interaction effects. An experimentally convenient way to construct very homogeneously disordered samples is the transmutation doping technique [19–21] which uses the homogeneous properties of neutron rays. Recent scaling results then suggest $\nu = 1.6 \pm 0.2$ [22].

As the localization phenomenon in disordered solids is intrinsically due to the wave nature of the electrons, it can also be observed in other systems exhibiting wave motion [1]. Localization has been studied, e.g. for water waves [23] in shallow basins with random obstacles, for light waves [24–26] in the presence of a fine dust of semiconductor material, for microwaves [27,28] in microwave cavities with random scatterers, and also for surface plasmon polariton waves [29] on rough semiconducting surfaces.

3 Scaling and the Anderson Model of Localization

In order to describe a disordered system, let us consider the Anderson model of localization [4],

$$H = \sum_{j\alpha,k\beta} t_{j\alpha,k\beta} |j\alpha\rangle\langle k\beta| + \sum_{j\alpha} \epsilon_{j\alpha} |j\alpha\rangle\langle j\alpha| . \tag{7}$$

The off-diagonal matrix $t_{j\alpha,k\beta}$ denotes the hopping integrals between the states $\{\alpha\}$ at sites $\{j\}$ with the states $\{\beta\}$ at sites $\{k\}$ and represents the discretization of the kinetic energy. For simplicity, one usually assumes that $\alpha = \beta = 1$ such that there is only one state per site. Moreover, the hopping is usually restricted to nearest-neighbor sites. The disorder is incorporated into the diagonal matrix $\epsilon_{j\alpha}$, whose elements are random numbers usually taken from a uniform distribution $[-W/2, W/2]$ with W parameterizing the strength of the disorder. Other distributions such as Gaussian and Lorentzian have also been investigated [30–34].

This model has been used extensively in conjunction with powerful numerical methods in order to study the localization problem [1]. MacKinnon and Kramer [35,36] have numerically verified the scaling hypothesis by showing that one can find the scaling behavior outlined in Sect. 1. For 3D the corresponding scaling curves have two branches corresponding to the metallic and insulating phases, whereas in 1D and 2D only the insulating branch exists. Recent numerical results – some of which shall be presented in the coming sections – indicate that $\nu = 1.58 \pm 0.02$ [37,38] which is in excellent agreement with the newer experiments reviewed above.

4 Numerical Methods and Finite-Size Scaling for Disordered Systems

The preferred numerical method for accurately computing localization lengths in disordered quantum systems is the transfer-matrix method (TMM) [35,36,39,40]. The TMM is based on a recursive reformulation of the Schrödinger equation such that, e.g. in a 2D strip of width M, length $N \gg M$ and uniform hopping $t_{j,k} = 1$ between nearest neighbors only,

$$\psi_{n+1,m} = (E - \epsilon_{n,m})\psi_{n,m} - \psi_{n,m-1} - \psi_{n,m+1} - \psi_{n-1,m} \tag{8}$$

where $\psi_{n,m}$ is the wave function at site (n,m). Equation (8) can be reformulated into a matrix equation as

$$\begin{pmatrix} \psi_{n+1} \\ \psi_n \end{pmatrix} = \begin{pmatrix} E - \epsilon_n - H_\perp & -1 \\ 1 & 0 \end{pmatrix} \begin{pmatrix} \psi_n \\ \psi_{n-1} \end{pmatrix} = T_n \begin{pmatrix} \psi_n \\ \psi_{n-1} \end{pmatrix}, \tag{9}$$

where $\psi_n = (\psi_{n,1}, \ldots, \psi_{n,M})^T$ denotes the wave function at all sites of the nth slice, $\epsilon_n = \mathrm{diag}(\epsilon_{n,1}, \ldots, \epsilon_{n,M})$, and $H_\perp$ represents the hopping terms in the transverse direction. The evolution of the wave function is given by the product of the transfer matrices $\tau_N = T_N T_{N-1} \ldots T_2 T_1$. Strong fluctuations, which increase exponentially with the system size, govern the evolution of the wave function and thus the behavior of the transmission coefficient through the sample [41–46]. Only the logarithm of the transmission coefficient [35,36,39–42] and the logarithm of the conductance [46–48] are statistically well-behaved self-averaging quantities. According to Oseledec's theorem [49] the eigenvalues $\exp[\pm\gamma_i(M)]$ of $\Gamma = \lim_{N\to\infty}(\tau_N^\dagger \tau_N)^{1/2N}$ exist and the smallest Lyapunov exponent $\gamma_{\min} > 0$ determines the localization length $\lambda(M) = 1/\gamma_{\min}$ at energy E. The accuracy of the λ's is determined as outlined in [35,36] from the variance of the changes of the exponents in the course of the iteration. Usually the method is performed with a complete and orthonormal set of initial vectors $(\psi_1, \psi_0)^T$. In order to preserve this orthogonality, the iterated vectors have to be reorthogonalized during the iteration process.

For small disorders, the $\lambda(M)$ values are of the same order of magnitude as the strip width M and thus subject to finite-size modifications. In order to avoid simple extrapolation schemes, a finite-size scaling (FSS) technique had been developed in [35,36,40] based on real space renormalization arguments for systems with finite size M and intimately related to the original scaling approach [2,50]. The connection to the experimentally perhaps more relevant finite-temperature scaling as in (6) is based on the idea that a finite system size M may be assumed to be equivalent to a measurement at finite temperature T since a finite temperature induces an effective length scale beyond which the electrons will scatter inelastically and thus lose the phase coherence necessary for quantum interference [3,36]. Scaling the $\lambda(M)/M$ data for various values of W onto a common scaling curve, i.e.,

$$\lambda(M)/M = f(\xi/M). \tag{10}$$

is the analogue of (6). One determines the FSS function f and the values of the scaling parameter ξ by a non-linear Levenberg-Marquardt fit [38], see also T. Ohtsuki's contribution in this volume. For diagonal disorder in 3D, this scaling hypothesis of localization has been shown to be valid with very high accuracy, and the ξ values of the extended (localized) branch are equal to the correlation (localization) length in the infinite system. A similar method based on the recursive Green's function technique is discussed by A. MacKinnon in this volume.

A number of more indirect numerical approaches to the MIT have been developed that either only require selected parts of the spectrum – so-called energy-level statistics (ELS) – or a few selected eigenvectors in the spectrum – so-called wave-function statistics (WFS) [51–53]. These methods are based on the connection of Anderson localization to random matrix theory (RMT) [54,55].

The (inverse) participation number [56] represents another possibility to distinguish localized states [57], its scaling with the system size yields a characteristic fractal dimension [58–64]. A generalization to higher moments of the spatial distribution of the wave function leads to the multifractal analysis (MFA) [65,66], where one computes a spectrum of exponents to describe scaling properties of the wave function [66–76]. For a given disorder, one can then read off from the system-size dependence whether the spectrum tends towards the metallic, insulating or truly critical behavior. At the MIT the singularity spectrum of the MFA is independent of the system size [75–77]. Thus we again have a means of studying the MIT. Note that WFS and MFA are not independent of each other, but one may be derived from the other [78]. The fractal characteristics of the eigenstates, i.e their scale incariance, can be beautifully visualized [79,80] by displaying the curdling of the wave functions at the MIT.

5 Scaling for Non-interacting, Disordered Systems

5.1 Early Scaling Results for the Isotropic Anderson Model

Following the seminal papers of MacKinnon and Kramer [35,36] the TMM and FSS approach was used to determine the phase diagram of localization i.e. the MIT in the entire (E, W) plane for the isotropic Anderson model with diagonal box, Gaussian and Lorentzian disorders [31–34,81] as well as at the band center for a binary [82] and a triangular distribution [83]. In 2D, not only the usual square lattice, but also a triangular and honeycomb lattice was studied [84]. The derivation of the critical exponents in these early studies was, however, impeded by the relatively large error of 1 % of the raw data and the limited cross section size of typically up to $M^2 = 13^2$ sites [83–86]. Much effort was needed to determine reliable values of ν [82,85].

We note that FSS has also been successfully employed for the analysis of ELS data and the derivation of critical exponents from the cumulative level-spacing distribution and Dyson-Metha statistics [87,88].

The scaling of the participation number has further been investigated for the Anderson model on 2D and 3D quasiperiodic lattices, too [90,89]. By means of FSS of the participation number, the MIT and the critical exponent could be computed [89].

The entire phase diagram of localization has also been established by MFA of the isotropic Anderson model with box, Gaussian and binary disorder [75], determining the parameter combinations of E and W for which the singularity spectrum is scale-invariant.

5.2 The Anderson Model with Anisotropic Hopping

As shown in Sect. 1, there is no MIT in 2D in the absence of many-body interactions, magnetic field and spin-orbit interactions. Furthermore, the $2+\epsilon$ expansion within the non-linear σ model [7] and numerical studies based on TMM data for bi-fractals [91] suggest that the critical exponent in $2+\epsilon$ dimensions changes continuously as $\epsilon \to 0$ for ϵ between 0 and 1. Thus one is led to ask the question whether a similarly continuous change does also happen, if we vary the hopping elements anisotropically (see C. Soukoulis in this volume). E.g., we decrease the hopping homogeneously in one or two directions yielding weakly coupled planes or chains. This then might model a transition from 3D to 2D or 1D, respectively.

We have investigated this problem using MFA [92,88], TMM [93,88] and ELS [94,95,88] together with FSS. We find that the critical disorder changes continuously with decreasing hopping strength t_a such that $W_c \sim t_a^\alpha$, where α is close to $\frac{1}{4}$ for planes and $\frac{1}{2}$ for chains. Here a represents z for coupled planes and y and z for chains. In Fig. 1 we show examples of FSS in this model. Note that the small relative error 0.07% of the raw data as well as the large system cross sections up to $M^2 = 46^2$ [94] (not shown here) for TMM make it possible – besides taking into account non-linear deviations from $|1 - \frac{W}{W_c}|$ – to determine estimates for *irrelevant* scaling exponents [38]. The value of the critical exponent ν is not affected by the anisotropy [96] and retains its usual value 1.6 ± 0.1 as in 3D [37,38]. Thus the 2D and 1D cases are reached only for $t_a = 0$ and we observe the 3D MIT at any finite t_a.

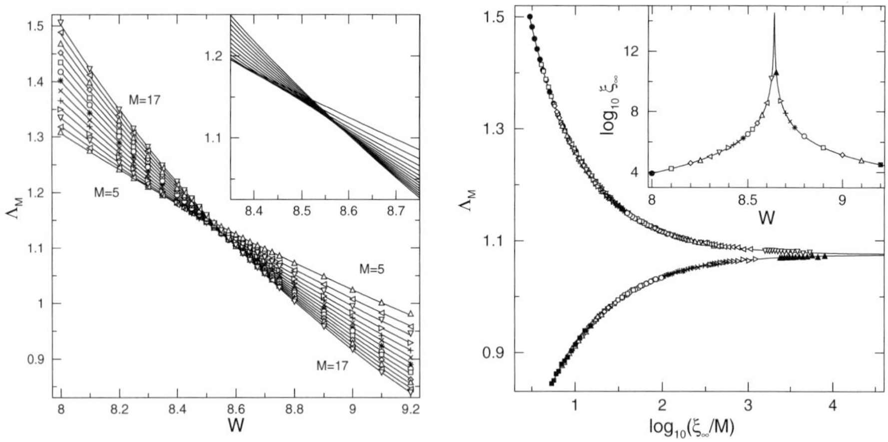

Fig. 1. *Left:* Reduced localization length for coupled planes with $t_z = 0.1$ and $M = 5, 6, \ldots, 17$. The lines are fits of the data according to [38] with $n_r = 3$ and $m_r = 2$. In the inset we enlarge the central region without the data points to show the shift of the crossing point. *Right:* Scaling function and scaling parameter, shown in the inset, corresponding to the fit in the left panel. The symbols distinguish different W values of the scaled data points

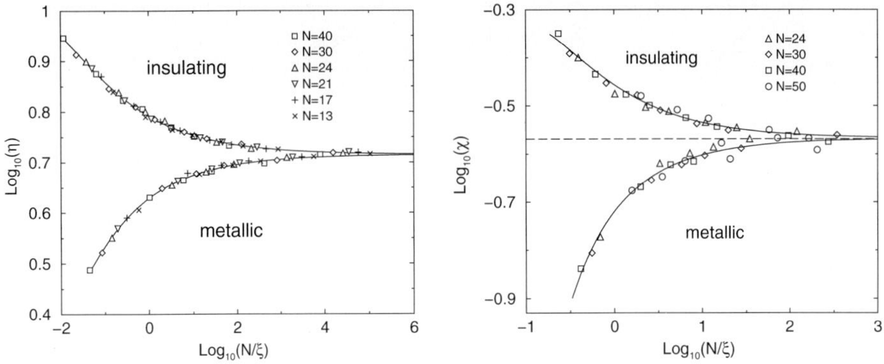

Fig. 2. *Left:* The one-parameter scaling dependence of η on ξ for different system sizes N and disorders $W \in [6,12]$. *Right:* The one-parameter scaling dependence of χ. The dashed line indicates the value $\chi_c = 0.27$ at the MIT obtained from this fit

The ELS and the singularity spectrum of the MFA at the MIT are independent of the system size and this size independence can be used to identify the MIT [97,98,87,68–70,99]. However, both ELS and MFA are influenced by the anisotropy and change considerably in comparison to the isotropic case. Also, the eigenfunctions are different from the isotropic case. Therefore ELS and the MFA singularity spectrum at the MIT are not universal, i.e., not independent of microscopic details of the system. This dependence on microscopic details is similar to the dependence on boundary conditions established recently for the ELS at the MIT [100,101]. However, the critical exponent should, of course, be universal. As it turned out [94,95], FSS can be applied successfully to various statistics of the spectrum, most accurately to the integrated Σ_2 and Δ_3 statistics. In Fig. 2, we give scaling results for the integrated Σ_2 statistics (η) and its derivative, the so-called level compressibility χ, both of which have been computed from spectral data with error 0.2 – 0.4%. The critical exponents $\nu_\eta = 1.43 \pm 0.13, \nu_\chi = 1.47 \pm 0.10$ derived from these data [95], although less accurate, are in agreement with the above-mentioned values.

5.3 The Anderson Model with Random Hopping

Let us change the model (7) such that all nearest-neighbor t values can be chosen randomly [102] with, e.g., $t \in [c-w/2, c+w/2]$ with c and w denoting the center and width of the distribution. In this parameterization, the ordered tight-binding model is recovered in the limit $c \to \infty$ after a suitable rescaling. The DOS has a peak at energy $E = 0$ for any strength of hopping disorder – known in 1D as *Dyson singularity* [103] – and the localization length at $E = 0$ diverges even in 1D [104].

We have studied the random hopping model in 2D [105–107] by TMM and by direct diagonalization of the Hamiltonian matrix and have shown that the singularity in the DOS still exists for bipartite square lattices where the energy

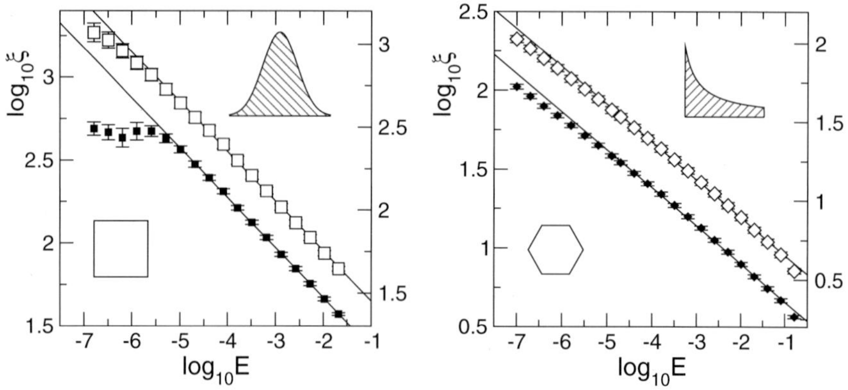

Fig. 3. Scaling parameter ξ vs. energy. *Left panel:* square lattice, Gaussian t distribution, *right panel:* honeycomb lattice, logarithmic t distribution; *filled symbols:* results for $M = 50$–100, *open symbols:* results for $M = 110$–160 (Gaussian distribution) or $M = 100$–150 (logarithmic distribution)

spectrum is strictly symmetric around $E = 0$. Furthermore, the localization length is also diverging at $E = 0$ [108–111], see also S. Evangelou's contribution in the volume.

For sufficiently large energies, it has been suggested [111] that the divergence of the localization length at the band center may be described by a power law, whereas it takes more complicated form below a certain crossover energy E^*. Our results at 0.1% error in the raw data for 2D suggest that the localization lengths exhibit power-law behaviour in a wide energy range with lower bound $E_{\min} \approx 10^{-7}$ and non-universal exponents $\nu \approx 0.25$ [112], see Fig. 3. For smaller energies we observe some deviations, however, there is also a possibility that this may be an effect of pronounced convergence problems which appear for strong hopping disorder.

In 3D, the hopping disorder alone is no longer sufficient to localize all states [102] as happens for (uniform) diagonal disorder at $W_c = 16.5$. Results of FSS for the 3D system at 0.1% error indicate that the critical exponent ν is the same regardless whether we study the MIT as a function of E or as a function of additional diagonal disorder W. Taking into account irrelevant scaling terms, we find that $\nu = 1.59 \pm 0.05$. Thus the results are again in agreement with the usual 3D case and the scaling hypothesis [114,113].

5.4 Thermoelectric Transport Coefficients in the Anderson Model

The conductivity σ is the quantity which is most often studied in transport measurements of disordered systems [9–13,17,18,115]. However, other transport properties such as the thermopower S, the thermal conductivity K and the Lorenz number L_0 have also been measured [116–118]. In [119–121], we have studied the behavior of S, K and L_0 by straightforwardly calculating the integrals in the linear response formulation of Chester-Greenwood-Kubo-Thellung [122–124].

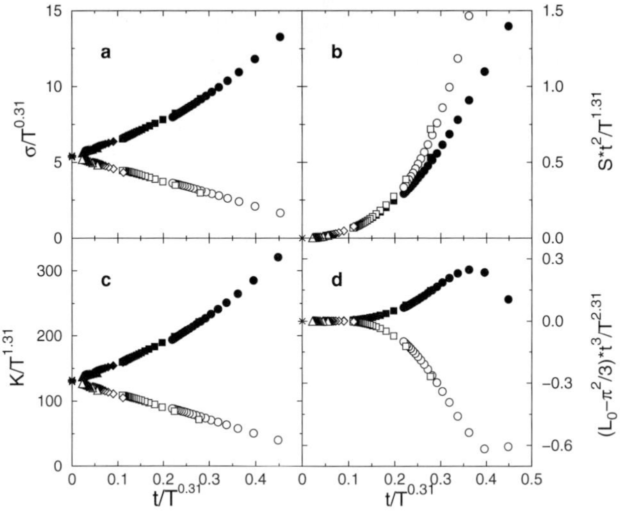

Fig. 4. Scaling of thermoelectric transport properties where $t = |1 - E_F/E_c|$. The different symbols denote the relative positions of various values of the Fermi energy E_F with respect to the mobility edge E_c

The only additional ingredients in our study were an averaged DOS and (a) the assumption of $\sigma(E)$ as in (4) [30,75] or (b) an energy-dependent conductivity σ obtained experimentally. For (a), we can show that the previous analytical considerations [125–127] apply in limited regimes of validity. Thus S/T diverges as $T \to 0$ when the MIT is approached from the metallic side, but S itself remains constant at the MIT. For (b), we show that the temperature-dependent σ, the thermoelectric power S, the thermal conductivity K and the Lorenz number L_0 obey scaling as shown in Fig. 4.

6 Scaling for Interacting, Disordered Systems

The research presented in the last section clearly supports the scaling hypothesis of localization for *non-interacting* electrons. However, real electrons of course interact [128], and their interaction is of relevance for the transport properties of disordered systems [129–131], especially in 2D and 1D [132] where screening [5] is much less efficient than in 3D. Recently, these theoretical considerations received a lot of renewed attention due to the experimental discovery of the 2D MIT [133,131].

In order to theoretically study the effects of the interplay between disorder and interactions, in principle one has to solve a problem with an exponentially growing number of states in the Hilbert space with increasing system size. At present, this can be achieved only for a few particles in 1D and very few particles in 2D [134–136]. However, in 1994 Shepelyansky [137] proposed to simply look at *two* interacting particles (TIP) in a random environment. He showed that two particles in 1D would form pair states even for repulsive interactions such that the TIP pairs would have a larger localization length

$$\lambda_2 \propto U^2 {\lambda_1}^2, \tag{11}$$

than the two seperate single particles (SP) leading to an enhanced possibility of transport through the system [138]. In (11) the pair energy is $E = 0$, U represents the onsite interaction strength and λ_1 is the SP localization length. Subsequent works have established that an enhancement due to interaction indeed exists, although the details are somewhat different from (11) [139].

6.1 Using Decimation to Study TIP in Random Environments

The obvious failure of the TMM approach to the TIP problem in a random potential [140–143] has led us to search for and apply another well tested method of computing localization lengths for disordered system: the decimation method [144]. We computed the TIP Green function in 1D at selected energies for 26 disorders between 0.5 and 9, 24 system sizes between 51 and 251, as well as 11 interaction strengths $U = 0, 0.1, \dots, 1$. For each such set of parameters, we averaged over at least 1000 samples. Furthermore, we constructed FSS curves (see Fig. 5) and from these curves computed scaling parameters which are the infinite-sample-size estimates $\xi_2(U)$ [145] of the localization lengths. We found

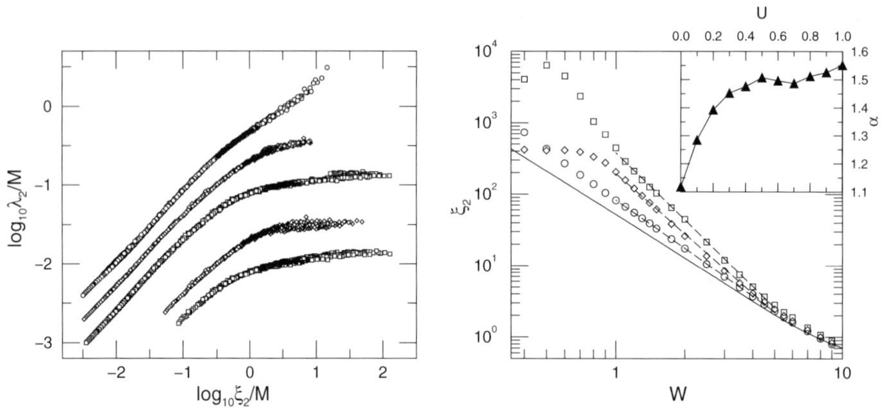

Fig. 5. *Left:* Finite-size scaling plot of the reduced TIP localisation lengths λ_2/M for $U = 0$ (○), $U = 0.2$ (◇) and $U = 1$ (□). The data for $U = 0.2$ ($U = 1$) have been divided by 2 (4) for clarity. Data corresponding to $W = 1$ are indicated by filled symbols. The two curves at the bottom show the same data for $U = 0.2$ and 1 and $W < 2.5$, shifted downward by one order of magnitude for clarity, but here the data for $W < 1$ are scaled with scaling parameters obtained from the power-law fits in the right panel. *Right:* TIP localisation lengths ξ_2 after FSS for $U = 0$ (○), $U = 0.2$ (◇) and $U = 1$ (□). The solid line represents 1D TMM data for SP localisation lengths $\lambda_1/2$, the dashed lines indicate power-law fits. *Inset:* Exponent α obtained by the fit of $\xi_2 \propto W^{-2\alpha}$ to the data for $U = 0, 0.1, \dots, 1$

[146,147] that onsite interaction in 1D indeed leads to a TIP localization length which is *larger* than the SP localization length at $E = 0$. However, the functional dependence is not simply given by (11). Our data follow $\xi_2(U) \sim \xi_2(0)^\beta$ with an exponent β which increases with increasing $|U|$ at $E = 0$. The best fit was obtained when the enhancement $\xi_2(U)/\xi_2(0)$ depends on an exponent β which is a function of U [148]. For values of $U > 1.5$ we observe that the enhancement decreases again; the position of the maximum depends upon W reflecting the expected duality between the behavior at small and large U [149].

6.2 The TIP Effect in a 2D Random Environment

In [150] we have employed the decimation method for TIP in quasi-1D strips of fixed length L and small cross-section $M < L$ at $E = 0$. Analytical considerations for 2D [138] predict that the enhancement of λ_2 at $E = 0$ should be

$$\lambda_2 \propto \lambda_1 \exp\left[\frac{U^2\lambda_1^2}{t^2}\right], \tag{12}$$

with the SP localization length $\lambda_1 \propto \exp(t^2/W^2)$ in 2D [36]. We found [150] that the enhancement is even stronger and as shown in Fig. 6 the scaling curves for $U \geq 0.5$ have two branches indicating a transition of TIP states from localized to delocalized behavior [151]. The scaling curves for $U \leq 0.2$ show a single branch corresponding to localized behavior. The quality of the scaling curves is not as good as in the 1D TIP analysis [146], due to the smaller samples and smaller number of configurations. Nevertheless, our data for 51 interaction strengths and 36 disorder values allow us to map the (U, W) phase diagram of the TIP delocalization-localization transition and we can study how the critical exponent of the localization length changes with U in Fig. 6. We find that for all $U \in (0, 2]$, the exponent is systematically larger than the critical exponent of the usual Anderson transition for non-interacting electrons in 3D. Let us emphasize that

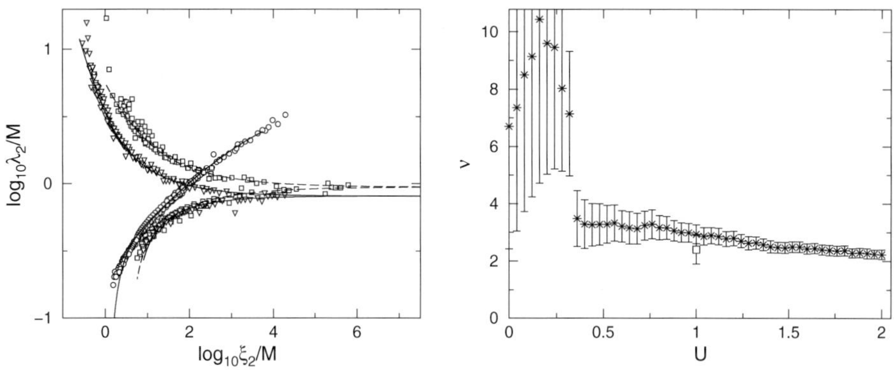

Fig. 6. *Left:* FSS scaling curves (lines) and reduced localization lengths λ_2/M for TIP in 2D as a function of the scaling parameter ξ_2 for $U = 0$ (○), 1 (□), and 2 (∇). *Right:* Critical exponent ν. The data point (□) for $U = 1$ represents the result of [151]

this transition is not a metal-insulator transition in the standard sense since only the TIP states show the delocalization transition. The majority of non-paired states remains localized.

6.3 The TIP Effect Close to an MIT

The numerical effort to study the influence of interaction directly at the MIT in 3D is currently prohibitive. This is true even for TIP, since the problem is equivalent to a six-dimensional SP system with correlated disorder. Fortunately, there is a 1D model which exhibits an MIT driven by increasing a local potential. This model is known as the Aubry-André model [9] which we extend by an interaction term, i.e.,

$$H = \sum_{\sigma,n=1}^{M} (c^{\dagger}_{\sigma,n+1} c_{\sigma,n} + h.c.) + \sum_{\sigma,n=1}^{M} \mu_n c^{\dagger}_{\sigma,n} c_{\sigma,n} + \sum_{n,m=1}^{M} U_{nm} c^{\dagger}_{n\downarrow} c_{n\downarrow} c^{\dagger}_{m\uparrow} c_{m\uparrow}, \tag{13}$$

where $\mu_n = 2\mu \cos(\alpha n + \beta)$ with $\alpha/2\pi$ an irrational number chosen as the inverse of the golden mean $\alpha/2\pi = (\sqrt{5}-1)/2$. β is an arbitrary phase shift. The $c^{\dagger}_{\sigma,n}$ and $c_{\sigma,n}$ are the creation and annihilation operators for an electron at site n with spin $\sigma = \uparrow, \downarrow$. U_{nm} denotes the interaction between particles: $U_{nm} = U\delta_{nm}$ for Hubbard onsite interaction or $U_{nm} = U/(|n-m|+1)$ for long-range interaction. For $\mu < 1$, all SP states in the model with $U = 0$ have been proven rigorously to be extended, whereas for $\mu > 1$ all SP states are localized [9,153–158]. Directly at $\mu = \mu_c = 1$ the SP states are critical. Thus the MIT is similar to the MIT in the 3D Anderson model, but there are no mobility edges.

The model has been previously considered at $U > 0$ from the TIP point of view in [159–161]. It has been shown that on the localized side, the TIP effects persists, i.e., the TIP localization length $\lambda_2 > \lambda_1$. On the extended side, it was argued that the interaction leads to a localization of the TIP states. In [162,163], we have used the TMM and decimation together with FSS to study the problem. We use up to $M = 377$ sites for the decimation method with at least 1000 samples. Let us emphasize that contrary to the problem with TMM for the TIP situation in finite systems, together with FSS the TMM approach can be used to give meaningful results. However, the computed localization lengths are no longer directly the localization lengths of a TIP pair, but rather measure the influence of the presence of the second particle on the transport properties of the first. In addition to investigating the onsite interacting case, we have also studied long-range interactions in [162]. We find that whereas onsite interaction does not shift the MIT from $\mu_c = 1$, long-range interaction might change the MIT towards smaller values $\mu_c \approx 0.92$.

For the quasiperiodic many-body system with nearest-neighbor interaction at finite particle density [164,165] we have recovered the transition at $\mu_c = 1$ independent of interaction strength, provided we consider densities like $\rho = 1/2$ which are incommensurate compared to the wave vector of the quasiperiodic potential – an irrational multiple of π. Thus, the low-density TIP case is comparable to finite but incommensurate densities. On the other hand, for commensurate densities, we find that the system can be completely localized even for $\mu \ll 1$, due to a Peierls resonance between the degrees of freedom of the electronic system and the quasiperiodic potential. Whereas for repulsive interactions the ground state remains localized, we find a region of extended states for attractive interaction due to the interplay between interaction and quasiperiodic potential. Thus, the physics of the model at finite densities is dominated by whether the density is commensurate or incommensurate and only in the latter case by interaction effects.

7 Conclusions

In the preceding sections, we have presented results of transport studies in disordered systems, ranging from modifications of the standard Anderson model of localization to effects of a two-body interaction. Of paramount importance in these studies was always the highest possible accuracy of the raw data combined with the careful subsequent application of the FSS technique. In fact, it is this scaling method that has allowed numerical studies to move beyond simple extrapolations and reliably construct estimates of quantities as if one were studying an infinite system. Of course, this statement is only a short and perhaps too short summary of the seminal paper by MacKinnon and Kramer [35], in which the FSS technique was first applied to the localization problem.

Acknowledgements

We gratefully acknowledge financial support by the DFG via SFB393 and priority research program "Quasikristalle" as well as by the DAAD. This paper is dedicated to Bernhard Kramer on the occasion of his 60th birthday. Both authors are grateful to him for many encouraging discussions and stimulating interactions over many years.

References

1. P.W. Anderson, Phys. Rev. **109**, 1492 (1958)
2. D. Belitz and T.R. Kirkpatrick, Rev. Mod. Phys. **66**, 261 (1994)
3. B. Kramer and A. MacKinnon, Rep. Prog. Phys. **56**, 1469 (1993)
4. P.A. Lee and T.V. Ramakrishnan, Rev. Mod. Phys. **57**, 287 (1985)
5. N.W. Ashcroft and N.D. Mermin, *Solid State Physics* (Saunders College, New York, 1976)
6. E. Abrahams, P.W. Anderson, D.C. Licciardello, and T.V. Ramakrishnan, Phys. Rev. Lett. **42**, 673 (1979)
7. F. Wegner, Z. Phys. B **44**, 9 (1981)
8. G.A. Thomas, in *Localisation and Interactions in Disordered and Doped Semiconductors*, edited by D.M. Finlayson (SUSSP, Edinburgh, 1986), p. 172
9. M.A. Paalanen and G.A. Thomas, Helv. Phys. Acta **56**, 27 (1983)

10. S. Waffenschmidt, C. Pfleiderer, and H. v. Löhneysen, Phys. Rev. Lett. **83**, 3005 (1999)
11. T.F. Rosenbaum, G.A. Thomas, and M.A. Paalanen, Phys. Rev. Lett. **72**, 2121 (1994)
12. H. Stupp et al., Phys. Rev. Lett. **71**, 2634 (1993)
13. H. Stupp et al., Phys. Rev. Lett. **72**, 2122 (1994)
14. P. Häussler, Phys. Rep. **222**, 65 (1992)
15. A. Möbius and C.J. Adkins, Current Opinion in Solid State and Materials Science **4**, 303 (1999)
16. U. Thomanschefsky and D. F. Holcomb, Phys. Rev. B **45**, 13356 (1988)
17. S. Bogdanovich, M.P. Sarachik, and R.N. Bhatt, Phys. Rev. Lett. **82**, 137 (1999)
18. S. Bogdanovich et al., Phys. Rev. B **60**, 2286 (1998)
19. K.M. Itoh et al., Phys. Rev. Lett. **77**, 4058 (1996)
20. B. Sandow et al., phys. stat. sol. (b) **205**, 281 (1998)
21. M. Watanabe, Y. Ootuka, K.M. Itoh, and E.E. Haller, Phys. Rev. B **58**, 9851 (1998)
22. T. Hayashi, Y. Hashimoto, S. Katsumoto, and Y. Iye, J. Phys. Chem. Solids **63**, 1315 (2002)
23. P.E. Lindelof, J. Nørregard, and J. Hanberg, Phys. Scr. T **14**, 317 (1986)
24. G. Maret, Phys. Blätter **48**, 161 (1992)
25. D.S. Wiersma, P. Bartolini, A. Lagendjik, and R. Righini, Nature **390**, 671 (1997)
26. P.-E. Wolf and G. Maret, Phys. Rev. Lett. **55**, 2696 (1985)
27. M. Barth, U. Kuhl, and H.-J. Stöckmann, Phys. Rev. Lett. **82**, 2026 (1999)
28. H.-J. Stöckmann, *Quantum Chaos – An Introduction* (Cambridge University Press, New York, 1999)
29. S.I. Bozhevolnyi, Ann. Phys. (Leipzig) **8**, 717 (1999)
30. B. Bulka, B. Kramer, and A. MacKinnon, Z. Phys. B **60**, 13 (1985)
31. B. Bulka, M. Schreiber, and B. Kramer, Z. Phys. B **66**, 21 (1987)
32. B. Kramer and M. Schreiber, in *Fluctuations and Stochastic Phenomena in Condensed Matter*, edited by L. Garrido, Lecture Notes in Physics (Springer, Berlin, Heidelberg) **268**, 351 (1987)
33. B. Kramer and M. Schreiber, in *Localization in Disordered Systems*, edited by W. Weller and P. Ziesche, Teubner Texte zur Physik (BSB Teubner, Leipzig) **16**, 96 (1988)
34. M. Schreiber, B. Bulka, and B. Kramer, in Proc. 18th Int. Conf. *Physics of Semiconductors*, Stockholm 1986, edited by O. Engström (World Scientific, Singapore 1987), pp. 1269–1272
35. A. MacKinnon and B. Kramer, Phys. Rev. Lett. **47**, 1546 (1981)
36. A. MacKinnon and B. Kramer, Z. Phys. B **53**, 1 (1983)
37. A. MacKinnon, J. Phys.: Condens. Matter **6**, 2511 (1994)
38. K. Slevin and T. Ohtsuki, Phys. Rev. Lett. **82**, 382 (1999)
39. B. Kramer and M. Schreiber, in *Computational Physics*, edited by K.H. Hoffmann and M. Schreiber (Springer, Berlin, 1996), pp. 166–188.
40. J.-L. Pichard and G. Sarma, J. Phys. C **14**, L127 (1981)
41. M. Schreiber, H. Grussbach, and M. Ottomeier, Mol. Cryst. Liq. Cryst. **216**, 67 (1992)
42. M. Schreiber and B. Kramer, in Proc. 19th Int. Conf. *Physics of Semiconductors*, Warschau 1988, edited by W. Zawadzki (Institute of Physics, Polish Academy of Sciences, Warsaw 1988), pp. 87–90
43. M. Schreiber and K. Maschke, Phil. Mag. B **65**, 639 (1992)

44. M. Schreiber and K. Maschke, in *Photonic Band Gaps and Localization*, edited by C.M. Soukoulis, NATO ARW (Plenum, New York) B **308**, 439 (1993)
45. M. Schreiber and K. Maschke, in *Quantum Coherence in Mesoscopic Systems*, edited by B. Kramer, NATO ASI (Plenum, New York) B **254**, 105 (1991)
46. J. Sak, B. Kramer, Phys. Rev. B **24**, 1761 (1981)
47. B. Kramer and M. Schreiber, J. Non-Cryst. Sol. **114**, 330 (1989)
48. B. Kramer, A. Kawabata, and M. Schreiber, Phil. Mag. B **65**, 595 (1992)
49. V.I. Oseledec, Trans. Moscow Math. Soc. **19**, 197 (1968)
50. F. Wegner, in *Localization and Metal-Insulator Transitions*, edited by H. Fritzsche and D. Adler (Plenum, New York, 1985), p. 337
51. K. Müller, B. Mehlig, F. Milde, and M. Schreiber, Phys. Rev. Lett. **78**, 215 (1997)
52. V. Uski, B. Mehlig, R.A. Römer, and M. Schreiber, Phys. Rev. B **62**, R7699 (2000)
53. V. Uski, B. Mehlig, and M. Schreiber, Phys. Rev. B **63**, 241101 (2001)
54. T. Guhr, A. Müller-Groeling, and H.A. Weidenmüller, Phys. Rep. **299**, 189 (1998)
55. F. Haake, *Quantum Signatures of Chaos*, 2nd ed. (Springer, Berlin, 1992)
56. D.J. Thouless, Phys. Rep. **13**, 93 (1974)
57. M. Schreiber and Y. Toyozawa, J. Phys. Soc. Japan **51**, 1537 (1982)
58. M. Schreiber, J. Phys. C **18**, 2493 (1985)
59. M. Schreiber, Phys. Rev. B **31**, 6146 (1985)
60. M. Schreiber, J. Non-Cryst. Sol. **97 & 98**, 221 (1987)
61. M. Schreiber, in *Localisation 1990*, edited by J.T. Chalker, Inst. Phys. Conf. Ser. **108**, 65 (1991)
62. M. Schreiber, in Proc. Int. Conf. *Localization, Interaction and Transport Phenomena in Impure Metals (LITPIM)*, edited by L. Schweitzer and B. Kramer, Suppl. PTB-PG-1 (PTB Braunschweig 1984), pp. 74–77
63. M. Schreiber and B. Kramer, in *Overview of Cyber 205 Projects in Bochum*, edited by H. Ehlich, K. H. Schlosser, and B. Wojcieszynski, Bochumer Schriften zur parallelen Datenverarbeitung (Rechenzentrum der Ruhruniversität, Bochum) **11**, 439 (1986)
64. M. Schreiber, Physica A **167**, 188 (1990)
65. H. Aoki, J. Phys. C **16**, L205 (1983)
66. M. Schreiber, in *Computational Physics*, edited by K.H. Hoffmann and M. Schreiber (Springer, Berlin, 1996), pp. 147–165
67. H. Grussbach, Ph.D. thesis, Johannes Gutenberg-Universität Mainz, 1995
68. M. Schreiber and H. Grussbach, Phys. Rev. Lett. **67**, 607 (1991)
69. M. Schreiber and H. Grussbach, Phil. Mag. B **65**, 704 (1992)
70. M. Schreiber and H. Grussbach, Mod. Phys. Lett. B **16**, 851 (1992)
71. H. Grussbach and M. Schreiber, J. Lumin. **53**, 137 (1992)
72. M. Schreiber and H. Grussbach, Phys. Rev. Lett. **68**, 2857 (1992)
73. M. Schreiber, in *New Horizons in Low-Dimensional Electron Systems*, edited by H. Aoki, M. Tsukuda, M. Schlüter, and F. Levy (Kluwer, Dordrecht 1992), pp. 167–178
74. H. Grussbach and M. Schreiber, Phys. Rev. B **48**, 6650 (1993)
75. H. Grussbach and M. Schreiber, Phys. Rev. B **51**, 663 (1995)
76. H. Grussbach and M. Schreiber, Chem. Phys. **177**, 733 (1993)
77. M. Schreiber and H. Grussbach, J. of Fractals **1**, 1037 (1993)
78. A.D. Mirlin and Y.V. Fyodorov, J. Phys. A: Math. Gen. **26**, L551 (1993)
79. H. Grussbach and M. Schreiber, Physica A **191**, 394 (1992)
80. F. Milde, PhD thesis, Technische Universität Chemnitz (2000), `http://archiv.tu-chemnitz.de/pub/2000/0060/index.html`

81. M. Schreiber, B. Kramer, and A. MacKinnon, Physica Scripta **T25**, 67 (1989)
82. E. Hofstetter and M. Schreiber, Europhys. Lett. **21**, 933 (1993)
83. M. Schreiber, in *Large Scale Molecular Systems - Quantum and Stochastic Aspects. Beyond the Simple Molecular Picture*, edited by W. Gans, A. Blumen, and A. Amann, NATO ASI (Plenum, New York) B **258**, 385 (1991)
84. M. Schreiber and M. Ottomeier, J. Phys.: Condens. Matter **4**, 1959 (1992)
85. B. Kramer, K. Broderix, A. MacKinnon, and M. Schreiber, Physica A **167**, 163 (1990)
86. M. Schreiber and B. Kramer, in *Anderson Localization*, edited by T. Ando, H. Fukuyama, Proceedings in Physics (Springer, Berlin, Heidelberg) **28**, 92 (1988)
87. E. Hofstetter and M. Schreiber, Phys. Rev. B **49**, 14726 (1994)
88. M. Schreiber and F. Milde, in *Computational Statistical Physics - From Billards to Monte Carlo*, edited by K.H. Hoffmann and M. Schreiber (Springer, Berlin, Heidelberg 2002), pp. 259–278
89. T. Rieth and M. Schreiber, Z. Phys. B **104**, 99 (1997)
90. T. Rieth and M. Schreiber, in Proc. 5th Int. Conf. *Quasicrystals*, edited by C. Janot and R. Mosseri (World Scientific, Singapore 1995), pp. 514–517
91. M. Schreiber and H. Grussbach, Phys. Rev. Lett. **76**, 1687 (1996)
92. F. Milde, R.A. Römer, and M. Schreiber, Phys. Rev. B **55**, 9463 (1997)
93. F. Milde, R.A. Römer, M. Schreiber, and V. Uski, Eur. Phys. J. B **15**, 685 (2000)
94. F. Milde, R.A. Römer, and M. Schreiber, Phys. Rev. B **61**, 6028 (2000)
95. M.L. Ndawana, R.A. Römer, and M. Schreiber, Eur. Phys. J. B **27**, 399 (2002)
96. F. Milde and R.A. Römer, Ann. Phys. (Leipzig) **7**, 452 (1998)
97. E. Hofstetter and M. Schreiber, Phys. Rev. B **48**, 16979 (1993)
98. E. Hofstetter and M. Schreiber, Phys. Rev. Lett. **73**, 3137 (1994)
99. I.K. Zharekeshev and B. Kramer, Phys. Rev. Lett. **79**, 717 (1997)
100. D. Braun, G. Montambaux, and M. Pascaud, Phys. Rev. Lett. **81**, 1062 (1998)
101. H. Potempa and L. Schweitzer, J. Phys.: Condens. Matter **10**, L431 (1998)
102. E.N. Economou and P.D. Antoniou, Solid State Commun. **21**, 285 (1977)
103. F.J. Dyson, Phys. Rev. **92**, 1331 (1953)
104. R.H. McKenzie, Phys. Rev. Lett. **77**, 4804 (1996)
105. A. Eilmes, R.A. Römer, and M. Schreiber, Eur. Phys. J. B **1**, 29 (1998)
106. A. Eilmes, R.A. Römer, and M. Schreiber, phys. stat. sol. (b) **205**, 229 (1998)
107. A. Eilmes, R.A. Römer, and M. Schreiber, Physica B **296**, 46 (2001)
108. F. Wegner, Z. Phys. B **35**, 207 (1979)
109. R. Gade and F. Wegner, Nucl. Phys. B **360**, 213 (1991)
110. R. Gade, Nucl. Phys. B **398**, 499 (1993)
111. M. Fabrizio and C. Castellani, Nucl. Phys. B **583**, 524 (2000)
112. A. Eilmes and R. A. Römer, J. Phys. Soc. Japan (2002), in press
113. P. Cain, Diploma thesis, Technische Universität Chemnitz (1998)
114. P. Biswas, P. Cain, R.A. Römer, and M. Schreiber, phys. stat. sol. (b) **218**, 205 (2000)
115. G.A. Thomas, Phil. Mag. B **52**, 479 (1985)
116. M. Lakner and H. v. Löhneysen, Phys. Rev. Lett. **70**, 3475 (1993)
117. C. Lauinger and F. Baumann, J. Phys.: Condens. Matter **7**, 1305 (1995)
118. G. Sherwood, M.A. Howson, and G.J. Morgan, J. Phys.: Condens. Matter **3**, 9395 (1991)
119. C. Villagonzalo and R.A. Römer, Ann. Phys. (Leipzig) **7**, 394 (1998)
120. C. Villagonzalo, R.A. Römer, and M. Schreiber, Eur. Phys. J. B **12**, 179 (1999)
121. C. Villagonzalo, R.A. Römer, M. Schreiber, and A. MacKinnon, Phys. Rev. B **62**, 16446 (2000)

122. G.V. Chester and A. Thellung, Proc. Phys. Soc. **77**, 1005 (1961)
123. D.A. Greenwood, Proc. Phys. Soc. **71**, 585 (1958)
124. R. Kubo, J. Phys. Soc. Japan **12**, 570 (1957)
125. C. Castellani, C.D. Castro, M. Grilli, and G. Strinati, Phys. Rev. B **37**, 6663 (1988)
126. U. Sivan and Y. Imry, Phys. Rev. B **33**, 551 (1986)
127. J.E. Enderby and A. C. Barnes, Phys. Rev. B **49**, 5062 (1994)
128. C.A. d. Coulomb, French physicist; discoverer and eponym of Coulomb's law, 1736–1806
129. A.L. Efros and B. Shklovskii, J. Phys. C **8**, L49 (1975)
130. N.F. Mott, *Metal-Insulator Transitions* (Taylor & Francis, London, 1990)
131. M.P. Sarachik and S.V. Kravchenko, Proc. Natl. Acad. Sci. USA **96**, 5900 (1999)
132. T. Giamarchi and B. S. Shastry, Phys. Rev. B **51**, 10915 (1995)
133. S.V. Kravchenko et al., Phys. Rev. B **50**, 8039 (1994)
134. P.H. Song and D.L. Shepelyansky, Phys. Rev. B **61**, 15546 (2000)
135. T. Vojta, F. Epperlein, and M. Schreiber, Phys. Rev. Lett. **81**, 4212 (1998)
136. F. Epperlein, M. Schreiber, and T. Vojta, phys. stat. sol. **205**, 233 (1998)
137. D.L. Shepelyansky, Phys. Rev. Lett. **73**, 2607 (1994)
138. Y. Imry, Europhys. Lett. **30**, 405 (1995)
139. R.A. Römer, M. Schreiber, and T. Vojta, Physica E **9**, 397 (2001)
140. K. Frahm, A. Müller-Groeling, J.L. Pichard, and D. Weinmann, Europhys. Lett. **31**, 169 (1995)
141. R.A. Römer and M. Schreiber, Phys. Rev. Lett. **78**, 515 (1997)
142. K. Frahm, A. Müller-Groeling, J.-L. Pichard, and D. Weinmann, Phys. Rev. Lett. **78**, 4889 (1997)
143. R.A. Römer and M. Schreiber, Phys. Rev. Lett. **78**, 4890 (1997)
144. C.J. Lambert and D. Weaire, phys. stat. sol. (b) **101**, 591 (1980)
145. P.H. Song and D. Kim, Phys. Rev. B **56**, 12217 (1997)
146. M. Leadbeater, R.A. Römer, and M. Schreiber, Eur. Phys. J. B **8**, 643 (1999)
147. R.A. Römer, M. Leadbeater, and M. Schreiber, Physica A **266**, 481 (1999)
148. I.V. Ponomarev and P. G. Silvestrov, Phys. Rev. B **56**, 3742 (1997)
149. X. Waintal, D. Weinmann, and J.-L. Pichard, Eur. Phys. J. B **7**, 451 (1999)
150. R.A. Römer, M. Leadbeater, and M. Schreiber, Ann. Phys. (Leipzig) **8**, 675 (1999)
151. M. Ortuño and E. Cuevas, Europhys. Lett. **46**, 224 (1999)
152. S. Aubry and G. André, Ann. Israel Phys. Soc. **3**, 133 (1980)
153. M. Kohmoto, Phys. Rev. Lett. **51**, 1198 (1983)
154. M. Kohmoto, L.P. Kadanoff, and C. Tang, Phys. Rev. Lett. **50**, 1870 (1983)
155. S. Östlund et al., Phys. Rev. Lett. **50**, 1873 (1983)
156. S. Östlund and R. Pandit, Phys. Rev. B **29**, 1394 (1984)
157. S.D. Sarma, S. He, and X.C. Xie, Phys. Rev. B **41**, 5544 (1990)
158. I. Varga, J. Pipek, and B. Vasvári, Phys. Rev. B **46**, 4978 (1992)
159. A. Barelli, J. Bellisard, P. Jacquod, and D. L. Shepelyansky, Phys. Rev. Lett. **77**, 4752 (1996)
160. A. Barelli, J. Bellisard, P. Jacquod, and D. L. Shepelyansky, Phys. Rev. B **55**, 9524 (1997)
161. D. L. Shepelyansky, Phys. Rev. B **54**, 14896 (1996)
162. A. Eilmes, U. Grimm, R.A. Römer, and M. Schreiber, Eur. Phys. J. B **8**, 547 (1999)
163. A. Eilmes, R. A. Römer, and M. Schreiber, Eur. Phys. J. B **23**, 229 (2001)
164. C. Schuster, R.A. Römer, and M. Schreiber, Phys. Rev. B **65**, 115114 (2002)
165. C. Schuster, R. A. Römer, and M. Schreiber, J. Phys. Soc. Japan (to appear)

Transfer Matrices and Disordered Systems

Angus MacKinnon

The Blackett Laboratory, Imperial College London, London SW7 2BW, UK

1 Introduction

Of the many approaches which have been developed for numerical simulations of disordered systems in general, and the Anderson Metal-Insulator transition (MIT) in particular, the transfer matrix has proved the most productive. In this chapter I will present several related approaches that might broadly be considered as *transfer matrix* methods. As a definition of a transfer matrix we may consider

> An operator which uses a set of quantities defined at one position to generate the same set of quantities at a subsequent position.

This concept may be expressed in terms of a recursive equation such as

$$\boldsymbol{a}(z+\delta z) = \hat{\mathbf{T}}(z)\boldsymbol{a}(z) \; , \tag{1}$$

or as a simple differential equation

$$\frac{\mathrm{d}}{\mathrm{d}z}\boldsymbol{a}(z) = \hat{\mathbf{D}}\boldsymbol{a}(z) \; . \tag{2}$$

The components of the vector $\boldsymbol{a}(z)$ may represent various quantities depending on the particular problem being considered. The important factor is that the set is *complete* in the sense that the number of components does not change as a z increases. Note in passing that the operators $\hat{\mathbf{T}}$ and $\hat{\mathbf{D}}$ are not necessarily linear (see Sect. 4 below).

The simplest examples may be derived from 1-dimensional Hamiltonians, which may be recast as simple differential or difference equations: the 1D Schrödinger equation

$$-\frac{\hbar^2}{2m}\frac{\mathrm{d}^2\psi}{\mathrm{d}z^2} + V(z)\psi = E\psi \; , \tag{3a}$$

$$\frac{\mathrm{d}}{\mathrm{d}z}\begin{pmatrix}\psi' \\ \psi\end{pmatrix} = \begin{pmatrix}0 & \frac{2m}{\hbar^2}\left(V(z)-E\right) \\ 1 & 0\end{pmatrix}\begin{pmatrix}\psi' \\ \psi\end{pmatrix} = \hat{\mathbf{D}}(z)\begin{pmatrix}\psi' \\ \psi\end{pmatrix} \; , \tag{3b}$$

or the Anderson [1] Hamiltonian

$$Va_{n+1} + \varepsilon_n a_n + Va_{n-1} = Ea_n \; , \tag{4a}$$

$$\begin{aligned}\begin{pmatrix}a_{n+1} \\ Va_n\end{pmatrix} &= \begin{pmatrix}V^{-1} & 0 \\ 0 & V\end{pmatrix}\begin{pmatrix}E-\varepsilon_n & -1 \\ 1 & 0\end{pmatrix}\begin{pmatrix}a_n \\ Va_{n-1}\end{pmatrix} \\ &= \hat{\mathbf{T}}_n\begin{pmatrix}a_n \\ Va_{n-1}\end{pmatrix} \; .\end{aligned} \tag{4b}$$

For linear operators $\hat{\mathbf{D}}(z)$ and $\hat{\mathbf{T}}_n$ it is possible to write formal solutions of (3b) & (4b) in the form

$$\begin{pmatrix} \psi'(z) \\ \psi(z) \end{pmatrix} = \hat{P} \exp\left(\int_0^z \hat{\mathrm{D}}(z')\mathrm{d}z' \right) \begin{pmatrix} \psi'(0) \\ \psi(0) \end{pmatrix} , \tag{5a}$$

$$\begin{pmatrix} a_{n+1} \\ V a_n \end{pmatrix} = \hat{P} \prod_{i=1}^{n} \hat{\mathrm{T}}_i \begin{pmatrix} a_1 \\ V a_0 \end{pmatrix} , \tag{5b}$$

where $\hat{P}$ represents a spatial-ordering operator analogous to the time-ordering operator that appears in most discussions of many-body theory. As we shall see, these formal solutions tend not to be useful in practice due to numerical instabilities and various techniques have been developed to overcome these difficulties.

2 The Standard Transfer Matrix

2.1 General Properties

Starting from (4a) we may consider a quasi-1D system as illustrated in Fig. 1 with a finite cross-section described by a Hamiltonian

$$V_{i,n;j,n+1} a_{j,n+1} + H_{i,n;j,n} a_{j,n} + V_{i,n;j,n-1} a_{j,n-1} = E a_{i,n} \tag{6a}$$

$$\mathbf{V}_{n,n+1} \boldsymbol{a}_{n+1} + \mathbf{H}_{n,n} \boldsymbol{a}_n + \mathbf{V}_{n,n-1} \boldsymbol{a}_{n-1} = E \boldsymbol{a}_n , \tag{6b}$$

where i and j run across the strip and n along it. In the vector representation (6b), $\boldsymbol{a}_n$ has components for each basis state in the nth slice and the matrices $\mathbf{H}_{n,n}$ and $\mathbf{V}_{n,n\pm1}$ are defined in the subspace of a single slice with their subscripts denoting slices. This Hamiltonian may be recast in terms of a transfer matrix as

$$\begin{bmatrix} \boldsymbol{a}_{n+1} \\ \mathbf{V}_{n+1,n} \boldsymbol{a}_n \end{bmatrix} = \begin{bmatrix} \left[\mathbf{V}^{\dagger}_{n+1,n} \right]^{-1} & \mathbf{0} \\ \mathbf{0} & \mathbf{V}_{n+1,n} \end{bmatrix} \begin{bmatrix} E - \mathbf{H}_{n,n} & -\mathbf{I} \\ \mathbf{I} & \mathbf{0} \end{bmatrix} \begin{bmatrix} \boldsymbol{a}_n \\ \mathbf{V}_{n,n-1} \boldsymbol{a}_{n-1} \end{bmatrix} . \tag{7}$$

In this form we can see one major restriction on this simple transfer matrix, namely that $\mathbf{V}$ may not be singular. In particular, problems in which off-diagonal elements may be zero, such as quantum percolation [2] cannot be treated in this way.

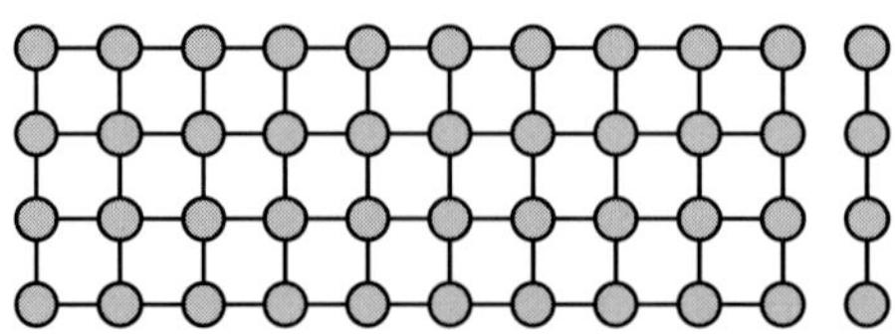

Fig. 1. A long strip of width $M(=4)$ and length $N(=10)$ with an additional slice about to be added on the right

Another important property is illustrated by considering the case $\mathbf{V} = \mathbf{I}$. In this case the transfer matrix $\mathbf{T}$ has the symmetry property

$$\mathbf{JTJ} = \mathbf{T}^{-1} \quad \text{where} \quad \mathbf{J} = \begin{bmatrix} \mathbf{0} & \mathbf{I} \\ \mathbf{I} & \mathbf{0} \end{bmatrix} , \tag{8}$$

from which it follows that the eigenvalues occur in pairs, (α, α^{-1}), with eigenvectors $(\boldsymbol{a}, \mathbf{J}\boldsymbol{a})$. This property is more general than appears from this simple derivation, although it sometimes applies only to the magnitude, but not the phases, of the eigenvalues.

2.2 Numerical Problems and Stabilisation

In the typical situation, where $\mathbf{H}$ and $\mathbf{V}$ are random matrices, the solutions of (7) rise exponentially. Often we require the longest length scale or smallest exponent [3–6] as this may be identified as the localisation length in the system. We consider an initial vector which contains approximately equally weighted contributions from every eigenvector, $\boldsymbol{a}_\alpha$ of $\mathbf{T}$. Then after n iterations the iterated vector will be approximately

$$\sum_\alpha \left[\exp(\alpha n)\boldsymbol{a}_\alpha + \exp(-\alpha n)\mathbf{J}\boldsymbol{a}_\alpha\right] , \tag{9}$$

in which the fractional weight of the eigenvector for the smallest (positive) exponent $\alpha_{\min}$ is roughly

$$w_n = \exp\left((\alpha_{\min} - \alpha_{\max})n\right) > \varepsilon_{\text{mach}} , \tag{10}$$

which must be larger than the accuracy $\varepsilon_{\text{mach}}$ of the computer; otherwise the information about $\alpha_{\min}$ will be lost. Unfortunately this is precisely the information required to calculate the localisation length.

In order to overcome this numerical problem and calculate the localisation length it is necessary to introduce a stabilisation procedure. We start by noting that to find the smallest eigenvalue of a $2N \times 2N$ transfer matrix it will be necessary to work with at least N vectors and to try to make sure these span an N-dimensional subspace. We write

$$\begin{bmatrix} \mathbf{A}_n^{\text{t}} \\ \mathbf{A}_n^{\text{b}} \end{bmatrix} = \prod_{i=1}^{n} \mathbf{T}_i \begin{bmatrix} \mathbf{A}_0^{\text{t}} \\ \mathbf{A}_0^{\text{b}} \end{bmatrix} , \tag{11}$$

where $\mathbf{A}^{\text{t}}$ and $\mathbf{A}^{\text{b}}$ are $N \times N$ matrices. To stabilise the system we multiply from the right by an appropriate $N \times N$ matrix. The most efficient is an upper triangular matrix $\mathbf{U}_n$ which orthonormalises the columns of $\mathbf{A}_n$. Thus

$$\begin{bmatrix} \mathbf{A}_n'^{\text{t}} \\ \mathbf{A}_n'^{\text{b}} \end{bmatrix} = \begin{bmatrix} \mathbf{A}_n^{\text{t}} \\ \mathbf{A}_n^{\text{b}} \end{bmatrix} \mathbf{U}_n . \tag{12}$$

Note also that if the process is repeated after another n iterations we have

$$\begin{bmatrix} \mathbf{A}''^{\mathrm{t}}_{2n} \\ \mathbf{A}''^{\mathrm{b}}_{2n} \end{bmatrix} = \begin{bmatrix} \mathbf{A}'^{\mathrm{t}}_{2n} \\ \mathbf{A}'^{\mathrm{b}}_{2n} \end{bmatrix} \mathbf{U}'_{2n} = \begin{bmatrix} \mathbf{A}^{\mathrm{t}}_{2n} \\ \mathbf{A}^{\mathrm{b}}_{2n} \end{bmatrix} \mathbf{U}_n \mathbf{U}'_{2n} = \begin{bmatrix} \mathbf{A}^{\mathrm{t}}_{n} \\ \mathbf{A}^{\mathrm{b}}_{n} \end{bmatrix} \mathbf{U}_{2n} , \tag{13}$$

where $\mathbf{U}_{2n}$ is the matrix required to orthonormalise the columns after $2n$ iterations and $\mathbf{U}'_{2n}$ is the corresponding matrix when the columns have already been orthonormalised after n iterations using $\mathbf{U}_n$. In this way we can maintain the orthogonality of the columns of $\mathbf{A}$ and thus the information required to extract the smallest positive eigenvalue of the transfer matrix.

This orthogonalisation process is invariably the time limiting step in any transfer matrix calculation. It scales as N^3 whereas the transfer matrix itself scales as N for each column, i.e. as N^2. Note, however, that it is not necessary to perform this after every iteration but only as often as is required to maintain a significant contribution from all N vectors. For the normal Anderson [1] this would typically be about every 12th iteration. In anisotropic models or those with correlated disorder the necessary orthogonalisation frequency may be very different. Traditionally this has been determined empirically, but it is possible to calculate the optimum value by noting the weight of the 1st $N-1$ already orthogonalised vectors in the Nth vector during the Gram–Schmidt orthogonalisation process.

As a further refinement the normalisation coefficients (or their logarithms) for each vector may be stored. These behave (on average) as $c_i = \exp(\alpha_i n)$ and by storing their logarithms we can determine α_i and hence the localisation length $\lambda = \alpha_{\min}^{-1}$.

2.3 Calculation of Transmission and Reflection Matrices

The transfer matrix may be used for the calculation of transmission and reflection coefficients where a sample is considered to be embedded between semi-infinite ordered wires [7,8]. To do so we start with an initial matrix $\mathbf{A_0}$ whose columns represent a complete set of N right-going waves. These are right-handed eigenvectors of the ordered transfer matrix (7). It is important to note that the transfer matrix is not Hermitian and that the corresponding left-handed eigenvectors take the form

$$\mathbf{A}^{\mathrm{L}}_0 = \begin{bmatrix} (\mathbf{A}^{\mathrm{t}}_0)^T & -(\mathbf{A}^{\mathrm{b}}_0)^T \end{bmatrix} \qquad \text{where} \qquad \mathbf{A}^{\mathrm{R}}_0 = \begin{bmatrix} \mathbf{A}^{\mathrm{t}}_0 \\ \mathbf{A}^{\mathrm{b}}_0 \end{bmatrix} . \tag{14}$$

The transmission coefficient may be calculated by noting that after operating with the transfer matrix on the initial vectors we can project out a set of incident right-going vectors on the left-hand side using the left-handed eigenvectors. Thus

$$\mathbf{t}^{-1}_{\boldsymbol{k},\boldsymbol{k}'} = \mathbf{A}^{\mathrm{L}}_0 \mathbf{T} \mathbf{A}^{\mathrm{R}}_0 \qquad \text{where} \qquad \mathbf{A}^{\mathrm{L}}_0 \mathbf{A}^{\mathrm{R}}_0 = \mathbf{I} , \tag{15}$$

where $\boldsymbol{k}$ labels the modes of the ordered leads. Unfortunately we cannot simply invert $\mathbf{t}^{-1}$ to obtain $\mathbf{t}$. As discussed above, Sect. 2.2, the contributions from the largest eigenvalues of $\mathbf{t}$ rapidly become negligible in $\mathbf{t}^{-1}$. To overcome this we note that the stabilisation procedure leaves us instead of $\mathbf{t}^{-1}$ with $\mathbf{B} = \mathbf{t}^{-1}\mathbf{U}$ and with $\mathbf{U}$ itself. Hence we can evaluate $\mathbf{t}$ from

$$\mathbf{t}^{-1} = \mathbf{B}\mathbf{U}^{-1} \quad \Rightarrow \quad \mathbf{t} = \mathbf{U}\mathbf{B}^{-1}\,, \tag{16}$$

without having to invert $\mathbf{t}^{-1}$ explicitly. The information about the largest eigenvalue is contained in $\mathbf{B}$ and $\mathbf{U}$.

Care is also required to distinguish between the amplitude and current transmission matrices. The above calculation results in the former whereas the latter are more commonly required. These are related by

$$\mathbf{t}^{\mathrm{c}}_{\boldsymbol{k}',\boldsymbol{k}} = \begin{cases} \mathbf{t}^{\mathrm{a}}_{\boldsymbol{k}',\boldsymbol{k}} v_{\mathrm{g}}(\boldsymbol{k}')/v_{\mathrm{g}}(\boldsymbol{k}) & \boldsymbol{k}' \,\&\, \boldsymbol{k} \text{ real} \\ 0 & \text{otherwise} \end{cases}\,, \tag{17}$$

where $v_{\mathrm{g}}(\boldsymbol{k})$ is the longitudinal group velocity of mode $\boldsymbol{k}$ of the leads.

The conductance of the system [8,6] is given by

$$G = \frac{e^2}{h} \mathrm{Tr}\, \mathbf{t}^{\mathrm{c}} \mathbf{t}^{\mathrm{c}\dagger}. \tag{18}$$

3 Generalisations of the Simple Transfer Matrix Method

There have been various attempts to generalise the method either to facilitate analytical solutions or to accelerate the subsequent analysis of the data.

3.1 Pendry's Generalisation

In a series of papers [9] Pendry discussed a generalisation of the transfer matrix by considering direct products of simple transfer matrices. Thus starting from

$$\boldsymbol{a}_{n+1} = \mathbf{T}_n \boldsymbol{a}_n, \tag{19}$$

it may be generalised to the form

$$[\boldsymbol{a}_{n+1} \otimes \boldsymbol{a}_{n+1}] = [\mathbf{T}_n \otimes \mathbf{T}_n]\, [\boldsymbol{a}_n \otimes \boldsymbol{a}_n]\,, \tag{20}$$

where $\otimes$ symbolises a direct or outer product. Thus the $2N \times 2N$ matrices become $(2N)^2 \times (2N)^2$, but it is possible to average over the disorder and obtain non-trivial results as ε^2 appears in the expression. In fact this can be generalised even further to arbitrary integer powers,

$$\left[\boldsymbol{a}_n^{\otimes m}\right] = \left[\mathbf{T}_n^{\otimes m}\right] \left[\boldsymbol{a}_n^{\otimes m}\right], \tag{21}$$

and, less rigorously, analytically continued to non positive integer powers. These generalised transfer matrices may be averaged and the considerable symmetry of the resulting matrix used to obtain analytical results.

3.2 Derivatives with Respect to Energy or Disorder

Often, such as when calculating critical behaviour it is useful to be able to focus in directly to the derivative with respect to energy or disorder, as this is the quantity which determines the critical exponents [10]. Starting from an expression such as (19) we can write

$$\left[\frac{\mathrm{d}}{\mathrm{d}\tau}\boldsymbol{a}_{n+1}\right] = \left[\frac{\mathrm{d}}{\mathrm{d}\tau}\mathbf{T}_n\right]\boldsymbol{a}_n + \mathbf{T}_n\left[\frac{\mathrm{d}}{\mathrm{d}\tau}\boldsymbol{a}_n\right], \tag{22}$$

where τ represents energy, disorder, or some other parameter. This may be rewritten in terms of a new transfer matrix of the form

$$\begin{bmatrix}\boldsymbol{a}_{n+1}\\ \boldsymbol{a}'_{n+1}\end{bmatrix} = \begin{bmatrix}\mathbf{T}_n & \mathbf{0}\\ \mathbf{T}'_n & \mathbf{T}_n\end{bmatrix}\begin{bmatrix}\boldsymbol{a}_n\\ \boldsymbol{a}'_n\end{bmatrix}, \tag{23}$$

where the primes refer to differentiation with respect to τ. This matrix has proven difficult to iterate for very long systems, but it has been possible to use it to study short systems embedded between ideal wires and to derive direct expressions for the critical exponent which do not require a fit over a range of disorder whose width may be uncertain.

4 The Green's Function Recursion Method

A less efficient, but potentially more powerful, approach is the Green's function recursion method. We start by defining the time-independent Green's function for the system as

$$(Z - \mathbf{H})\,\mathbf{G} = \mathbf{I}, \tag{24}$$

where $Z = E + \mathrm{i}\gamma$ is the complex energy. The Dyson equation

$$\mathbf{G} = \mathbf{G}_0 + \mathbf{G}_0 \varDelta\mathbf{H}\mathbf{G}, \tag{25}$$

provides a more useful starting point for the derivation of the method, where $\mathbf{G}_0$ corresponds to the unperturbed Hamiltonian and $\varDelta\mathbf{H}$ is a perturbation.

As with the simple transfer matrix (2) we start by considering a strip or bar (Fig. 1) of length N with an additional slice disconnected from the rest. The perturbation will consist of the off-diagonal matrix elements of $\mathbf{H}$ required to couple the extra slice to the bar. Using a notation where $\mathbf{G}_{ij}^{(N)}$ is the submatrix of $\mathbf{G}$ for a system of length N which couples the ith and jth slices. Using (25) we can write

$$\begin{aligned}
\mathbf{G}^{(N+1)}_{N+1,N+1} &= \mathbf{G}^{(N)}_{N+1,N+1} + \mathbf{G}^{(N)}_{N+1,N+1}\mathbf{V}_{N+1,N}\mathbf{G}^{(N+1)}_{N,N+1} \\
&= \mathbf{G}^{(N)}_{N+1,N+1} + \mathbf{G}^{(N)}_{N+1,N+1}\mathbf{V}_{N+1,N}\mathbf{G}^{(N)}_{NN}\mathbf{V}_{N,N+1}\mathbf{G}^{(N+1)}_{N+1,N+1} \\
&= \left[Z - \mathbf{H}_{N+1,N+1} - \mathbf{V}_{N+1,N}\mathbf{G}^{(N)}_{NN}\mathbf{V}_{N,N+1}\right]^{-1}, \qquad (26a) \\
\mathbf{G}^{(N+1)}_{i\le N,N+1} &= \mathbf{G}^{(N)}_{iN}\mathbf{V}_{N,N+1}\mathbf{G}^{(N+1)}_{N+1,N+1}, \qquad (26b) \\
\mathbf{G}^{(N+1)}_{N+1,j\le N} &= \mathbf{G}^{(N+1)}_{N+1,N+1}\mathbf{V}_{N+1,N}\mathbf{G}^{(N)}_{N,j}, \qquad (26c) \\
\mathbf{G}^{(N+1)}_{i\le N,j\le N} &= \mathbf{G}^{(N)}_{ij} + \mathbf{G}^{(N)}_{iN}\mathbf{V}_{N,N+1}\mathbf{G}^{(N+1)}_{N+1,j} \\
&= \mathbf{G}^{(N)}_{ij} + \mathbf{G}^{(N)}_{iN}\mathbf{V}_{N,N+1}\mathbf{G}^{(N+1)}_{N+1,N+1}\mathbf{V}_{N+1,N}\mathbf{G}^{(N)}_{Nj}\,. \qquad (26d)
\end{aligned}$$

Before proceeding we note that by inverting the matrices in (26b) we obtain

$$\begin{aligned}
\mathbf{B}_{N+1} &= \left[\mathbf{G}^{(N)}_{1N}\mathbf{V}_{N,N+1}\right]^{-1}, \\
\left[\mathbf{G}^{(N+1)}_{1,N+1}\right]^{-1} &= \left[\mathbf{G}^{(N+1)}_{N+1,N+1}\right]^{-1}\mathbf{B}_{N+1} \\
&= \left[Z - \mathbf{H}_{N+1,N+1} - \mathbf{V}_{N+1,N}\mathbf{G}^{(N)}_{NN}\mathbf{V}_{N,N+1}\right]\mathbf{B}_{N+1} \\
&= \left[Z - \mathbf{H}_{N+1,N+1}\right]\mathbf{B}_{N+1} - \mathbf{V}_{N+1,N}\mathbf{B}_N \\
&= \mathbf{V}_{N+1,N+2}\mathbf{B}_{N+2}\,, \qquad (27)
\end{aligned}$$

which should be compared with (7) to establish the relationship between the simple transfer matrix and Green's function recursion methods. It can also be used to relate the transmission matrix $\mathbf{t}$ with $\mathbf{G}_{1N}$. This is the form in which the method is most often applied. A particularly important case arises in situations when the transfer matrix fails: when the number of channels varies from slice to slice on the long strip or bar. In such cases the simple Green's function recursion method applying (26a) and (26b) may still be used successfully [2].

The method is much more powerful, however [12]. To illustrate this we consider a calculation of the density of states using

$$\rho(E) = -\frac{1}{\pi}\,\mathrm{Im}\,\mathrm{Tr}\,\mathbf{G}(E+\mathrm{i}\gamma) = -\frac{1}{\pi}\sum_i \mathrm{Tr}\,\mathrm{Im}\,\mathbf{G}_{ii}(E+\mathrm{i}\gamma)\,. \qquad (28)$$

Using (26d) we can write

$$\begin{aligned}
\sum_{i=1}^{N+1}\mathrm{Tr}\,\mathbf{G}^{(N+1)}_{ii} &= \sum_{i=1}^{N}\mathrm{Tr}\left(\mathbf{G}^{(N)}_{ii} + \mathbf{G}^{(N)}_{iN}\mathbf{V}_{N,N+1}\mathbf{G}^{(N+1)}_{N+1,N+1}\mathbf{V}_{N+1,N}\mathbf{G}^{(N)}_{Ni}\right) \\
&\quad + \mathrm{Tr}\,\mathbf{G}^{(N+1)}_{N+1,N+1}\,. \qquad (29)
\end{aligned}$$

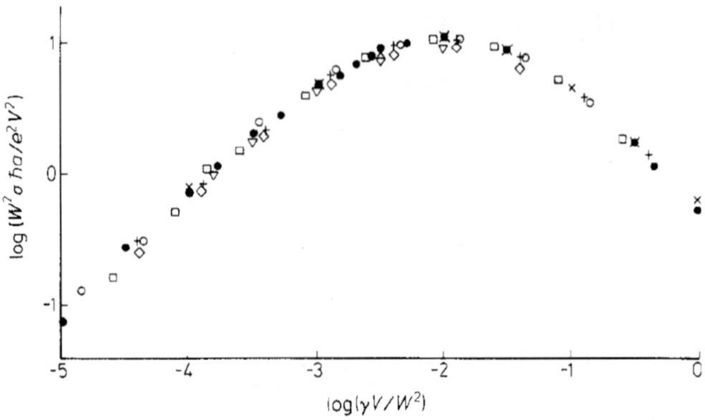

Fig. 2. $\log(W^2\sigma h a/e^2V^2)$ against $\log(\gamma V/W^2)$ for various values of disorder $W/V = 0.1, 0.5, 1.0, 15, 2.0, 5.0, 10.0$ in a 1D disordered system. The error bars are in all cases smaller than the sizes of the symbols [11]. Here W is the width of a box distribution for ε_i in the Anderson Hamiltonian (4a), σ the conductivity

By defining

$$s_\rho^N = \sum_{i=1}^{N} \mathrm{Tr}\, \mathbf{G}_{ii}^{(N)}, \tag{30a}$$

$$\mathbf{F}_N = \mathbf{V}_{N+1,N} \sum_{i=1}^{N} \mathbf{G}_{Ni}^{(N)} \mathbf{G}_{iN}^{(N)} \mathbf{V}_{N,N+1}, \tag{30b}$$

we can rewrite (29) as

$$s_\rho^{(N+1)} = s_\rho^N + \mathrm{Tr}\left[(\mathbf{F}_N + \mathbf{I})\, \mathbf{G}_{N+1,N+1}^{(N+1)}\right] . \tag{31}$$

Similarly, we can write a recursion relation for $\mathbf{F}_N$ by using (26b) and (26c):

$$\mathbf{F}_{N+1} = \mathbf{V}_{N+2,N+1} \mathbf{G}_{N+1,N+1}^{(N+1)} (\mathbf{F}_N + \mathbf{I})\, \mathbf{G}_{N+1,N+1}^{(N+1)} \mathbf{V}_{N+1,N+2} \, . \tag{32}$$

Hence, we can iterate (31) and (32) together with (26a) as often as required to calculate the density of states of a strip or bar of any length. The time required will be proportional to the length but also to the cube of the cross section.

It is important to note that this technique makes it possible, not only to calculate the Green's function itself, but also various quantities which may be expressed in terms of combinations of such Green's functions.

The normal and Hall conductivity [11,12] may be calculated similarly by starting from the Kubo–Greenwood formula [13,14]. Figure 2 shows an early example of the calculation of the conductivity in a 1D system. Note the role of the imaginary part of the energy, γ, in these results in which it may be interpreted as an inverse lifetime or inelastic scattering rate. In this case it was possible to calculate sufficiently long systems such that there were always around 10^4 states

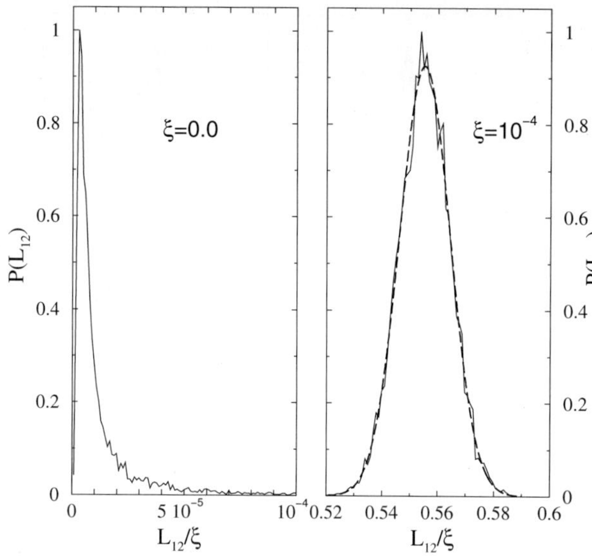

Fig. 3. Distribution function $P(L_{12})$ for zero and finite value of $\xi(\equiv \gamma)$ at $E = \mu = -1$, $W = 2$ and $T = 0$. Data is computed from 10000 samples and length $Na = 3 \times 10^6 a$. P is normalized to 1 at the maximum. The *dashed line* corresponds to a Gaussian (*right*) fit [16]

within the energy width γ. This resulted in an error of about 1% on each data point. By using the scaling relationship illustrated in the figure one can deduce that the conductivity is always zero in the limit of small γ, something which had previously defied numerical verification.

More recently the method has been extended to the calculation of thermoelectric quantities [15,16], including the kinetic coefficients L_{11}, L_{12}, L_{22}, by starting from the Chester–Thelling–Kubo–Greenwood formula [17]. From these it is possible to derive expressions for the thermoelectric power, the Peltier coefficient and the thermal conductivity. In fact it turns out that the dimensionless thermopower τ may be approximated by the real part of s_ρ as defined above (30a). Initial results are illustrated in Figs. 3 and 4. Again we see the crucial role played by the imaginary part of the energy γ (labelled ξ in the figures) in the physics. In fact the kinetic coefficient L_{12} has a prefactor of γ and is therefore zero in the $\gamma \to 0$ limit. In both figures we see that the distribution for small γ, when the inelastic scattering length is large, is Lorentzian in form, whereas it becomes Gaussian for larger γ.

5 Summary

In this paper I have discussed various transfer matrices together with some advantages and disadvantages of each. In Sect. 2 I have given a full discussion of the implementation of the method concentrating on some of the pitfalls commonly

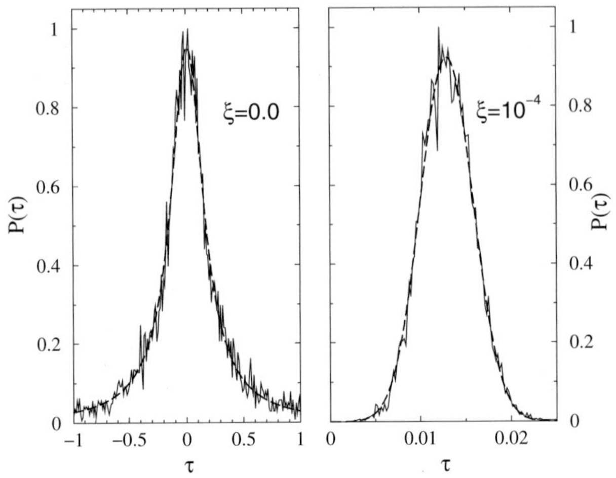

Fig. 4. Distribution function $P(\tau)$ for zero and finite value of $\xi(\equiv \gamma)$ and the same parameters as Fig. 3. The *dashed lines* correspond to a Lorentzian (*left*) and a Gaussian (*right*) fit [16]

encountered by beginners. The subsequent sections have discussed various generalisations more in outline than in detail. This should provide potential users with useful information and references which will enable them to make use of the methodology in their own problems.

References

1. P.W. Anderson: Phys. Rev. **109**, 1492 (1958)
2. J.P.G. Taylor and A. MacKinnon: J.Phys: Condensed Matter **1** 9963 (1989)
3. A. MacKinnon and B. Kramer: Phys. Rev. Lett., **47**, 1546 (1981)
4. A. MacKinnon and B. Kramer: Z. Phys. B **53**, 1 (1983)
5. J.L. Pichard and G. Sarma,G: J. Phys. C **14**, L127 (1981)
6. B. Kramer and A. MacKinnon: Reports on Progress in Physics **56** 1469 (1993)
7. R. Landauer: Phil. Mag. **21**, 863 (1970)
8. M. Büttiker, Y. Imry, R. Landauer and S. Pinhas: Phys. Rev. B **31**, 6207 (1985)
9. J.B. Pendry: J. Phys. C **17**, 5317 (1984)
10. A.P. Taylor and A. MacKinnon: J. Phys. Condensed Matter: **14** 8663 (2002)
11. A. MacKinnon: J. Phys. C **13**, L1031 (1980)
12. A. MacKinnon: Z. Phys. B **59**, 385 (1985)
13. R. Kubo, J. Phys. Soc. Japan **12**, 570 (1957)
14. D.A. Greenwood, Proc. Phys. Soc. **71**, 585 (1958)
15. C. Villagonzalo: Dissertation, Chemnitz University of Technology (2001) at `http://archiv.tu-chemnitz.de/pub/2001/0060`.
16. R.A. Römer, A. MacKinnon and C. Villagonzalo: J. Phys. Soc. Jap. (to be published)
17. G.V. Chester and A. Thellung, Proc. Phys. Soc. **77**, 1005 (1961)

Corrections to Single Parameter Scaling at the Anderson Transition

Tomi Ohtsuki[1] and Keith Slevin[2]

[1] Department of Physics, Sophia University, Kioi-cho 7-1, Chiyoda-ku, Tokyo 102-8554, Japan
[2] Department of Physics, Graduate School of Science, Osaka University, 1-1 Machikaneyama, Toyonaka, Osaka 560-0043, Japan

1 Introduction

The Anderson transition [1], which is a continuous quantum phase transition induced by disorder, can be described by the single parameter scaling [2,3]. One of the conclusion of the scaling theory is that the Anderson transition is universal, i.e., it does not depend on the details of the models but depends only on the basic symmetry of the system. The field theoretic approach made clear that the transitions are classified into three universality classes, namely, orthogonal, unitary and symplectic classes [4]. However, analytic approaches fail to estimate quantitatively the critical exponents. Numerical investigation, therefore, has been playing an important role to discuss quantitatively the critical behaviour of the Anderson transition [5–7].

Two main purposes of the numerical simulations are; i) to demonstrate the single parameter scaling, and ii) to evaluate quantitatively the critical behaviour. The first one is not at all trivial, since there are doubts that there might be a lot of relevant scaling variables [8,9]. Distinguishing one universality class from the other is categorised into the second purpose.

The critical exponents of different universality classes in three dimensional (3D) systems are close to each other. To estimate critical exponents precisely enough to distinguish one universality class from the others requires high precision data obtained by large scale numerical simulations [10]. Such high precision data reveal that there are deviations from the single parameter scaling, i.e., the corrections to scaling [11–14].

In this paper, we review recent developments of the numerical simulations for the Anderson transition. We will review how to treat the corrections to scaling [14], and show that there is only a relevant scaling variable and others are irrelevant. The boundary condition effects can also be treated within the framework of the corrections to scaling method [15]. We will also review a model for two dimensional systems with spin-orbit coupling, where the corrections to scaling is minimised [16].

2 Critical Behaviour in 3D Systems

The tight-binding Hamiltonian for 3D non-interacting electrons on a simple cubic lattice with nearest neighbour interactions is given by

$$\begin{aligned} < \boldsymbol{r}|H|\boldsymbol{r} > &= V(\boldsymbol{r}), \\ < \boldsymbol{r}|H|\boldsymbol{r} - \hat{x} > &= -1, \\ < \boldsymbol{r}|H|\boldsymbol{r} - \hat{y} > &= -1, \\ < \boldsymbol{r}|H|\boldsymbol{r} - \hat{z} > &= -1. \end{aligned} \tag{1}$$

Here $\hat{x}$, $\hat{y}$ and $\hat{z}$ are the lattice basis vectors. The potential V is independently and identically distributed with probability density $p(V)$. Typical probability densities are; i) the box distribution

$$\begin{aligned} p(V) &= 1/W \ \ |V| \leq W/2, \\ &= 0 \qquad \text{otherwise}, \end{aligned}$$

ii) the Gaussian distribution

$$p(V) = \frac{1}{\sqrt{2\pi\sigma^2}} \exp\left(-\frac{V^2}{2\sigma^2}\right),$$

with $\sigma^2 = W^2/12$, and iii) the Cauchy distribution

$$p(V) = \frac{W}{\pi \left(W^2 + V^2\right)}.$$

For the last distribution all moments higher than the mean are divergent and the parameter W is proportional to the full width at half maximum of the distribution. The Hamiltonian corresponding to this potential is called the Lloyd model.

For these three models we analyse the localisation length λ for electrons on a quasi-1D dimensional bar of cross section whose linear dimension is L. The length λ was determined to within a specified accuracy using a standard transfer matrix technique [17,5–7]. Then we define the renormalised quasi 1D localisation length Λ as

$$\Lambda = \frac{\lambda}{L}, \tag{2}$$

which we call here MacKinnon-Kramer parameter.

The MacKinnon-Kramer parameter Λ is a function of the Fermi energy E,the strength of disorder W, the size L and so on. The single parameter scaling means that these parameters change Λ only through single scaling variable and Λ reads

$$\Lambda = f(\chi L^{1/\nu}), \tag{3}$$

where χ is a function of physical parameters except L. The length scale (which is the localisation length ξ_- in the insulating regime and the correlation length ξ_+ in the metallic regime) $\xi_\pm$ diverges as

$$\xi_\pm \sim |\chi|^{-\nu}. \tag{4}$$

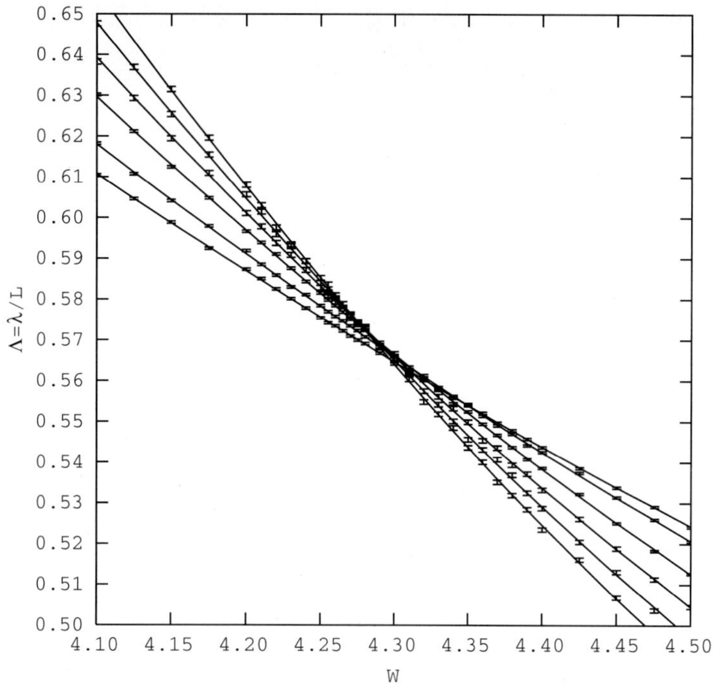

Fig. 1. Λ as a function of disorder for the three dimensional Lloyd model. From [14]

Generally, by tuning one of the arguments of χ we can set $\chi = 0$, i.e., reach the critical point.

From (3), we see that Λ does not depend on the system size L when $\chi = 0$. In Fig. 1, we plot Λ as a function of W. The accuracy of raw data is typically 0.1%. We see that the common crossing does not occur, but there is a systematic deviation of the crossing point as we change the system size.

This deviation is due to the corrections to scaling [18]. Λ is not only a function of single parameter $\chi_{\mathrm{R}} L^{1/\nu}$ but a function of $\chi_{\mathrm{I}_1} L^{y_1}, \chi_{\mathrm{I}_2} L^{y_2}, \cdots$. If these $\chi_{\mathrm{I}_i} L^{y_i}$'s are irrelevant ($y_i < 0$), i.e., with the increase of system size they decrease and vanish in the limit of infinite system size, the single parameter scaling is recovered. In Fig. 2, we show schematic view of the renormalisation flow for 2 scaling variables, one is relevant and the other is irrelevant. The origin of the figure is the critical point, which we do not know. When we change a physical parameter such as the strength of disorder, the initial values of scaling variable change, which are indicated by the solid and dashed curves. The former corresponds to the situation where the information close to the critical point is hard to obtain, while we can investigate the critical behaviour accurately in the latter situation (dashed curve).

Corrections to scaling behaviour are taken into account as follows [14]. Our starting point is the equation which expresses the MacKinnon-Kramer parameter

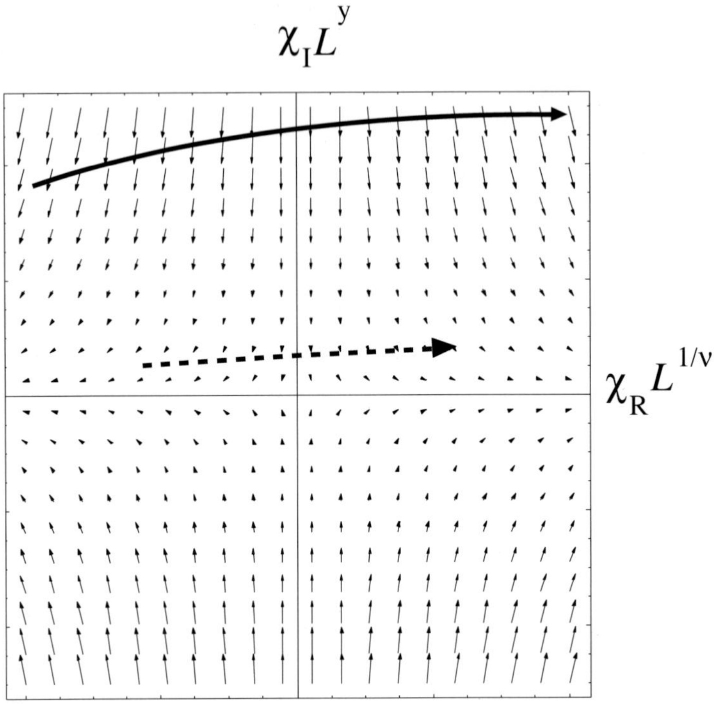

Fig. 2. Schematic view of the renormalisation flow of scaling variables $\chi_R L^{1/\nu}$ and $\chi_I L^y$ about the critical point. The solid line indicates the situation where we change the physical parameter far from the critical point, while the dashed line corresponds to the situation where the corrections to scaling happen to be small

Λ as a function of the scaling variables

$$\Lambda = F\left(\chi_R L^{1/\nu}, \chi_I L^y\right) . \tag{5}$$

For L finite there is no phase transition and F is a smooth function of its arguments. Assuming the irrelevant scaling variable is not dangerous, we make a Taylor expansion up to order n_I

$$\Lambda = \sum_{n=0}^{n_I} \chi_I^n L^{ny} F_n\left(\chi_R L^{1/\nu}\right) , \tag{6}$$

and obtain a series of functions F_n. Each F_n is then expanded as a Taylor series up to order n_R

$$F_n(\chi_R L^{1/\nu}) = \sum_{m=0}^{n_R} \chi_R^m L^{m/\nu} F_{nm} . \tag{7}$$

To take account of non-linearities in the scaling variables we expand both of them in terms of the dimensionless disorder $w = (W_c - W)/W_c$ where W_c is

Table 1. The best fit estimates of the critical disorder and the critical exponent and their 95% confidence intervals. The quantity $\Lambda_c = F_0(0)$ is expected to be universal

	W_c	Λ_c	ν	y
Box	16.54(52,56)	0.576(73,78)	1.57(55,60)	-2.8(3.4,2.3)
Gau.	21.29(27,32)	0.576(73,78)	1.58(55,61)	-3.9(5.9,2.6)
Lor.	4.27(25,28)	0.579(76,88)	1.58(46,65)	-2.5(3.2,1.2

the critical disorder separating the insulating ($w < 0$) and conducting phases ($w > 0$).

$$\chi_{\mathrm{R}} = \sum_{n=1}^{m_{\mathrm{R}}} b_n w^n \ , \ \chi_{\mathrm{I}} = \sum_{n=0}^{m_{\mathrm{I}}} c_n w^n . \tag{8}$$

The orders of the expansions are m_{R} and m_{I} respectively. Notice that $\chi(w = 0) = 0$. The absolute scale of the arguments in (5) are undefined so we fix them by setting $F_{01} = F_{10} = 1$ in (7). The total number of parameters introduced is then $N_p = (n_{\mathrm{I}} + 1)(n_{\mathrm{R}} + 1) + m_{\mathrm{R}} + m_{\mathrm{I}} + 2$.

The critical exponent ν, the irrelevant exponent y and the scaling function F_0 are expected to be universal. Though we have explicitly considered corrections due to the leading irrelevant scaling variable only, the analysis can be easily extended to two or more such variables.

To exhibit scaling it is necessary to subtract the corrections due to the irrelevant scaling variable. Setting $n_{\mathrm{I}} = 1$ we define

$$\Lambda_{\mathrm{corrected}} = \Lambda - \chi_{\mathrm{I}} L^y F_1 \left(\chi_{\mathrm{R}} L^{1/\nu}\right) , \tag{9}$$

with the obvious generalisation when $n_{\mathrm{I}} > 1$. We then have

$$\Lambda_{\mathrm{corrected}} = F_{\pm} \left(\frac{L}{\xi_{\pm}}\right) . \tag{10}$$

The functions $F_{\pm}$ are defined by $F_{\pm}(x) = F_0(\pm(\xi_{\pm} L)^{1/\nu})$. By subtracting the corrections, the single parameter scaling behaviour is recovered as shown in the scaling plot, Fig. 3. The best fit estimate of critical disorder W_c, critical exponent ν as well as $\Lambda_c = F_0(0)$ are summarised in Table 1.

3 Boundary Condition Effect

The above calculation has been done with the periodic boundary conditions (pbc) imposed on transverse directions to the quasi-1D bar. Another possibility is to impose fixed boundary conditions (fbc). We can also impose pbc in one direction, and the fixed boundary condition on the other direction that we call here mixed boundary condition (mbc). With these boundary conditions, Λ's as a function of W are shown in Fig. 4. No common crossing seems to exist in the case of mbc and fbc, and one might even get the wrong impression that the critical disorder depends on the boundary condition. After corrections to

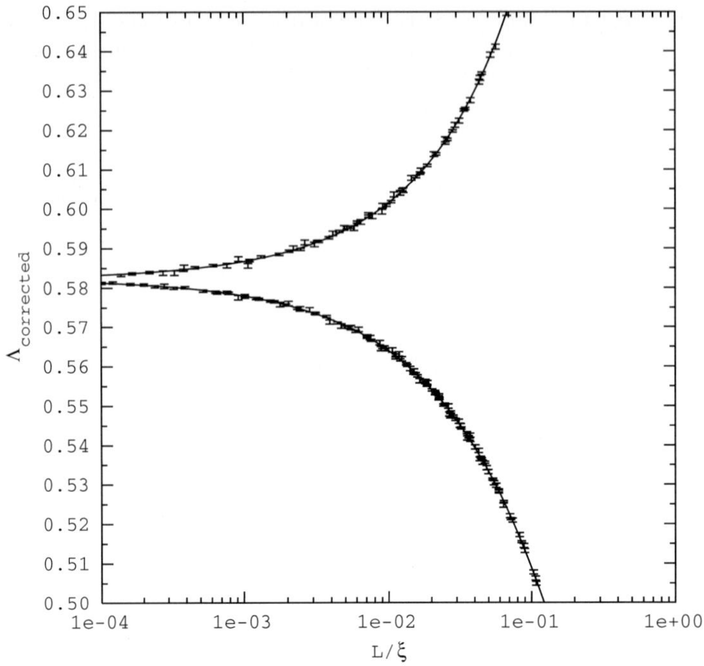

Fig. 3. $\Lambda_{\rm corrected}$ defined by (9) versus L/ξ. *Solid lines* are the scaling functions F_+ and F_- defined by (10). From [14]

Table 2. The best fit estimates of the critical exponent, the critical disorder and Λ_c together with their 95% confidence intervals as well as the two terminal conductance, its average and variance, extrapolated to the infinite system size

	ν	W_c	Λ_c	$< g(\infty) >$	$\mathrm{var}(g(\infty))$
pbc	1.56(55,58)	16.54(53,55)	0.576(.574,.577)	$0.89 \pm .02$	$0.472 \pm .01$
mbc	1.60(56,64)	16.47(42,52)	0.502(.494,.509)	$0.71 \pm .01$	$0.409 \pm .003$
fbc	1.54(41,61)	16.49(39,64)	0.427(.403,.442)	$0.560 \pm .002$	$0.349 \pm .001$

scaling is removed, however, the common crossing is recovered (Fig. 5), and the critical disorders as well as the critical exponents estimated for different boundary conditions become consistent.

What is unexpected is that the critical value of MacKinnon- Kramer parameter Λ_c strongly depends on the boundary condition. The boundary condition dependence of the critical quantity, such as the energy level statistics was noticed in [19]. What is to be stressed is that the conductance, which plays the central role in the scaling theory of localisation, does depend on the boundary condition [20,21]. Table 2 summarises the results of ν, W_c, Λ_c as well as the statistics of two terminal conductance defined by $2\mathrm{Tr}t^\dagger t$, 2 coming from electron spin and t is the transmission matrix.

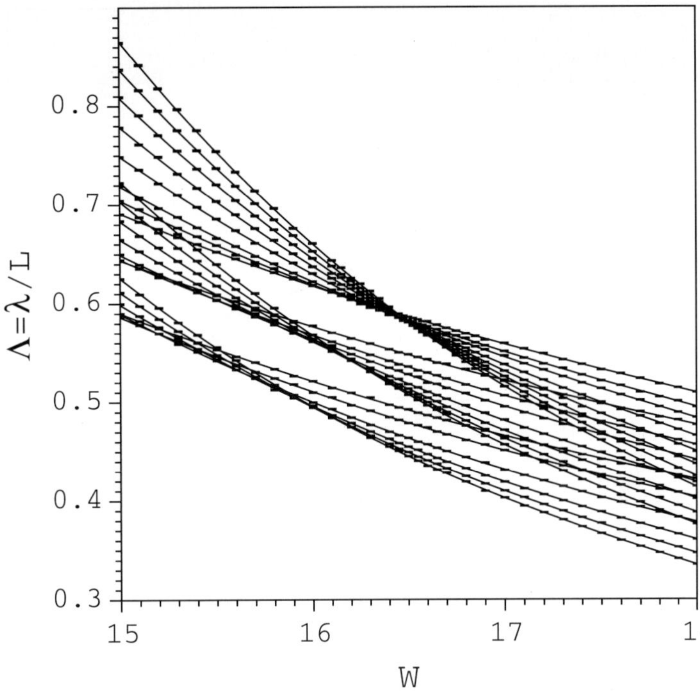

Fig. 4. Λ vs. W. for periodic (*upper curves*), mixed (*middle curves*) and fixed (*lower curves*) boundary conditions. From [15]

4 2D Symplectic Ensemble

Now we pay attention to the Anderson transition in two dimensional (2D) systems. According to the scaling theory of localisation, all the systems are driven to be insulator in 2D. Exceptions are quantum Hall systems (unitary class) and the systems with strong spin-orbit scattering (symplectic class).

One might think that 2D Anderson transition is more easily analysed numerically than 3D one. As was noted by Huckestein [11,12], in the case of quantum Hall effect, we encounter a large corrections to scaling which make the numerical analysis of the critical exponent very difficult. This is also true of the Ando model [22] which describes the system belonging to the symplectic class. Evangelou-Ziman model [23,24] and network model [25,26] also suffers from the corrections to scaling [22–29].

To estimate the critical exponent accurately, the smaller the corrections to scaling are the better. Here we propose the SU(2) model [16],

$$H = \sum_{i,\sigma} \epsilon_i c^{\dagger}_{i,\sigma} c_{i,\sigma} - V \sum_{\langle i,j\angle,\sigma,\sigma'} R(i,j)_{\sigma,\sigma'} c^{\dagger}_{i,\sigma} c_{j,\sigma'}, \tag{11}$$

where $R(i,j)$ incorporates the spin-orbit coupling between the nearest neighbours. This belongs to the group SU(2) of 2×2 unitary matrices and can be

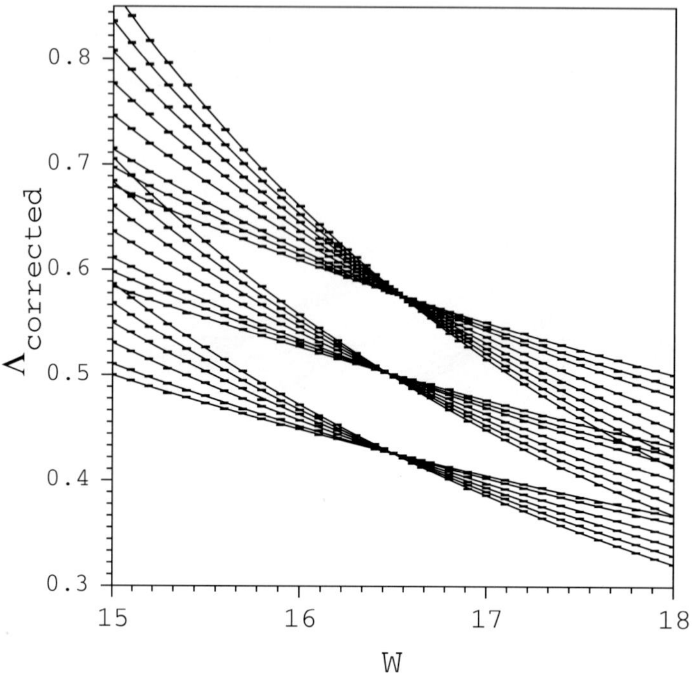

Fig. 5. Λ vs. W after the surface corrections are removed. From [15]

parametrised as

$$R(i,j) = \begin{pmatrix} e^{i\alpha_{i,j}} \cos \beta_{i,j} & e^{i\gamma_{i,j}} \sin \beta_{i,j} \\ -e^{-i\gamma_{i,j}} \sin \beta_{i,j} & e^{-i\alpha_{i,j}} \cos \beta_{i,j} \end{pmatrix} . \tag{12}$$

α and γ are distributed uniformly in the range $[0, 2\pi)$ and β according to the probability density,

$$P(\beta) = \begin{cases} \sin(2\beta) & 0 \leq \beta \leq \frac{\pi}{2} \\ 0 & \text{otherwise.} \end{cases} \tag{13}$$

This uniform distribution on SU(2) space guarantees that the spin relaxation time is shortest possible, and the corrections to scaling are expected to be small compared to the other models.

In Fig. 6 we plot $\ln \Lambda$ as a function of W for E=1. Even with the high accuracy of 0.1% an 0.05%, the raw data fit well to the fitting formula without corrections to scaling terms. From this fitting, the critical exponent is estimated to be $\nu = 2.73$ with 95% confidence interval [2.71, 2.75].

5 Summary

We have reviewed in this article how to handle the corrections to single parameter scaling that appear in MacKinnon-Kramer method. The appearance of

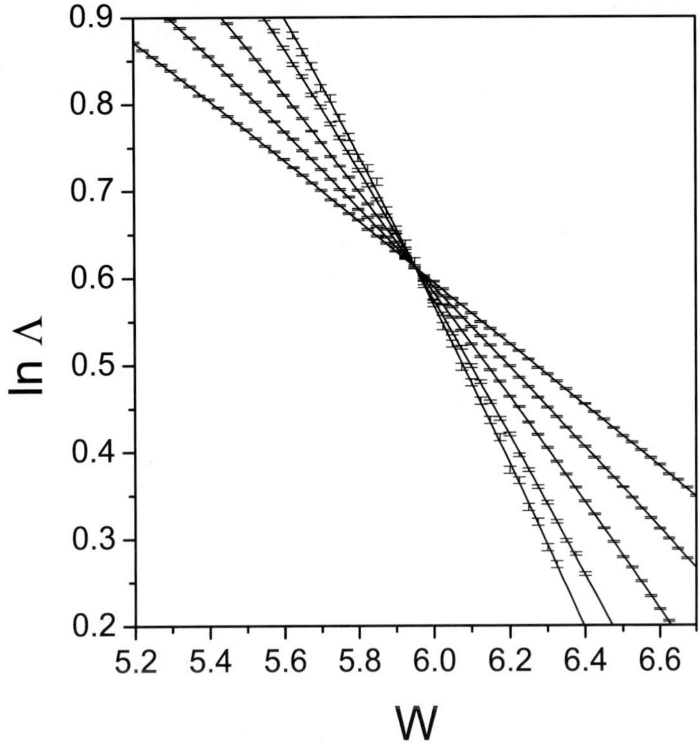

Fig. 6. Λ vs. W for two dimensional SU(2) model. From [16]

the corrections does not mean that MacKinnon-Kramer method is not efficient. It is so efficient that it detects the corrections while other numerical methods often fail to find them. Either by expanding the scaling function with respect to the corrections to scaling (Sects. 2 and 3) or by analysing a proper model which minimises the corrections (Sect. 4) we have shown that the critical exponent can be estimated up to 1% accuracy.

The origin of the corrections to scaling remains unclear. In Sect. 3, it was shown that the large corrections to scaling originate from the boundary condition. The exponent y is close to -1 for fbc and mbc, which suggests that the surface corrections exist. However, if we analyse the scaling of conductance, y is close to -1 for mbc and pbc, and somewhat smaller (≈ -2) for fbc [15].

In the case of symplectic ensemble, the Ando model gives larger corrections to scaling than SU(2) model, which implies that the dominant contribution comes from the spin relaxation length. If we adopt a model for QHE where the corrections to scaling is minimised, we will be able to analyse the QH critical behavior more precisely, which is a problem left in the future.

References

1. P. Anderson, Phys. Rev. **109**, 1492 (1958)
2. F. Wegner, Z. Phys. B **25**, 327 (1976)
3. E. Abrahams et al., Phys. Rev. Lett. **42**, 673 (1979)
4. S. Hikami, A. Larkin, Y. Nagaoka, Prog. Theor. Phys. **57** (1980)
5. A. MacKinnon and B. Kramer, Phys. Rev. Lett. **47**, 1546 (1981)
6. A. MacKinnon and B. Kramer, Z. Phys. B **53**, 1 (1983)
7. B. Kramer and A. MacKinnon, Rep. Prog. Phys. **56**, 1469 (1993)
8. V.E. Kravtsov, I.V. Lerner, V.I. Yudson, Sov-Phys. JETP **67**, 1441 (1988)
9. H. Mall and F. Wegner, Nucl. Phys. B **393**, 495 (1993)
10. K. Slevin and T. Ohtsuki, Phys. Rev. Lett. **78**, 4083 (1997)
11. B. Huckestein, Phys. Rev. Lett. **72**, 1080 (1994)
12. B. Huckestein, Rev. Mod. Phys. **67**, 357 (1995)
13. A. MacKinnon, J. Phys.: Condens. Matt. **6**, 2511 (1994)
14. K. Slevin, T. Ohtsuki, Phys. Rev. Lett. **82**, 669 (1999)
15. K. Slevin, T. Ohtsuki, T. Kawarabayashi, Phys. Rev. Lett. **84**, 3915 (1999)
16. Y. Asada, K. Slevin and T. Ohtsuki, Phys. Rev. Lett. **89**, 9256601 (2002)
17. J.-L. Pichard and G. Sarma, J. Phys. C **14**, L127 (1981)
18. Ch. 3, *Scaling and Renormalization in Statistical Physics*, J. Cardy (Cambridge University Press, 1996)
19. D. Braun, G. Montambaux and M. Pascaud, Phys. Rev. Lett. **81**, 1062 (1998)
20. C.M. Soukoulis et al., Phys. Rev. Lett. **82**, 668 (1999)
21. K. Slevin and T. Ohtsuki, Phys. Rev. Lett. **82**, 669 (1999)
22. T. Ando, Phys. Rev. B **40**, 5325 (1989)
23. S.N. Evangelou and T. Ziman, J. Phys. C **20**, L235 (1987)
24. S.N. Evangelou, Phys. Rev. Lett. **75**, 2550 (1995)
25. R. Merkt, M. Janssen and B. Huckestein, Phys. Rev. B **58**, 4394 (1998)
26. K. Minakuchi Phys. Rev. B **58**, 9627 (1998)
27. U. Fastenrath et al., Physica **172A**, 302 (1998)
28. L. Schweitzer and I.K. Zarekeshev, J. Phys. Condens. Matter **9**, L441 (1997)
29. K. Yakubo and M. Ono, Phys. Rev. B **58**, 9767 (1998)

On the Critical Exponent of the Anderson Transition

Arisato Kawabata

Department of Physics, Gakushuin University, 1-5-1 Mejiro, Toshima-ku, Tokyo 171-8588, Japan

1 Introduction

In the elementary course of quantum mechanics, we learn that there are two kinds of eigenstates, i.e., bound states and scattering states. In 1958 Anderson predicted the existence of *localised states* [1]. A localised state is different from a bound state, although its wave function vanishes at large distance, too.

For a bound state, the eigenenergy is smaller than the potential energy except in a restricted region and the Schrödinger equation has only exponentially growing or damping solutions at large distance. On the other hand, a localised state exists only in a potential with spatial randomness, and the envelope of its wave function damps exponentially at large distance although there are infinite number of regions where the eigenenergy is larger than the potential energy. Moreover, under certain conditions, an eigenstate can be localised even if the eigenenergy is larger than the potential energy in the whole space.

In fact in one and two-dimensional systems, all the eigenstates are localised irrespective of the value of the random potential (the theoretical proof is less rigorous for two-dimensions). On the other hand, in three-dimensional systems, there can be eigenstates of which the wave functions do not vanish at large distance. Such states in a random potential are called *extended states*. Generally, there exists such an energy E_C that a state is localised or extended according as the eigenenergy is smaller or larger than E_C: This energy is called *mobility edge*.

When the Fermi energy E_F is smaller than E_C, all the occupied states are localised at 0K and the system is an insulator. If we increase E_F, a transition from insulator to conductor take places at $E_F = E_C$ because the electrons in the extended states can carry the current: This transition is called *Anderson transition*.

Although the existence of localised states was predicted in 1958, there were no appropriate methods to describe them mathematically, until the renormalization group theory was introduced to this problem by Wegner [2] and Abrahams et al. [3].

The renormalization group method predict that for $E_F \gtrsim E_C$ the conductivity σ behaves like

$$\sigma = \sigma_0 (E_\mathrm{F} - E_\mathrm{C})^s \, , \tag{1}$$

and s is called *the critical exponent* of the Anderson transition.

The transition may be controlled by some other parameters, for example, the density n_d of the dopants in doped semiconductors. In this case, both E_F and E_C change if n_d changes. Nevertheless, provided that E_F and E_C are smooth functions of n_d, the behavior of the conductance is characterized by the same exponent s as

$$\sigma = \sigma_1 |n_\mathrm{d} - n_\mathrm{C}|^s , \tag{2}$$

where n_C is the critical value of n_d.

Moreover, it is widely accepted that s is universal and that it depends only on the fundamental symmetry of the system, for example, whether or not time reversal symmetry or/and spin-orbit coupling exists, as was pointed out by Hikami et al. [4].

Therefore, the critical exponent is one of the most fundamental quantities in Anderson localisation and it has been studied by many investigators. In spite of their efforts, a reliable value of the critical exponents has not yet been obtained by an analytic method. In the following we review the renormalization group method and the results obtained by it, and we will propose an improvement of it.

2 Renormalization Group Theory

In the renormalization group theory of Anderson localisation, we introduce the idea of the system-size-dependent conductivity $\sigma(L)$. In the elementary course of electricity, we learn that the conductivity σ is independent of the system size, while the resistance R depends on the shape and the size of the system, i.e., $R = 1/(\sigma L)$ if the system is a cube of $L \times L \times L$. This is correct, however, only for good conductors.

Supposed that $E_\mathrm{F} < E_\mathrm{C}$, and that a wave function at the Fermi level behaves like

$$|\psi(\boldsymbol{r})| \propto \mathrm{e}^{-|\boldsymbol{r}-\boldsymbol{r}_0|/\xi} . \tag{3}$$

Here ξ is called *localisation length.* At 0K, the current is carried by the electrons at the Fermi level, and if we put an electron into the system in the form of a wave packet composed of such wave functions, the electron can diffuse only over a distance $\sim \xi$. Therefore, the system behaves like a conductor or an insulator according to $L \ll \xi$ or $L \gg \xi$, and the conductivity is dependent on the system size.

In the renormalization group theory, the conductance (reciprocal of the resistance) rather than the conductivity is regarded as the fundamental quantity to characterized the electron conduction. The renormalization group theory of Anderson transition is based on the scaling assumption:

Let $G(L)$ be the conductance of a system of size L, i.e.,

$$G(L) = L\sigma(L) . \tag{4}$$

Then, $G(L + \delta L)$ with the same microscopic structure, e.g., the same electron density and the same impurity density, is determined only by $G(L)$ and $\delta L/L$, namely,

$$G(L + \delta L) = f(G(L), \delta L/L) \ , \tag{5}$$

in terms of a universal function $f(G, b)$. Although a mathematically rigorous proof of this assumption is not yet given, it is accepted by most people in this field and also is supported by numerical analyses [5–7]. From this equation we obtain

$$\frac{\mathrm{d}G(L)}{\mathrm{d}\log L} = \left.\frac{\partial f(G(L), b)}{\partial b}\right|_{b=0} \ . \tag{6}$$

It is to be noted that the right hand side of the above equation is dependent only on $G(L)$, and this type of equations are called renormalization group equations. Usually we define the dimensionless conductance $g(L)$,

$$g(L) \equiv \frac{2\pi^2 \hbar G(L)}{e^2} \ , \tag{7}$$

and write (6) in the form

$$\frac{\mathrm{d}\log g(L)}{\mathrm{d}\log L} = \beta(g(L)) \equiv \frac{1}{G(L)} \left.\frac{\partial f(G(L), b)}{\partial b}\right|_{b=0} \ . \tag{8}$$

Critical exponent s is obtained from the *beta-function* $\beta(g)$:

$$s = \frac{1}{g_c \beta'(g_c)} \ , \tag{9}$$

where g_c is the fixed point of $\beta(g)$, i.e., $\beta(g_c) = 0$ [3,8]. Thus the calculation of the critical exponent reduces to that of the beta-function.

In order to calculate the beta-function, we have to calculate the conductance for an arbitrary system size. On the other hand, the critical exponent is defined only in terms of the conductivity of an infinite system. Therefore one might think that the calculation of the critical exponent using the beta-function is more difficult than the direct calculation using (1). This is, however, not the case. The conductivity is dependent on the details of the system, while the beta-function is universal. Therefore, the characteristics of the system do not show up in the beta-function, and we can expect that the calculation of the beta-fuction is easier than it seems.

3 Calculation of the Beta-Function

3.1 $1/g$-Expansion

At present we know only one systematic method to calculate the beta-function, i.e., a series expansion in terms of $1/g$. Many years ago, Langer and Neal have

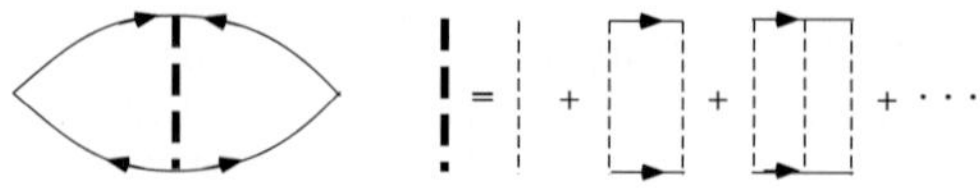

Fig. 1. Feynman graph for the Langer-Neal term: The *solid lines* and the *dotted lines* represent the electrons and impurities, respectively

pointed out that a contribution to the conductivity due to multiple scatterings of the electrons by the impurities is divergent for two-dimensional systems [9]. This contribution is represented by the Feynman graph in Fig. 1, and later it has been pointed out by Abrahams et al. that this term gives rise to the first order term in the $1/g$ expansion of the beta-function [3]. In fact this term represents the effects of the interference of the multiply scattered electronic waves toward the localisation [10,11].

3.2 Model System

We assume a system of an electron gas with randomly placed impurities, of which the potential is assumed to be of δ-function type for simplicity. Then the Hamiltonian is of the form:

$$H = \frac{\boldsymbol{p}^2}{2m} + \sum_i u\,\delta(\boldsymbol{r} - \boldsymbol{R}_i)\,, \tag{10}$$

where $\boldsymbol{R}_i$ is the position of the i-th impurity.

Here we remark that in the presence of the magnetic field or the spin-orbit coupling the system belongs to a different universality class and that the beta-function changes essentially [4].

Without Langer-Neal term, within the lowest order Born approximation, the conductivity is given by

$$\sigma = \sigma_0 \equiv \frac{n_{\rm e} e^2 \tau_0}{m}\,, \tag{11}$$

where $n_{\rm e}$ is the electron density and the mean scattering time τ_0 is given by

$$\frac{1}{\tau_0} \equiv \frac{2\pi}{\hbar} u^2 N_0\,, \tag{12}$$

N_0 being the density of states at the Fermi level. The conductivity can also be written in the form

$$\sigma_0 = 2e^2 N_0 D_0 \tag{13}$$

where $D_0 \equiv v_{\rm F}^2 \tau_0/3$ is the diffusion coefficient, $v_{\rm F}$ being the Fermi velocity.

The system-size-dependent conductivity with Langer-Neal term is given by [8]

$$\sigma(L) = \sigma_0 \left(1 - \frac{1}{\pi\hbar N_0} \int_{\alpha/L<q<q_{\rm c}} \frac{1}{D_0 q^2} \frac{{\rm d}^3\boldsymbol{q}}{(2\pi)^3}\right)\,. \tag{14}$$

Here we take into account the dependence on the system size as the lower cut off in the integral with $\alpha \approx 1$, and q_c is the upper cut off of order of the reciprocal of the mean free path of the electrons.

From eqs. (4), (7), and (14), we obtain

$$g(L) = L\frac{2\pi^2\sigma_0\hbar}{e^2}\left(1 - \frac{1}{\pi\hbar N_0}\int_{\alpha/L<q<q_c}\frac{1}{D_0 q^2}\frac{d^3\boldsymbol{q}}{(2\pi)^3}\right), \tag{15}$$

hence from (13)

$$g(L) = L\frac{2\pi^2\sigma_0\hbar}{e^2} - \frac{2Lq_c}{\pi} + \frac{2\alpha}{\pi}, \tag{16}$$

and from (8) we easily obtain

$$\beta(g) = 1 - \frac{2}{g}, \tag{17}$$

with $\alpha = \pi$ (the ambiguity in α does not affect the value of the critical exponent). The critical exponent is obtained from (9):

$$s = 1\,. \tag{18}$$

The higher order terms have been extensively investigated by Wegner [12] and Hikami [13]. Hikami calculated the beta-function up to fifth order:

$$\beta(g) = 1 - \frac{2}{g} - \frac{12\zeta(3)}{g^4} + \frac{27\zeta(4)}{2g^5}\,. \tag{19}$$

Using (9), we obtain $s = 0.67$ with $g_c = 2.53$ (Hikami obtained $s = 0.73$ using Borel-Padé analysis).

4 Numerical Analyses and Experiments

Along with the analytic methods, the numerical methods have been developed by MacKinnon and Kramer [5], and extensive studies based on their methods have been done by Ohtsuki, Slevin, and Kawarabayashi [6,7]. As for the details of the numerical methods the reader is referred to the article by T. Ohtsuki in this volume as well as the above references.

The numerical methods give

$$s = 1.54 \sim 1.58\,, \tag{20}$$

and those values seem very reliable.

On the other hand, the experimental value of the critical exponent are rather scattered, i.e., $s = 0.5 \sim 1.3$. It has been suspected that for doped semiconductors the exponent depends on the compensation. In fact, according to the recent analyses of the experimental data by Itoh et al. [14,15], it is almost settled that $s \approx 0.5$ without compensation and that $s \approx 1.2$ with compensation.

At present there is no theory to explain the effects of the compensation on the exponent. In fact, the compensation can not be incorporated in such a simple model hamiltonian as is given by (10), and in this article we will not be concerned about the experimental values of the exponent.

5 Self-Consistent Treatment

5.1 Beta-Function

In the preceding sections we have seen that the analytic methods do not reproduce the values of the exponent by the numerical methods. Since the results of the numerical methods are very reliable, the discrepancy is to be ascribed to the inadequacy of the approximation in the analytic method. In fact, the $1/g$ expansion method is equivalent to the ε expansion method, ε being the dimension of the system minus 2. Since $\varepsilon = 1$ for three dimension, we can not expect a good convergence of the series.

In this section we propose another method to calculate the beta-function based on self-consistent method. We have seen that (14), the lowest order interference correction to the conductance, gives the first order term of beta-function in $1/g$, and that it results in $s = 1$. Therefore, (14) is not a bad starting point to obtain a value of $s \approx 1.5 \sim 1.6$.

We introduce a q-dependence of the diffusion coefficient in the integral in (14). We write it in terms of the diffusion coefficient:

$$D(L) = D_0 \left(1 - \frac{1}{\pi \hbar N_0} \int_{\alpha/L<q<q_c} \frac{1}{\mathcal{D}(q) q^2} \frac{\mathrm{d}^3 \boldsymbol{q}}{(2\pi)^3} \right) . \tag{21}$$

where $D(L) \equiv \sigma(L)/(2e^2 N_0)$. Note that the two diffusion coefficients on the right hand side of this equation come from the different parts of the Feynman graph in Fig. 1 and that they have to be treated in different ways.

We assume a form

$$\mathcal{D}(q) = A q^\gamma , \tag{22}$$

for large q, A being a constant. Then, putting it to (21), we obtain

$$D(L) = D_0 - D_1 + C \left(\frac{\alpha}{L} \right)^{1-\gamma} , \tag{23}$$

where D_1 and C are some constants. Since $D(\infty) = D_0 - D_1$, within the present approximation the Anderson transition occurs for $D_0 = D_1$. Therefore, near the transition we can neglect $D_0 - D_1$ in (23).

Here we require a self-consistency relation:

$$D(L) = \mathcal{D}\left(\frac{\alpha}{L} \right) , \tag{24}$$

then, comparing (22) with (23), we obtain

$$\gamma = \frac{1}{2} \,. \tag{25}$$

Since we assumed (22) for large q, we assume an interpolation formula

$$\mathcal{D}(q) = D_0(1 + c\sqrt{\lambda_0 q}\,) \,, \tag{26}$$

where

$$\lambda_0 \equiv \frac{1}{2\pi\hbar N_0 D_0} \,, \tag{27}$$

and c is a constant to be determined later.

We put it into (21), and we obtain

$$D(L) = D_0 \left[1 + \frac{2}{\pi^2 c} \left\{ \sqrt{\frac{\alpha\lambda_0}{L}} - \sqrt{\lambda_0\, q_c} + \frac{1}{c} \log \frac{c\sqrt{\lambda_0\, q_c} + 1}{c\sqrt{\alpha\lambda_0/L} + 1} \right\} \right] \,. \tag{28}$$

Here we require the self-consistent relation (24). Although this requirement is not fulfilled for all the regions of L, the most important L-dependence of the right hand side of of this equation comes from the first term in the curly bracket. Then, comparing it with the second term of the right hand side of (26), we obtain

$$c = \frac{\sqrt{2}}{\pi} \,. \tag{29}$$

We define the dimensionless conductance with

$$g(L) \equiv \frac{bL}{\lambda(L)} \,, \tag{30}$$

where

$$\lambda(L) \equiv \frac{1}{2\pi\hbar N_0 D(L)} \,, \tag{31}$$

and b is a constant of the order of unity. Then (28) becomes

$$\begin{aligned} g(L) = b\frac{L}{\lambda_0}\left(1 - \frac{\sqrt{2}}{\pi}\sqrt{\lambda_0 q_c}\right) + b\frac{\sqrt{2}}{\pi}\sqrt{\alpha\frac{L}{\lambda_0}} \\ + b\frac{L}{\lambda_0}\log\frac{\sqrt{\lambda_0 q_c} + \pi/\sqrt{2}}{\sqrt{\alpha\lambda_0/L} + \pi/\sqrt{2}} \,, \end{aligned} \tag{32}$$

and the beta-function is obtained from (8):

$$\beta(g) = \frac{1}{g(L)}\left[g(L) - \eta\sqrt{b\frac{L}{\lambda_0}} + \frac{bL}{\lambda_0}\frac{1}{2 + \eta^{-1}\sqrt{bL/\lambda_0}} \right] \,, \tag{33}$$

with

$$\eta \equiv \frac{\sqrt{b\alpha}}{\sqrt{2\pi}} \,. \tag{34}$$

According to the scaling assumption, the right hand side of (33) must be a function only of $g(L)$, but it is not the case due to the inadequacy of the approximation. Here we replace λ_0 with $\lambda(L)$ defined by (31). Then, putting $\eta = 1$, we obtain a simple expression for the beta-function:

$$\beta(g) = 1 - \frac{1}{\sqrt{g}} + \frac{1}{2+\sqrt{g}} \,. \tag{35}$$

5.2 Critical Exponent

From the beta-function of (35), we can calculate the critical exponent easily using (9):

$$g_c = 4 - 2\sqrt{3} \,, \tag{36}$$

$$s = 1 + \frac{1}{\sqrt{3}} = 1.58 \,. \tag{37}$$

We find that the value of the critical exponent is very close to those of the numerical calculations i.e., $s = 1.54 \sim 1.58$. Here it should be remarked that such a good agreement does not necessarily mean that the approximations which have been adopted in deriving the beta-function are very good. We have intended only to obtain a value $s \gtrsim 1$, and very good agreement should be regarded somewhat accidental.

5.3 Properties of Beta-Function

It is clear that the beta-function obtained above can not be expanded in terms of $1/g$, and for large g it behaves like

$$\beta(g) = 1 - \frac{2}{g} + \frac{4}{g\sqrt{g}} - \cdots \tag{38}$$

Although it is incompatible with the idea of $1/g$-expansion, at present there is not a rigorous proof that the beta-function can be expanded in terms of $1/g$.

As for the behavior for small g, it is widely accepted that it should behave like $\beta(g) \approx \log g$ [3]. The beta-fucntion obtained above clearly does not show this behavior. It is not, however, a necessary condition to obtain a critical exponent with a good accuracy, for it is determined by the behavior of the beta-function at $g \approx 1$.

6 Phenomenological Theory

As we have seen in the previous section, it is rather difficult to derive analytically a beta-function which behaves reasonably in the whole regions of g. In the following we review a phenomenological theory to give such a beta-function by Janssen [16], based on the arguments by Shapiro [17].

Firstly, we assume that the conductance of the system of size L is characterized by a transmission coefficient $T(L)$ as in one-dimension. Suppose the system is b times elongated along the current. Then, we have a scaling relation:

$$T(bL) = [T(L)]^b . \tag{39}$$

Next we assume that the dimensionless conductance is given by the Landauer formula [18]

$$g(L) = \frac{T(L)}{1 - T(L)} . \tag{40}$$

Then, the conductance of the system of size bL (expanded to all the direction) is given by

$$g(bL) = \frac{b^2}{\{1/g(L) + 1\}^b - 1} , \tag{41}$$

where b^2 in the numerator of the right hand side is due to the increase in the cross section perpendicular to the current. Taking the derivative of the logarithm of both sides with respect to b and putting $b = 1$, we obtain

$$\beta(g) = 2 - (1 + g) \log \left(1 + \frac{1}{g}\right) . \tag{42}$$

We easily find that it behaves reasonably in both regions $g \gg 1$ and $g \ll 1$:

$$\beta(g) = \begin{cases} 1 - 1/(2g) , & (g \gg 1) \\ \log g , & (g \ll 1) \end{cases} . \tag{43}$$

The difference of the numerical factor of $1/g$ term from that in (17) is due to the different definition of g. The critical exponents is calculated numerically:

$$s = 1.65 , \tag{44}$$

and it is not a very bad value.

As we have seen, this theory is based on assumptions hard to justify from a microscopic point of view, and it is not clear how to incorporate the effects of magnetic field or spin-orbit coupling, which alter the fundamental behavior of beta-function [4]. Nevertheless, there are some things to be learned in this theory.

After looking at this theory, one might think that any beta-functions which behave properly in both limits give reasonable values for the critical exponent, because their behaviors at $g \approx 1$ can not be very much different from each other (see Fig. 2). The situation is, however, not as simple as that. For example, Vollhardt and Wölfle have derived such a beta-function, but it still gives $s = 1$ [19].

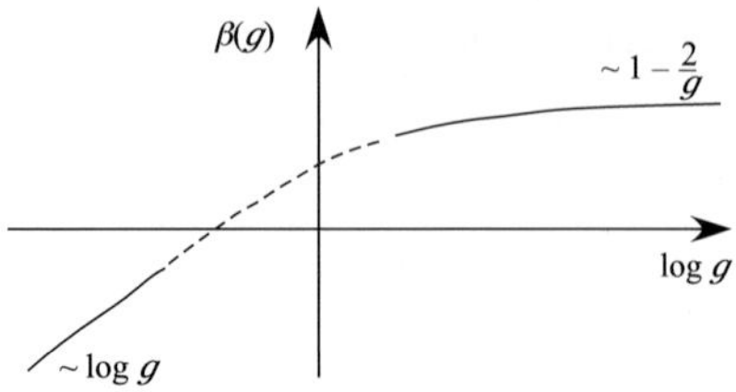

Fig. 2. The expected behavior of beta-function in the regions $g \gg 1$ and $g \ll 1$

7 Summaries

We have proposed a self-consistent method to calculate the beta-function of the Anderson transition. We introduced the wave number dependence of the diffusion coefficient in the first order interference correction term. The critical exponent obtained from the beta-function is in very good agreement with those of numerical analyses. Such a good agreement is beyond the expectation, and should be regarded somewhat accidental. The author believes, however, he is proceeding to a right direction.

As for the experiments, it is almost established that the critical exponent is dependent on the compensation for doped semiconductors. So far no idea has been proposed to incorporate the effects of the compensation, and we need investigations on more realistic models.

Acknowledgments

The author is grateful to Prof. T. Ohtsuki and Prof. Y. Ono for their valuable discussions. This work is partly supported by the "High Technology Research Center Project" of Ministry of Education, Culture, Sports, Sciences and Technology.

References

1. P.W. Anderson: Phys. Rev. **109**, 1492 (1958)
2. F.J. Wegner: Z. Phys. B **25**, 327 (1976)
3. E. Abrahams, P.W. Anderson, D.C. Licciardello, and T.V. Ramakrishnan: Phys. Rev. Lett. **42**, 673 (1979)
4. S. Hikami, A.I. Larkin and Y. Nagaoka: Prog. Theor. Phys. **63**, 707 (1980)
5. A. MacKinnon and B. Kramer: Z. Phys. B **53**, 1 (1983)
6. T. Ohtsuki, K. Slevin, and T. Kawarabayashi: *Proc. Int. Conf. Localisation 1999, ed. M. Schreiber, 1999 Hamburg*, Ann. Phys. (Leipzig) **8**, 655 (1999)
7. K. Slevin, P. MarKoš, and T. Ohtsuki: Phys. Rev. Lett. **86**, 3597 (2001)
8. A. Kawabata: Prog. Theor. Phys. Suppl. No. 84, 16 (1985)
9. J.S. Langer and T.Neal: Phys. Rev. Lett. **16**, 984 (1966)
10. G. Bergmann: Phys. Rep. **107**, 1 (1984)
11. Y. Nagaoka: Prog. Theor. Phys. Suppl. No. 84, 1 (1985)
12. F.J. Wegner: Nucl. Phys. B **316**, 663 (1989)

13. S. Hikami: *Proc. Low Dimensional Field Theories and Condensed Matter Physics, 4th Yukawa International Seminar, 1991 Kyoto*, Prog. Theor. Phys. Suppl. No. 107, 213 (1992)
14. K.M. Itoh, M. Watanabe, Y. Ohtsuka, and E.E. Haller: *Proc. Int. Conf. Localisation 1999, ed. M. Schreiber, 1999 Hamburg*, Ann. Phys. (Leipzig) **8**, 631 (1999)
15. K.M. Itoh, M. Watanabe, Y. Ohtsuka, and E.E. Haller: *Proc. Int. Conf. Localisation 2002, ed. S. Komiyama, 2002 Tokyo*, J. Phys. Soc. Jpn. Suppl. Vol. 72 (2003), in press
16. M. Janssen: Phys. Rep. **295**, 1 (1998)
17. B. Shapiro: Phil. Mag. B **56**, 1031 (1987)
18. R. Landauer: Z. Phys. B, Condens. Matter **68**, 217 (1987)
19. D. Vollhardt and P. Woelfle: Phys. Rev. Lett. **45**, 842 (1980)

Conductance Statistics near the Anderson Transition

Peter Markoš

Institute of Physics, Slovak Academy of Sciences, Dubravska cesta 9, 842 28 Bratislava, Slovakia

1 Introduction

Since the pioneering work of Anderson [1] we know that disorder localizes electrons. At zero temperature any small amount of disorder localizes all electrons in one-dimensional systems [2]. In higher dimensions ($d > 2$ for systems with orthogonal symmetry) weak disorder does not destroy the metallic regime. Only when the strength of the disorder increases beyond a certain specific value (critical disorder), the electron becomes localized. This phenomenon – transition from the metallic to the insulating regime due to an increase of disorder – is called Anderson transition [3].

Scaling theory of the Anderson transition uses the conductance g [4] as the order parameter. It is supposed [5–7] that the system size dependence of the conductance is determined only by the value of the conductance itself:

$$\frac{\partial \ln g}{\partial \ln L} = \beta(g) \tag{1}$$

where $\beta(g)$ is an analytical function of g. β is positive (negative) in the metallic (localized) regimes, respectively. For dimension $d > 2$ the function $\beta(g)$ changes its sign, being positive for $g \gg 1$ and negative in the limit $\ln g \to -\infty$. There is an unstable fixed point g_c defined as the solution of $\beta(g = g_c) = 0$. The system-size independent critical conductance g_c represents the critical point of the Anderson transition.

Relation (1) contains no information about the *microscopic* structure of the model. This means that the Anderson transition is universal. The form of the β function is determined only by the physical symmetry and dimension of the system [8].

Soon after the formulation of the scaling theory of localization it became clear that the conductance g is not a self-averaged quantity. Reproducible fluctuations of the conductance were found both in the metallic and in the insulating regimes [9,10]. The knowledge of the mean value $\langle g\rangle$ is therefore not sufficient for a complete description of the transport properties. One has to deal with the conductance distribution $P(g)$ [11,12] or, equivalently, with all cumulants of the conductance. This is easier in the metallic regime, where $P(g)$ is Gaussian and the conductance fluctuations are universal [13–21] and independent on the value of the mean conductance and/or the system size. The width of the distribution depends only on the dimension, the physical symmetry of the system [13] and on the boundary conditions [22,23]. In the insulator, the conductance wildly fluctuates within the ensemble of macroscopically equivalent ensembles. It is the logarithm of the conductance which is distributed normally in the limit of large system size [11,12,17,24].

At present we have no complete analytical theory able to describe the conductance statistics at the critical point. Analytical results are known only for systems of dimension $d = 2 + \epsilon$ ($\epsilon \ll 1$) [12,25,26]. The expression for conductance cumulants [25] enabled one to estimate the shape of the critical conductance distribution $P_c(g)$ [26]. $P_c(g)$ is system size independent and Gaussian in the neighbourhood of the mean value $\langle g \rangle \sim \epsilon^{-1}$. The distribution possesses a power law tail $P_c(g) \sim g^{-1-2/\epsilon}$ for $g \to \infty$, and the delta-function peak $\delta(g)$. These results can not be applied to three dimensional (3D) systems ($\epsilon = 1$) where $\langle g \rangle \sim 1$ [27]. Numerical simulations are therefore crucial in 3D systems. The first systematic numerical analysis of the conductance statistics in 3D was done in [27] and was followed by a series of papers [28–34]. In 2D systems, the critical conductance distribution was numerically studied in the regime of the quantum Hall effect [35] and in systems with the spin-orbit interaction [28,36,37].

This paper reviews our recent numerical data for the conductance distribution. We address the question of the shape, universality and the scaling of the critical conductance distribution.

In the numerical calculation of the conductance we suppose that two opposite sites of the sample are connected to semi-infinite perfect leads and use the multichannel Landauer formula [38] which relates the conductance g (in units $2e^2/\hbar$) to the transmission matrix t:

$$g = \mathrm{Tr}\ t^\dagger t = \sum_{i=1}^{N_{\mathrm{open}}} \cosh^{-2} \frac{z_i}{2}. \tag{2}$$

In (2) we introduced the variables z_i, $i = 1, 2 \dots N_{\mathrm{open}}$, ($z_1 < z_2 < \dots$) which parametrize the eigenvalues of the matrix $t^\dagger t$. In the limit $L_z \gg L$, z_i converges to $2L_z/\lambda_i$ where λ_i is the ith localization length of the quasi-one dimensional (quasi-1D) system [17,27]. N_{open} is the number of open channels.

Owing to relation (2), the analysis of the conductance can be reduced to the calculation of the eigenvalues of the matrix $t^\dagger t$. The general formula for t was derived in [39,40].

According to the scaling theory, critical exponents of the Anderson transition as well as conductance statistics do not depend on the microscopic details of the model. In numerical simulations, we consider the Anderson Hamiltonian

$$\mathcal{H} = W \sum_n \varepsilon_n c_n^\dagger c_n + \sum_{[nn']} \tau_{[nn']} c_n^\dagger c_{n'}. \tag{3}$$

In (9) n counts the sites of the d-dimensional lattice and $[nn']$ means nearest neighbor sites. The hopping term τ equals to unity for orthogonal systems, unless the anisotropy is considered. When spin orbit scattering is considered, τ is a

2×2 matrix [18,42]. The parameter W measures the strength of the disorder. For a box distribution of random energies ε_n ($P(\varepsilon) = 1$ for $|\varepsilon| \le 1/2$ and $P(\varepsilon) = 0$ otherwise), the 3D Anderson model (9) exhibits an Anderson transition at $W_c = 16.5$.

2 Finite Size Scaling

As it was discussed in Sect. 1, the conductance g is not a self-averaged quantity. To avoid statistical fluctuations, quasi-1D systems were introduced [43] and the localization length γ is calculated instead of the conductance. In the limit $L_z/L \to \infty$ (L_z and L are the length and the width of the system, respectively) γ is a self-averaged quantity which converges to its mean value. Finite size scaling is then used [44] for the analysis of the disorder and the system width dependence of γ. It is assumed that the variable $\Lambda = \gamma/L$ is a function of only one parameter: $\Lambda(L,W) = \Lambda(L/\xi(W))$. Here $\xi = \xi(W)$ is the correlation length which diverges in the vicinity of the critical point $\xi(W) = |W - W_c|^{-\nu}$. The critical exponents ν and $s = (d-2)\nu$ characterize the critical behavior of the localization length and of the conductance, respectively [5].

Finite size scaling analysis of the quasi-1D data enabled to test the universality of the Anderson transition [45] and provided us with the more accurate estimation of the critical exponent $\nu \approx 1.57$ [46].

2.1 Scaling of the Mean Conductance

Verification of the scaling theory of the localization requires the proof of the universal scaling of the mean conductance and of the entire conductance distribution in the critical regime. Single parameter scaling of various mean values, $\langle g\rangle$, $\exp\langle \ln g\rangle$, and $1/\langle g^{-1}\rangle$, was proved numerically for the 3D Anderson model [47]. Statistical ensembles of more than $N_{\rm stat} \gtrsim 10^6$ cubes of the size from 4^3 to 18^3 were collected for various values of the disorder W. Figure 1 shows typical data for the system size dependence of the mean conductance. In agreement with (1), $\langle g\rangle$ increases (decreases) with the system size in the metallic (localized) regime. By the use of the general fitting procedure [46], the critical disorder and the critical exponent $\nu \approx 1.57$ were obtained.

Data in Fig. 1 confirm that the variance, $\mathrm{var}\, g = \langle g^2\rangle - \langle g\rangle^2$ is an unambiguous function of the mean $\langle g\rangle$ in the critical regime. This supports, but still does not prove the single parameter scaling theory. A general proof of the theory requires verification of the single parameter scaling of all conductance cummulants. This is numerically impossible since higher cummulants are fully determined by rare events with very large values of the conductance.

2.2 Scaling of the Conductance Distribution

As higher cummulants are not treatable numerically, we test the scaling of the conductance distribution by the analysis of the scaling of percentiles g_α [34]. The

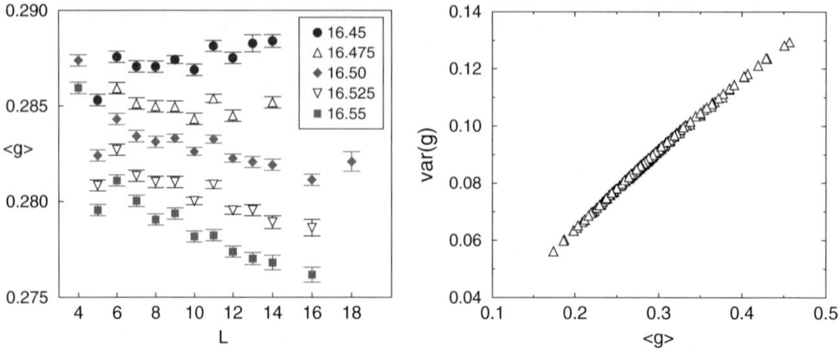

Fig. 1. *Left:* The L-dependence of the mean conductance $\langle g\rangle$ for different values of the disorder in the critical region. Data show that $16.45 < W_c < 16.55$. More exact estimation of the critical disorder was done in [47]. *Right:* The unambiguous dependence var g vs $\langle g\rangle$ in the critical region of the metal-insulator transition. This agrees with the single parameter scaling theory

percentile g_α is defined as

$$\alpha = \int_0^{g_\alpha} P(g)\mathrm{d}g. \tag{4}$$

Owing to (4), the probability to find $g < g_\alpha$ equals to α. Of course, the percentile g_α is a function of disorder and system size: $g_\alpha = g_\alpha(L, W)$. Single parameter scaling of percentiles has been proved for several values of α [34].

Suppose that g_α and g_β $(\alpha < \beta)$ obey the single parameter scaling. Then g_γ $(\alpha < \gamma < \beta)$ scales, too. Therefore, in contrast to the analysis of the conductance cumulants, it is enough to analyze only a few percentiles. Next, if g_α and g_β scale, then the difference $g_\beta - g_\alpha$ scales. Scaling of percentiles assures thus the scaling of the entire conductance distribution. Of course, this analysis is not applicable to the limit $\alpha \to 1$, because available statistical ensembles are never big enough to provide us with sufficient information about the tail of the distribution.

3 Critical Conductance Distribution

At present, we have no analytical description of the critical conductance distribution in 3D systems. Analytical results were obtained only for the conductance cumulants in the dimension $d = 2+\epsilon$ close to the lower critical dimension $(\epsilon \ll 1)$ [25]. In spite of the non-universality of higher order conductance cumulants

$$\langle \delta g^n\rangle = \begin{cases} \epsilon^{n-2} & n < n_0 = \epsilon^{-1} \\ \sim L^{\epsilon n^2 - n} & n > \epsilon^{-1} \end{cases} \tag{5}$$

the critical distribution $P_c(g)$ was shown to be universal and L-independent in the limit $L \to \infty$ [26]. However, theoretical analysis of the form of the critical distribution, is applicable only in the limit of very small ϵ [27].

3.1 The Form of the Critical Conductance Distribution

All what we know about the $P_c(g)$ in 3D is based on numerical data. In Fig. 2 we present $P_c(g)$ for the 3D Anderson model. Data confirm that the critical conductance distribution is system size independent, as required [12]. The shape of $P_c(g)$ differs considerably from the conductance distribution in the metallic and in the insulating regimes. To explain the typical properties of the critical conductance distribution, we use our knowledge about statistical properties of parameters z, as defined in (2).

The decrease of $P_c(g)$ to zero when $g \to 0$ is visible when very large statistical ensembles are studied (left inset of Fig. 2). Due to (2), small conductance means that z_1 is large. From numerical data we know that the distribution $P(z_1)$ is similar to a Wigner surmise and decreases as $\exp(-z_1^2)$ for large z_1. Consequently, $\ln P_c(\ln g)$ decreases as $-\ln^2 g$ (right figure in Fig. 2), and guarantees that $\lim_{g\to 0} P_c(g) = 0$ [27,30].

Large g behavior of $P_c(g)$ is determined by the chance that many parameters z_i are small. Statistical analysis of parameters z_i showed that for $i > 1$ the distribution $P(z_i)$ is Gaussian with mean value $\langle z_i\rangle \propto \sqrt{i}$ and var $z_i \propto \langle z_i\rangle^{-1}$ [30]. The probability to find a sample with small value of higher z_i is therefore very small: $P_c(g \approx i)$ decreases as $\exp -i^{3/2}$ [30]. Figure 2 indeed shows very fast decrease of the probability $P_c(g)$ for $g > 1$. The chance to have $g > 1$ is only 3%. Probability to find large values of g drastically decreases: we found that in the ensemble of 10^7 samples ($L = 10$) only 470 samples have $g > 2$ and only one sample has $g > 3$ [33]. The analysis of the contribution of the first two channels

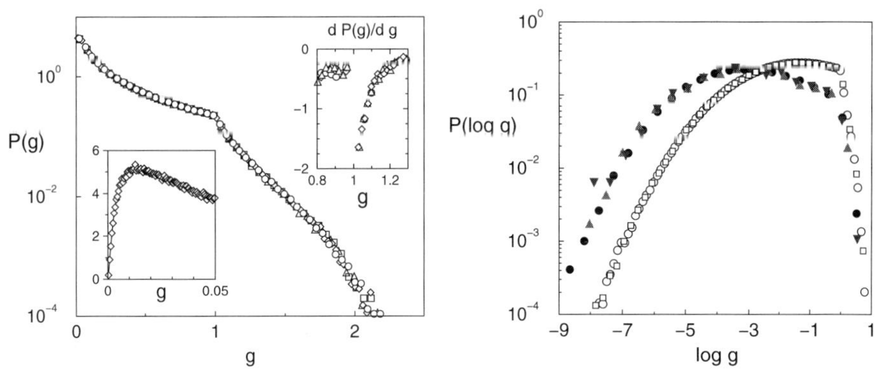

Fig. 2. *Left:* Critical conductance distribution $P_c(g)$ of the 3D Anderson model, obtained for statistical ensembles of $N_{\rm stat} = 10^6$ samples of the size 10^3–18^3. $P_c(g)$ is system size independent and decreases faster than exponentially when $g > 1$. The main properties of the critical distribution are shown in insets: *Left inset* shows in details the small-g behavior and proves that the distribution decreases to zero as $g \to 0^+$. *Right inset* shows the discontinuity of the derivative $\mathrm{d}P(g)/\mathrm{d}g$ at $g = 1$. *Right* figure presents the distribution $P_c(\ln g)$ at the critical point for the three dimensional ($W_c = 16.5$, *open symbols*) and four dimensional (4D) ($W_c = 34.3$, *full symbols*) systems of various system size

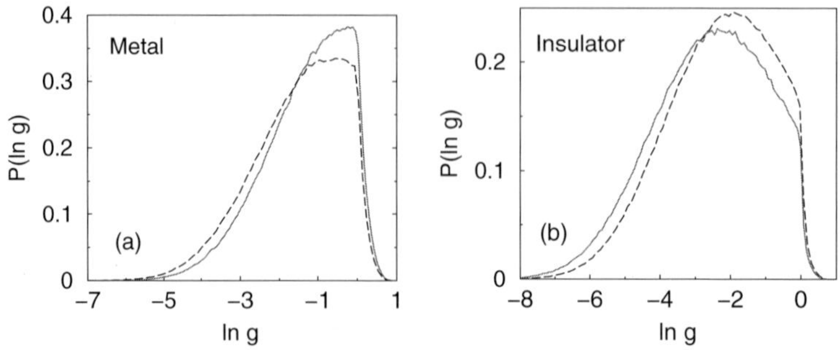

Fig. 3. Conductance distribution for ensembles of cubes 6^3 (*dashed lines*) and 12^3 (*solid line*): **a** The metallic regime $W = 15.4$ (the critical disorder is $W_c = 16.5$). Distribution $P(g)$ moves toward higher values of the conductance as the system size increases and $P(g)$ becomes Gaussian when $L \to \infty$. **b** Insulator $W = 17.6$; mean conductance decreases as the system size increases and the distribution of the $\ln g$ becomes Gaussian in the limit $L \to \infty$

is therefore sufficient for the understanding of the qualitative properties of the critical distribution.

Numerical data also show that the critical distribution is non-analytical at $g = 1$. The right inset of Fig. 2 shows the discontinuity of the first derivative $\mathrm{d}P_c(g)/\mathrm{d}g$. The same non-analyticity was found in 4D systems (right Fig. 2), in the unitary [29] and symplectic (Fig. 4) systems, and also in the weakly disordered quasi-1D systems [48,49].

The present description of the critical distribution is based on the analysis of statistical properties of parameters z. It is applicable to any system, for which the mean values of parameters z are of the order of unity. Then only a few (two, or three) channels contribute to the conductance. This analysis is, however, not applicable to systems close to the lower critical dimension $d = 2 + \epsilon$. Here, the mean conductance $\langle g\rangle \sim \epsilon^{-1}$, which means that the number of channels which contribute to the conductance, is large, $\sim \epsilon^{-1}$. It is therefore no surprise that the critical conductance distribution found in [26] differs from that shown in Fig. 2.

3.2 Conductance Distribution in the Critical Regime

Figure 3 shows that the typical properties of the critical conductance distribution hold also for $P(g)$ in the neighbourhood of the critical point. This is because the statistical properties of the parameters z depend continuously on the disorder in the critical regime. Critical properties of $P(g)$ survive until the system size L exceeds the correlation length ξ. Only when $\xi \ll L$ the distribution typical for the metallic or the insulating regime can be observed.

3.3 Universality

Single parameter scaling theory of localization supposes that the critical conductance distribution is universal. Its form does not depend on the microscopic

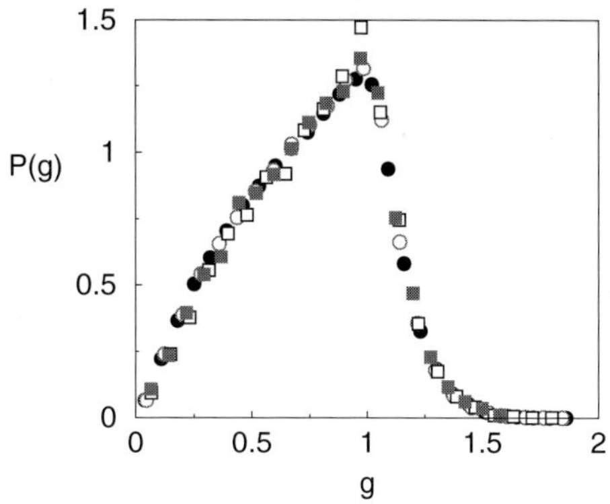

Fig. 4. Comparison of the critical conductance distribution of two different 2D models with spin-orbit coupling: *full symbols* Ando model [42] *open symbols* Evangelou-Ziman model [18]. Squares 40×40 and 80×80 were simulated to prove the system size independence of the critical distribution. Note also the non-analytic behavior for $g = 1$

details of the model. As an example, we present in Fig. 4 the critical conductance distribution for two 2D models with spin-orbit scattering: Evangelou-Ziman model [18] and Ando model [42]. In spite of the different microscopic definition of both models, $P_c(g)$ is universal [28]. Universality of $P_c(g)$ with respect to various distributions of random energies ε_n was confirmed in [28] and [32].

As was shown already in Fig. 2, the shape of $P_c(g)$ depends on the dimension of the model. Ref. [29] confirmed that also the physical symmetry influences the form of $P_c(g)$. Less expectable was the observation [22,31] that $P_c(g)$ as well as the spectral statistics [50] depend on the *boundary conditions* in the transversal direction. Nevertheless, this is consistent with the original definition of the conductance as a measure of the sensitivity of the energy spectrum of the system to the change of the boundary conditions [4].

For completeness, let us note that the critical conductance distribution depends also on other parameters of the model: lattice topology [51], anisotropy [37] and, of course, on the length of the system. We believe that these non-universalities could be compensated by the change of another model parameter (see, for instance [37], where the anisotropy is compensated by the length of the system).

3.4 Dimensional Dependence

The right figure in Fig. 2 compares the critical conductance distribution for 3D and 4D cubes. As supposed, the maximum of $P_c(g)$ for 4D is shifted toward smaller conductances, because the critical disorder increases as the spatial dimension increases and higher disorder means lower mean conductance [5,26]. Qualitatively, however, both distributions are very similar: $P_c(\ln g)$ decreases as $\exp[-\ln^2 g]$ for $\ln g \to -\infty$ and possesses the non-analyticity at $\ln g = 0$. This similarity is not surprising, because the form of the distribution is determined

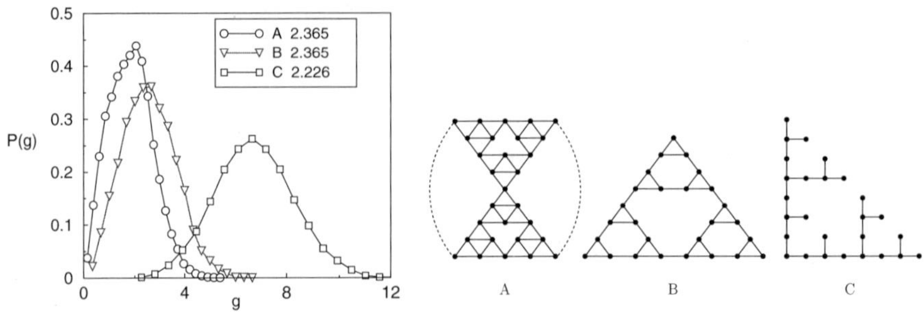

Fig. 5. Critical conductance distribution for bifractal lattices. Legend gives spectral dimension. 5th generation of fractals was used to calculate the distribution. Bifractal is linear along the z axis and fractal in a plane perpendicular to z. *Right* figure shows the structure of the fractal (the 3rd generation) cross section. All fractals have the same fractal dimension $d_{\rm f} = \ln 3/\ln 2$. Spectral dimension $d_{\rm s} = \ln 9/\ln 5$ for fractals A and B, and $= \ln 9/\ln 6$ for fractal C. Note that although bifractals A and B have both the same fractal and spectral dimensions, they posses different critical conductance distributions. $P_c(g)$ depends on the lattice topology [51]

mostly by the statistics of z_1 and z_2, which are qualitatively similar in 3D and 4D [30].

Surprisingly, the relation (5) seems to hold also for $\epsilon = 1$ and 2, at least for the first two cumulants. We obtained numerically that $\langle g\rangle = 0.285$ for 3D and 0.135 for 4D, so that $\langle g\rangle_{\rm 3D} \approx 2\langle g\rangle_{\rm 4D}$. For the second cumulants we found $\mathrm{var}\, g_{\rm 3D} \approx \mathrm{var} g_{\rm 4D} \approx 0.17$ [51].

More interesting is the investigation of the $P_c(g)$ in systems of dimension $2+\epsilon$ [51]. As we are not able to create d-dimensional hyper-cubes with non-integer d in computers, we simulated the transport on bifractal latices [52,51]. Bifractals are linear along the propagation direction and possess the fractal lattice in the cross section (Fig. 5). We proved that the critical exponent ν depends only on the *spectral* dimension of the lattice. Mean conductance, var g and the critical distribution $P_c(g)$ depend, however, on the lattice topology. For instance, Fig. 5 shows that bifractals A and B have the same spectral dimension, but different critical distribution. This is the reason why the obtained data can not be used for the verification of relations (5).

In Fig. 5 we present $P_c(g)$ for three different bifractals. As expected, $\langle g\rangle$ increases and the distribution converges to Gaussian when $\epsilon \to 0$. However, we found neither the δ-function peak at $g = 0$ nor the power-law tail of the distribution for $g \gg \langle g\rangle$, predicted by the theory [26].

4 Conductance Distribution in Non-critical Regime

Although we have no analytical theory of the conductance statistics in the critical regime, we can learn some typical properties of the conductance distribution from the analysis of the quasi-1D weakly disordered systems [48]. Starting from the

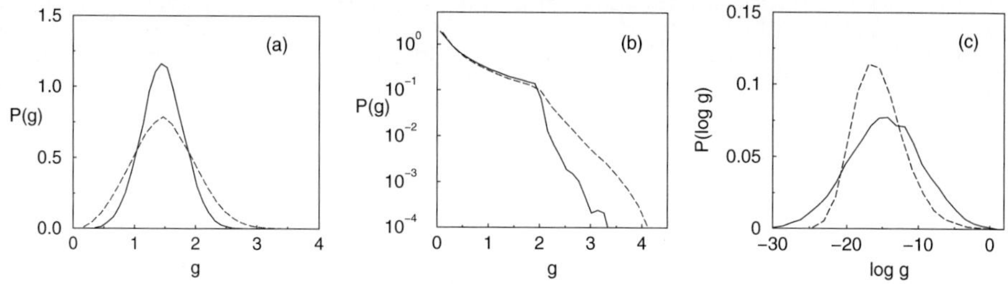

Fig. 6. Comparison of the conductance distribution of the 3D (*dashed lines*) and quasi-1D (*solid line*). **a** Metallic ($\langle g\rangle > 1$) **b** critical ($\langle g\rangle \approx 1$) and **c** localized ($\langle g\rangle \ll 1$) regimes are shown. In the quasi-1D systems, the strength of the disorder is fixed to $W = 4$ and the length of the system is tuned to obtain the same mean conductance as in the 3D system

Dorokhov-Mello-Pereyra-Kumar equation [15] for the probability distribution of parameters z, the conductance distribution $P(g)$ can be calculated. $P(g)$ depends on the length of the system. For short samples, $P(g)$ is Gaussian [17] (see also Fig. 6a). When the length of the system increases, the conductance decreases and the system passes from the metallic regime into the localized one. For the intermediate system length, where $\langle g\rangle \approx 1$, $P(g)$ is expected to be *qualitatively* similar to the critical conductance distribution [48]. This is shown in Fig. 6, where we compare the conductance distribution of quasi-1D and 3D systems of the same mean conductance.

Figure 6 also shows that the analogy between 3D and quasi-1D systems is not exact. To understand *quantitative* differences between the quasi-1D and 3D systems, we analyzed the spectrum of parameters z. In the metallic regime, the difference between 3D and quasi-1D systems is only in the value of var g [13]. $P(g)$ is Gaussian and the spectrum of parameters z is linear $\langle z_i\rangle \propto i$ [14,17] independently of the dimension of the system [54]. In contrast to the metallic regime, the spectrum of z becomes dimension dependent in the critical region. Both quasi-1D [53,54] and 3D-dimensional [49] numerical studies confirmed that at the critical point

$$\langle z_i\rangle^{d-1} \propto i, \tag{6}$$

($d > 2$). Owing to (6), the difference $\Delta = \langle z_2 - z_1\rangle$ is smaller that $\langle z_1\rangle$ in 3D, while it equals to $\langle z_1\rangle$ in the quasi-1D system. The contribution of the second channel is therefore more important in 3D than in the quasi-1D. This explains the longer tail of the distribution in the 3D system (Fig. 6b).

It is commonly believed that the distribution $P(\ln g)$ is Gaussian in the insulating phase, independently of the dimension of the system. This is,however, not true. The spectrum of z depends namely on the dimension of the system also in the localized regime. For the 3D systems it was proved numerically [49,54] that the difference $\Delta = \langle z_2 - z_1\rangle$ is *constant*, independent of the disorder and on the system size. Therefore the second channel influences always the statistics

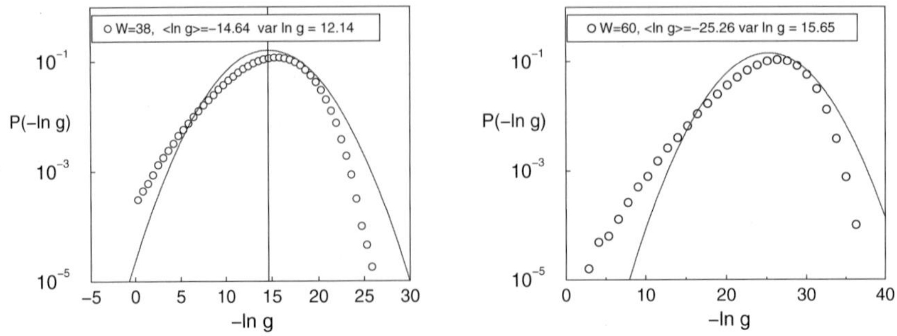

Fig. 7. Distribution of $-\ln g$ for strongly localized regime. Note that the distribution is not symmetric. It possesses the long tail for small values of the argument, and decreases much faster on the opposite side. *Solid lines* are a Gaussian distribution with the mean and variance found from numerical data. Size of the cubes is $L = 10$. *Left*: $W = 38$. Note that the sharp decrease of the distribution at $\ln g = 0$, discussed in Sect. 3.1, is still present [48]. *Right*: $W = 60$

of the conductance. Beyond its contribution to the value of the conductance it is important that a constant value of Δ prevents the distribution $P(z_1)$ to develop into the Gaussian form. While the values $z_1 \ll \langle z_1 \rangle$ are still possible, the probability to find systems with much higher values $z_1 \gg \langle z_1 \rangle$ is strongly suppressed. The distribution $P(z_1)$ is therefore not symmetric. The same is true for the distribution $P(-\ln g)$ (Fig. 7) which possesses a long tail for small values of $|\ln g|$ and decreases much faster than Gaussian for $|\ln g| \to \infty$.

Note that in weakly disordered quasi-1D systems $\Delta \sim \langle z_1 \rangle$. The distance $\langle z_2 - z_1 \rangle$ is much larger than the width of the distribution $P(z_1)$. Higher channels therefore do not influence the distribution of z_1 and $P \ln g$ is Gaussian.

5 Conclusion

We reviewed recent progress in numerical studies of the statistics of the conductance in the critical regime. Numerical analysis confirms that the conductance distribution in the 3D Anderson model obeys single parameter scaling. Analysis of the statistics of the eigenvalues of the transmission matrix enables us to understand the main features of the conductance distribution in the critical regime. Although we still have no analytical description of the conductance statistics in the critical regime, we hope that results of numerical experiments will inspire theoreticians to formulate the general analytical theory of the Anderson transition.

References

1. P.W. Anderson: Phys. Rev. **109**, 1492 (1958)
2. N.F. Mott, W. D. Twose: Adv. Phys. **10**, 107 (1961)
3. N.F. Mott, E.A. Davis: Electron processes in non-crystalline materials, Clarendon Press, Oxford 1979
4. J.T. Edwards, D.J. Thouless: J. Phys. C **5**, 807 (1972); D. C. Licciardello, D. J. Thouless: J. Phys. C **8**, 4157 (1975)
5. F. Wegner: Z. Phys. B **25**, 327 (1976)
6. E. Abrahams, P.W. Anderson, D.C. Licciardello, T.V. Ramakrishnan: Phys. Rev. Lett. **42**, 673 (1979)
7. L P. Gorkov, A.I. Larkin, D.E. Khmelnitskii: JETP Lett. **30**, 228 (1979)
8. B. Kramer, A. MacKinnon: Rep. Prog. Phys **56**, 1469 (1993)
9. S. Washburn, R.A. Webb: Adv. Phys. **35**, 375 (1986)
10. A.B. Fowler, J.J. Wainer, R.A. Webb: IBM J. Res. Dev. **32**, 372 (1988)
11. P.W. Anderson, D.J. Thouless, E. Abrahams, D.S. Fisher: Phys. Rev. B **22**, 3519 (1980)
12. B. Shapiro: Philos. Mag. B **56**, 1032 (1987); Phys. Rev. Lett. **65** 1510 (1990)
13. P.A. Lee, A.D. Stone: Phys. Rev. Lett. **55**, 1622 (1985); P.A. Lee, A.D. Stone, H. Fukuyama: Phys. Rev. B **35**, 1039 (1987)
14. Y. Imry: Europhys. Lett. **1**, 249 (1986)
15. O.N. Dorokhov: JETP Lett. **36**, 318 (1982); P.A. Mello, P. Pereyra, N. Kumar: Ann. Phys. (NY) **181**, 290 (1988)
16. C.W.J. Beenakker: Rev. Mod. Phys. **69**, 731 (1997)
17. J.-L. Pichard: 'Random transfer matrix theory and conductance fluctuations'. In: *Quantum Coherence in Mesoscopic Systems* NATO ASI Ser. B: Physics vol. **254** ed. by B. Kramer (Plenum Press, 1991)
18. P.M. Bell, A. MacKinnon: J. Phys.. Condens. Matt. **6**, 5423 (1994)
19. N. Giordano: Phys. Rev. B **38**, 4746 (1988)
20. J.-L. Pichard, N. Zannon, I. Imry, A.D. Stone: J. Phys. France **51**, 587 (1990)
21. Y. Avishai, J.-L. Pichard, K.A. Muttalib: J. Phys. France **3**, 1387 (1991)
22. D. Braun, E. Hofstetter, G. Montambaux, A. MacKinnon: Phys. Rev. B **64**, 155107 (2001)
23. M. Rühländer, P. Markoš, C.M. Soukoulis: Phys. Rev. B **64**, 172202 (2001)
24. P. Markoš, B. Kramer: Ann. Physik (Leipzig) **2**, 339 (1993)
25. B.L. Altshuler, V.E. Kravtsov, I.V. Lerner: Sov. Phys. JETP **64**, 1352 (1986); JETP Lett. **43**, 441 (1986)
26. A. Cohen, B. Shapiro: Int. J. Mod. Phys. B **6**, 1243 (1992)
27. P. Markoš, B. Kramer: Philos. Mag. B **68**, 357 (1993)
28. P. Markoš: Europhys. Lett. **26**, 431 (1994)
29. K. Slevin, T. Ohtsuki: Phys. Rev. Lett. **78**, 4083 (1997);
30. P. Markoš: Phys. Rev. Lett. **83**, 588 (1999)
31. K. Slevin, T. Ohtsuki, T. Kawarabayashi: Phys. Rev. Lett. **84**, 3915 (2000)
32. K. Slevin, T. Ohtsuki: Phys. Rev. B **63**, 45108 (2001)
33. M. Rühländer, P. Markoš, C. M. Soukoulis: Phys. Rev. B **64**, 212202 (2001)
34. K. Slevin, P. Markoš, T. Ohtsuki: cond-mat/0208157
35. Z. Wang, B. Jovanovic, D.H. Lee: Phys. Rev. Lett. **77**, 4426 (1996); V. Plerou, Z. Wang: Phys. Rev. B **58**, 1967 (1998); X. Wang, Q. Li, C.M. Soukoulis: Phys. Rev. B **58**, R3576 (1998)
36. P. Markoš: J. Phys. I France **4**, 551 (1994)

37. M. Rühländer, P. Markoš, C.M. Soukoulis: Phys. Rev. B **64**, 193103 (2001)
38. R. Landauer: IBM J. Res. Dev. **1**, 223 (1957); R. Landauer: Phil. Mag. **21**, 683 (1970); E.N. Economou, C. M. Soukoulis: Phys. Rev. Lett. **46**, 618 (1981)
39. T. Ando: Phys. Rev. B **44**, 8017 (1991)
40. J.B. Pendry, A. MacKinnon, P.J. Roberts: Proc. P. Soc. London A **437**, 67 (1992)
41. S.N. Evangelou, T. Ziman: J. Phys. C **20**, L235 (1987)
42. T. Ando: Phys. Rev. B **40**, 5325 (1989)
43. J.-L. Pichard, G. Sarma: J. Phys. C **14**, L127 (1981); *ibid*, L167
44. A. MacKinnon, B. Kramer: Phys. Rev. Lett. **47**, 1546 (1981)
45. B. Bulka, B. Kramer, A. MacKinnon: Z. Phys. B: Condens. Matt. **60**, 13 (1985); B. Kramer, K. Broderix, A. MacKinnon, M. Schreiber: Physica A **167**, 163 (1990); A. MacKinnon: J. Phys.: Condens. Matt **6**, 2511 (1994)
46. K. Slevin, T. Ohtsuki: Phys. Rev. Lett. **82**, 382 (1999)
47. K. Slevin, P. Markoš, T. Ohtsuki: Phys. Rev. Lett. **86**, 3594 (2001)
48. K.A. Muttalib, P.Wölfle: Phys. Rev. Lett. **83**, 3013 (1999); V.A. Gopar, K.A. Muttalib, P. Wölfle: cond-mat/0205643; K.A. Muttalib, P. Wölfle, A. García-Martín, V.A. Gopar: cond-mat/0210293
49. P. Markoš: Phys. Rev. B **65**, 104207 (2002)
50. D. Braun, G. Montambaux, M. Pascaud: Phys. Rev. Let. **81**, 1062 (1998)
51. I. Travěnec, P. Markoš: Phys. Rev. B **65**, 113109 (2002)
52. M. Schreiber, H. Grußbach: Phys. Rev. Lett. **76**, 1687 (1996)
53. P. Markoš, M. Henneke: J. Phys.: Condens. Matt. **6**, L765 (1994)
54. P. Markoš: J. Phys.: Condens. Matt. **7**, 8361 (1995)

Frequency Dependent Electrical Transport in the Integer Quantum Hall Effect

Ludwig Schweitzer

Physikalisch-Technischen Bundesanstalt, Bundesallee 100,
D-38116 Braunschweig, Germany

1 Introduction

It is well established to view the integer quantum Hall effect (QHE) as a sequence of quantum phase transitions associated with critical points that separate energy regions of localised states where the Hall-conductivity σ_{xy} is quantised in integer units of e^2/h (see, e.g., [1,2]). Simultaneously, the longitudinal conductivity σ_{xx} becomes unmeasurably small in the limit of vanishing temperature and zero frequency. To check the inherent consequences of this theoretical picture, various experiments have been devised to investigate those properties that should occur near the critical energies E_n assigned to the critical points. For example, due to the divergence of the localisation length $\xi(E) \propto |E - E_n|^{-\mu}$, the width Δ of the longitudinal conductivity peaks emerging at the transitions is expected to exhibit power-law scaling with respect to temperature, system size, or an externally applied frequency.

High frequency Hall-conductivity experiments, initially aimed at resolving the problem of the so-called low frequency breakdown of the QHE apparently observed at $\sim 1\,\mathrm{MHz}$, were successfully carried out at microwave frequencies ($\sim 33\,\mathrm{MHz}$) [3]. The longitudinal ac conductivity was also studied to obtain some information about localisation and the formation of Hall plateaus in the frequency ranges 100 Hz to 20 kHz [4] and 50–600 MHz [5]. Later, frequency dependent transport has been investigated also in the Gigahertz frequency range below 15 GHz [6,7] and above 30 GHz [8–10].

Dynamical scaling has been studied in several experiments, some of which show indeed power law scaling of the $\sigma_{xx}(\omega)$ peak width as expected, $\Delta \sim \omega^\kappa$ [6,11,12], whereas others do not [7]. The exponent $\kappa = (\mu z)^{-1}$ contains both the critical exponent μ of the localisation length and the dynamical exponent z which relates energy and length scales, $E \sim L^{-z}$. The value of $\mu = 2.35 \pm 0.03$ is well known from numerical calculations [13,14], and it also coincides with the outcome of a finite size scaling experiment [15]. However, it is presently only accepted as true that $z = 2$ for non-interacting particles, and $z = 1$ if Coulomb electron-electron interactions are present [16–20]. Therefore, a theoretical description of the ac conductivities would clearly contribute to a better understanding of dynamical scaling at quantum critical points.

Legal metrology represents a second area where a better knowledge of frequency dependent transport is highly desirable because the ac quantum Hall effect is applied for the realization and dissemination of the impedance standard and the unit of capacitance, the farad. At the moment the achieved relative uncertainty at a frequency of 1 kHz is of the order 10^{-7} which still is at least one order of magnitude too large [21–24]. It is unclear whether the observed deviations from the quantised dc value are due to external influences like capacitive and inductive couplings caused by the leads and contacts. Alternatively, the measured frequency effects that make exact quantisation impossible could be already inherent in an ideal non-interacting two-dimensional electron gas in the presence of disorder and a perpendicular magnetic field. Of course, applying a finite frequency ω will lead to a finite $\sigma_{xx}(\omega)$ and in turn will influence the quantisation of $\sigma_{xy}(\omega)$, but it remains to be investigated how large the deviation will be.

Theoretical studies of the ac conductivity in quantum Hall systems started in 1985 [25] when it was shown within a semiclassical percolation theory that for finite frequencies the longitudinal conductivity is not zero, thus influencing the quantisation of the Hall-conductivity. The quantum mechanical problem of non-interacting electrons in a 2d disordered system in the presence of a strong perpendicular magnetic field B was tackled by Apel [26] using a variational method. Applying an instanton approximation and confining to the high field limit, i.e., restricting to the lowest Landau level (LLL), an analytical solution for the real part of the frequency dependent longitudinal conductivity could be presented, $\sigma_{xx}(\omega) \propto \omega^2 \ln(1/\omega^2)$, a result that should hold if the Fermi energy lies deep down in the lower tail of the LLL.

Generalising the above result, both real and imaginary parts of the frequency dependent conductivities were obtained in a sequence of papers by Viehweger and Efetov [27–29]. The Kubo conductivities were determined by calculating the functional integrals in super-symmetric representation near non-trivial saddle points. Still, for the final results the Fermi energy was restricted to lie within the energy range of localised states in the lowest tail of the lowest Landau band and, therefore, no proposition for the critical regions at half fillings could be given. The longitudinal conductivity was found to be

$$\sigma_{xx}(\omega) = c\omega^2 \ln^b(1/\omega^2) - \mathrm{i}e^2\omega 2l_B^2 \varrho(E_\mathrm{F}) \,, \tag{1}$$

with the density of states $\varrho(E_\mathrm{F})$, and two unspecified constants b and c [27]. The real part of the Hall-conductivity in the same limit was proposed as

$$\sigma_{yx}(\omega) = e^2/h \, (\hbar\omega/\Gamma)^2 \, 8\pi l_B^2 n_e \,, \tag{2}$$

where $l_B = \sqrt{\hbar/(eB)}$ is the magnetic length, $\Gamma^2 = \lambda/2\pi l_B^2$ is the second moment of the white noise disorder potential distribution with disorder strength λ, and n_e denotes the electron density. The deviation due to frequency of the Hall-conductivity from its quantised dc plateau value can be perceived from an approximate expression proposed by Viehweger and Efetov [29] for the 2nd plateau, e.g., for filling factor $\nu \approx 2$,

$$\sigma_{yx}(\omega) = e^2/h \left[\frac{2}{1-\omega^2/\omega_c^2} - \left(\frac{2\hbar\omega}{\Gamma}\right)^2 (2-\nu) \right] . \tag{3}$$

Again, Γ is a measure of the disorder strength describing the width of the disorder broadened Landau band, and $\omega_c = eB/m_e$ is the cyclotron frequency with electron mass m_e. According to (3), due to frequency a deviation from the quantised value becomes apparent even for integer filling.

Before reviewing the attempts which applied numerical methods to overcome the limitations that had to be conceded in connection with the position of E_F in the analytical work, and to check the permissiveness of the approximations made, it is appropriate here to mention a result for the hopping regime. Polyakov and Shklovskii [30] obtained for the dissipative part of the ac conductivity a relation which, in contrast to (1), is linear in frequency ω,

$$\Re \sigma_{xx}(\omega) = K\, \epsilon \xi \omega \,. \tag{4}$$

This expression has recently been used to successfully describe experimental data [31]. Here, ξ is the localisation length, ϵ the dielectric constant, and the pre-factor is $K = 1/6$ in the limit $\hbar\omega \gg k_B T$. Also, the frequency scaling of the peak width $\Delta \sim \omega^\kappa$ was proposed within the same hopping model [30].

We now turn to the numerical approaches which were started by Gammel and Brenig who considered the low frequency anomalies and the finite size scaling of the real part of the conductivity peak $\sigma^p_{xx}(\omega, L_y)$ at the critical point of the lowest Landau band [32]. For these purposes the authors utilised the random Landau model in the high field limit (lowest Landau band only) [1,13] and generalised MacKinnon's recursive Green function method [33] for the evaluation of the real part of the dynamical conductivity. In contrast to the conventional quadratic Drude-like behaviour the peak value decayed linearly with frequency, $\sigma^p_{xx}(\omega) = \sigma^0_{xx} - \text{const.}\,|\omega|$, which was attributed to the long time tails in the velocity correlations which were observed also in a semiclassical model [34,35]. The range of this unusual linear frequency dependence varied with the spatial correlation length of the disorder potentials. A second result concerns the scaling of $\sigma^p_{xx}(\omega, L_y)$ at low frequencies as a function of the system width L_y, $\sigma^p_{xx}(\omega L^2_y) \propto (\omega L^2_y)^{2-\eta/2}$, where $\eta = 2 - D(2) = 0.36 \pm 0.06$ [32] is related to the multi-fractal wave functions [36–39], and to the anomalous diffusion at the critical point with $\eta \simeq 0.38$ [40] and $D(2) \simeq 1.62$ [41].

The frequency scaling of the $\sigma_{xx}(\omega)$ peak width was considered numerically for the first time in a paper by Avishai and Luck [42]. Using a continuum model with spatially correlated Gaussian disorder potentials placed on a square lattice, which then was diagonalised within the subspace of functions pertaining to the lowest Landau level, the real part of the dissipative conductivity was evaluated from the Kubo formula involving matrix elements of the velocity of the guiding centres [43]. This is always necessary in single band approximations because of the vanishing of the current matrix elements between states belonging to the same Landau level. As a result, a broadening of the conductivity peak was observed and from a finite size scaling analysis a dynamical exponent $z = 1.19 \pm 0.13$ could be extracted using $\nu = 2.33$ from Ref. [13]. This is rather startling because it is firmly believed that for non-interacting systems z equals the Euclidean dimension of space which gives $z = d = 2$ in the QHE case.

A different theoretical approach for the low frequency behaviour of the $\sigma_{xx}(\omega)$ peak has been pursued by Jug and Ziegler [44] who studied a Dirac fermion model with an inhomogeneous mass [45] applying a non-perturbative calculation. This model leads to a non-zero density of states and to a finite bandwidth of extended states near the centre of the Landau band [46]. Therefore, the ac conductivity as well as its peak width do neither show power-law behaviour nor do they vanish in the limit $\omega \to 0$. This latter feature of the model has been asserted to explain the linear frequency dependence and the finite intercept at $\omega = 0$ observed experimentally for the width of the conductivity peak by Balaban *et al.* [7], but, up to now, there is no other experiment showing such a peculiar behaviour.

2 Preliminary Considerations – Basic Relations

In the usual experiments on two-dimensional systems a current $I_x(\omega)$ is driven through the sample of length L_x and width L_y. The voltage drop along the current direction, $U_x(\omega)$, and that across the sample, $U_y(\omega)$, are measured from which the Hall-resistance $R_{\rm H}(\omega) = U_y(\omega)/I_x(\omega) = \rho_{xy}(\omega)$ and the longitudinal resistance $R_x(\omega) = U_x(\omega)/I_x(\omega) = \rho_{xx}(\omega)L_x/L_y$ are obtained, where ρ_{xy} and ρ_{xx} denote the respective resistivities. To compare with the theoretically calculated conductivities one has to use the relations below. Only in Corbino samples $\sigma_{xx}(\omega)$ can be experimentally detected directly.

The total current through a cross-section, $I_x(\omega) = \int_0^{L_y} j_x(\omega, \boldsymbol{r})\mathrm{d}y$, is determined by the local current density $j_x(\omega, \boldsymbol{r})$ which constitutes the response to the applied electric field

$$j_x(\omega, \boldsymbol{r}) = \int \sum_{u\in\{x,y\}} \sigma_{xu}(\omega, \boldsymbol{r}, \boldsymbol{r}')E_u(\omega, \boldsymbol{r}')\, \mathrm{d}^2\boldsymbol{r}' \,. \tag{5}$$

The nonlocal conductivity tensor is particularly important in phase-coherent mesoscopic samples. Usually, for the investigation of the measured macroscopic conductivity tensor one is not interested in its spatial dependence. Therefore, one relies on a local approximation and considers the electric field to be effectively constant. This leads to Ohm's law $\boldsymbol{j} = \widehat{\sigma}\boldsymbol{E}$ from which the resistance components are simply given by inverting the conductivity tensor $\widehat{\sigma}$,

$$\begin{pmatrix} \rho_{xx} & \rho_{xy} \\ \rho_{yx} & \rho_{yy} \end{pmatrix} = \frac{1}{\sigma_{xx}\sigma_{yy} - \sigma_{xy}\sigma_{yx}} \begin{pmatrix} \sigma_{yy} & \sigma_{yx} \\ \sigma_{xy} & \sigma_{xx} \end{pmatrix} . \tag{6}$$

For an isotropic system we have $\sigma_{xx} = \sigma_{yy}$ and $\sigma_{yx} = -\sigma_{xy}$ which in case of zero frequency gives the well known relations

$$\rho_{xx} = \frac{\sigma_{xx}}{\sigma_{xx}^2 + \sigma_{xy}^2}, \qquad \rho_{xy} = \frac{-\sigma_{xy}}{\sigma_{xx}^2 + \sigma_{xy}^2} . \tag{7}$$

From experiment one knows that whenever ρ_{xy} is quantised ρ_{xx} gets unmeasurably small which in turn means that $\sigma_{xx} \to 0$ and $\rho_{xy} = 1/\sigma_{yx}$. Therefore, to make this happen one normally concludes that the corresponding electronic states have to be localised.

In the presence of frequency this argument no longer holds because electrons in localised states do respond to an applied time dependent electric field giving rise to an alternating current. Also, both real and imaginary parts have to be considered now

$$\sigma_{xx}(\omega) = \sigma^R_{xx}(\omega) + \mathrm{i}\,\sigma^I_{xx}(\omega), \qquad \sigma_{xy}(\omega) = \sigma^R_{xy}(\omega) + \mathrm{i}\,\sigma^I_{xy}(\omega) \,. \tag{8}$$

Assuming an isotropic system, the respective tensor components of the ac resistivity $\rho_{uv}(\omega) = \rho^R_{uv}(\omega) + \mathrm{i}\,\rho^I_{uv}(\omega)$ with $u, v \in \{x, y\}$ can be written as

$$\rho^R_{uv}(\omega) = \frac{\sigma^R_{vu}(\delta\sigma^2_{xx} + \delta\sigma^2_{xy}) + 2\sigma^I_{vu}(\sigma^R_{xx}\,\sigma^I_{xx} + \sigma^R_{xy}\,\sigma^I_{xy})}{(\delta\sigma^2_{xx} + \delta\sigma^2_{xy})^2 + 4(\sigma^R_{xx}\,\sigma^I_{xx} + \sigma^R_{xy}\,\sigma^I_{xy})^2} \tag{9}$$

$$\rho^I_{uv}(\omega) = \frac{\sigma^I_{vu}(\delta\sigma^2_{xx} + \delta\sigma^2_{xy}) + 2\sigma^R_{uv}(\sigma^R_{xx}\,\sigma^I_{xx} + \sigma^R_{xy}\,\sigma^I_{xy})}{(\delta\sigma^2_{xx} + \delta\sigma^2_{xy})^2 + 4(\sigma^R_{xx}\,\sigma^I_{xx} + \sigma^R_{xy}\,\sigma^I_{xy})^2} \tag{10}$$

with the abbreviations $\delta\sigma^2_{xx} \equiv (\sigma^R_{xx})^2 - (\sigma^I_{xx})^2$ and $\delta\sigma^2_{xy} \equiv (\sigma^R_{xy})^2 - (\sigma^I_{xy})^2$.

3 Model and Transport Theory

We describe the dynamics of non-interacting particles moving within a two-dimensional plane in the presence of a perpendicular magnetic field and random electrostatic disorder potentials by a lattice model with Hamiltonian

$$H = \sum_{\boldsymbol{r}} w_{\boldsymbol{r}} |\boldsymbol{r}\rangle\langle\boldsymbol{r}| - \sum_{<\boldsymbol{r}\neq\boldsymbol{r}'>} V_{\boldsymbol{r}\boldsymbol{r}'} |\boldsymbol{r}\rangle\langle\boldsymbol{r}'| \,. \tag{11}$$

The random disorder potentials associated with the lattice sites are denoted by $w_{\boldsymbol{r}}$ with probability density distribution $P(w_{\boldsymbol{r}}) = 1/W$ within the interval $[-W/2, W/2]$, where W is the disorder strength, and the $|\boldsymbol{r}\rangle$ are the lattice base vectors. The transfer terms

$$V_{\boldsymbol{r}\boldsymbol{r}'} = V \exp\Big(- \mathrm{i}\, e^2/\hbar \int_{\boldsymbol{r}}^{\boldsymbol{r}'} \boldsymbol{A}(\boldsymbol{l})\, \mathrm{d}\boldsymbol{l} \Big) \,, \tag{12}$$

which connect only nearest neighbours on the lattice, contain the influence of the applied magnetic field via the vector potential $\boldsymbol{A}(\boldsymbol{r}) = (0, Bx, 0)$ in their phase factors. V and the lattice constant a define the units of energy and length, respectively.

The electrical transport is calculated within linear response theory using the Kubo formula which allows to determine the time dependent linear conductivity from the current matrix elements of the unperturbed system

$$\sigma_{uv}(E_{\mathrm{F}}, T, \omega) = \frac{\pi e^2}{\Omega} \int_{-\infty}^{\infty} \mathrm{d}E \frac{f(E) - f(E + \hbar\omega)}{\omega} \times \mathrm{Tr}\big\{\hat{v}_u \delta(E_{\mathrm{F}} - H) \hat{v}_v \delta(E_{\mathrm{F}} + \hbar\omega - H)\big\} \,, \tag{13}$$

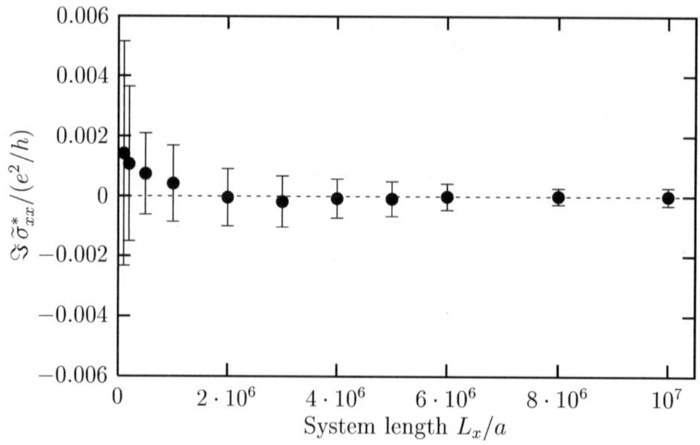

Fig. 1. Disorder averaged imaginary part of $\widetilde{\sigma}^*_{xx}(E,\omega,\varepsilon)$ at energy $E/V = -3.35$, frequency $\hbar\omega/V = 0.001$, and $\varepsilon/V = 0.0008$ as a function of sample length $L_x = Na$. The system width is $L_y/a = 32$ and the number of realisations amounts to 29 for $L_x/a \le 5\cdot 10^6$ and to 8 for larger lengths

where $f(E) = (\exp[(E - E_{\rm F})/(k_{\rm B}T)] - 1)^{-1}$ is the Fermi function. The area of the system is $\Omega = L_x L_y$, $\hat{v}_u = {\rm i}/\hbar[H,u]$ signifies the u-component of the velocity operator, and $\delta(E+\hbar\omega - H) = {\rm i}/(2\pi)(G^{\omega,+}(E) - G^{\omega,-}(E))$, where $G^{\omega,\pm}(E) = ((E+\hbar\omega\pm{\rm i}\varepsilon){\rm I} - H)^{-1}$ is the resolvent with imaginary frequency ${\rm i}\varepsilon$ and unit matrix I.

For finite systems at temperature $T = 0\,$K, ensuring the correct order of limits for size Ω and imaginary frequency ${\rm i}\varepsilon$, one gets with $\gamma = \omega + 2{\rm i}\varepsilon/\hbar$

$$\begin{aligned}\sigma_{uv}(E_{\rm F},\omega) = \lim_{\varepsilon\to 0}\lim_{\Omega\to\infty}\frac{e^2}{h}\frac{1}{\Omega\hbar\omega}\Bigg[\int_{E_F-\hbar\omega}^{E_F}{\rm d}E\,{\rm Tr}\Big\{(\hbar\gamma)^2\left[uG^{\omega,+}vG^-\right]\\ -(\hbar\omega)^2\left[uG^{\omega,+}vG^+\right] + 2{\rm i}\varepsilon\left[uv(G^{\omega,+}-G^-)\right]\Big\}\\ -\int_{-\infty}^{E_F-\hbar\omega}{\rm d}E\,{\rm Tr}\Big\{(\hbar\omega)^2\left[uG^{\omega,+}vG^+ - uG^{\omega,-}vG^-\right]\Big\}\Bigg]\end{aligned} \tag{14}$$

$$\begin{aligned}= \lim_{\varepsilon\to 0}\lim_{\Omega\to\infty}\frac{e^2}{h}\frac{1}{\Omega\hbar\omega}\Bigg[\int_{E_F-\hbar\omega}^{E_F}\widetilde{\sigma}_{uv}(E,\omega,\varepsilon,L_x,L_y)\,{\rm d}E\\ +\int_{-\infty}^{E_F-\hbar\omega}\widetilde{\sigma}^*_{uv}(E,\omega,\varepsilon,L_x,L_y)\,{\rm d}E\Bigg]\,.\end{aligned} \tag{15}$$

One can show that the second integral with the limits $(-\infty, E_{\rm F}-\hbar\omega)$ does not contribute to the real part of $\sigma_{uv}(E_{\rm F},\omega)$ because the kernel is identically zero, but we were not able to proof the same also for the imaginary part. Therefore, using the recursive Green function method explained in the next section, we numerically studied $\Im\,\widetilde{\sigma}^*_{uv}(E,\omega,\varepsilon,L_x,L_y)$ and found it to become very small only after disorder averaging. As an example we show in Fig. 1 the dependence of $\langle\Im\,\widetilde{\sigma}^*_{xx}\rangle$, averaged over up to 29 realisations, on the length of the system for

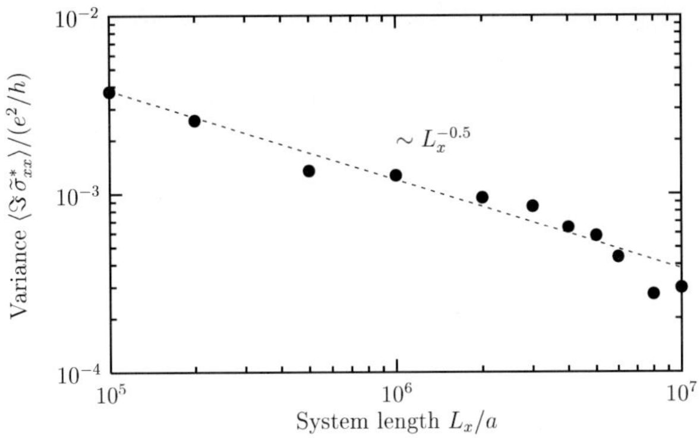

Fig. 2. The variance of $\langle \Im \tilde{\sigma}^*_{xx}(E, \omega, \varepsilon) \rangle$ versus system length L_x calculated for the averages shown in Fig. 1 exhibiting an empirical power-law $\sim L_x^{0.5}$

a particular energy $E/V = -3.35$ and width $L_y/a = 32$. Also the variance of $\langle \tilde{\sigma}^*_{xx}(E, \omega, \varepsilon) \rangle$ gets smaller with system length following an empirical power law $\sim (L_x/a)^{-0.5}$ (see Fig. 2). In what follows we neglect the second integral for the calculation of the imaginary parts of the conductivities and assume that only the contribution of the first one with limits $(E_{\mathrm{F}} - \hbar\omega, E_{\mathrm{F}})$ matters. Of course, one has to be particularly careful even if the kernel is very small because with increasing disorder strength the energy range that contributes to the integral (finite density of states) tends to infinity. Therefore, a rigorous proof for the vanishing of this integral kernel is highly desirable.

4 Recursive Green Function Method

A very efficient method for the numerical investigation of large disordered chains, strips and bars that are assembled by successively adding one slice at a time has been pioneered by MacKinnon [47]. This iterative technique relies on the property that the Hamiltonian $H^{(N+1)}$ of a lattice system containing $N + 1$ slices, each a lattice constant a apart, can be decomposed into parts that describe the system containing N slices, $H^{(N)}$, the next slice added, $H_{N+1,N+1}$, and a term that connects the last slice to the rest, $H'_N = H_{N,N+1} + H_{N+1,N}$. Then the corresponding resolvent is formally equivalent to the Dyson equation $G = G_0 + G_0\widehat{V}G$ where the 'unperturbed' G_0 represents the direct sum of $H^{(N)}$ and $H_{N+1,N+1}$, and $\widehat{V}$ corresponds to the 'interaction' H'_N. The essential advantage of this method is the fact that, for a fixed width L_y, the system size is increased in length adding slice by slice, whereas the size of the matrices to be dealt with numerically remains the same [47,48].

A number of physical quantities like localisation length [49–52], density of states [51,53], and some dc transport coefficients [33,54,55] have been calculated by this technique over the years. Also, this method was implemented for the evaluation of the real part of the ac conductivity in 1d [56,57] and 2d systems [32,58,59]. Further efforts to include also the real and imaginary parts of the

Hall- and the imaginary part of the longitudinal conductivity in quantum Hall systems were also successfully accomplished [60–62].

The iteration equations of the resolvent matrix acting on the subspace of such slices with indices $i, j \leq N$ in the N-th iteration step can be written as

$$\begin{aligned}
G_{i,j}^{\omega,\pm,(N+1)} &= G_{i,j}^{\omega,\pm,(N)} + G_{i,N}^{\omega,\pm,(N)} H_{N,N+1} G_{N+1,N+1}^{\omega,\pm,(N+1)} H_{N+1,N} G_{N,j}^{\omega,\pm,(N)} \\
G_{N+1,N+1}^{\omega,\pm,(N+1)} &= [(E + \hbar\omega \pm \mathrm{i}\varepsilon)\mathrm{I} - H_{N+1,N+1} - H_{N+1,N} G_{N,N}^{\omega,\pm,(N)} H_{N,N+1}]^{-1} \\
G_{i,N+1}^{\omega,\pm,(N+1)} &= G_{i,N}^{\omega,\pm,(N)} H_{N,N+1} G_{N+1,N+1}^{\omega,\pm,(N+1)} \\
G_{N+1,j}^{\omega,\pm,(N+1)} &= G_{N+1,N+1}^{\omega,\pm,(N+1)} H_{N+1,N} G_{N,j}^{\omega,\pm,(N)} \,. \qquad (16)
\end{aligned}$$

The calculation of the ac conductivities starts with the Kubo formula (15) by setting up a recursion equation for fixed energy E, width $L_y = Ma$, and imaginary frequency ε, which, e.g., for the longitudinal component reads

$$\begin{aligned}
\widetilde{\sigma}_{xx}(E, \omega, \varepsilon, N) = \frac{e^2}{hMNa^2} S_N^{xx} = \frac{e^2}{hMNa^2} \mathrm{Tr}\Big\{ \sum_{i,j}^{N} \Big[(\hbar\gamma)^2 x_i G_{ij}^{\omega,+} x_j G_{ji}^{-} \\
- (\hbar\omega)^2 x_i G_{ij}^{\omega,+} x_j G_{ji}^{+} + 2\mathrm{i}\varepsilon \delta_{ij} x_i^2 (G_{ij}^{\omega,+} - G_{ji}^{-}) \Big] \Big\} \,. \qquad (17)
\end{aligned}$$

The iteration equation for adding a new slice is given by

$$\begin{aligned}
S_{N+1}^{xx} = S_N^{xx} + \mathrm{Tr}\{ A_N R_{N+1}^{-} + D_N^1 R_{N+1}^{\omega,+} D_N^2 R_{N+1}^{-} \\
+ B_N R_{N+1}^{\omega,+} - D_N^3 R_{N+1}^{+} D_N^4 R_{N+1}^{\omega,+} - C_N R_{N+1}^{+} \} \,. \qquad (18)
\end{aligned}$$

with $R_{N+1}^{\omega,\pm} \equiv G_{N+1,N+1}^{\omega,\pm,(N+1)}$, and a set of auxiliary quantities as defined in the appendix. The coupled iteration equations and the auxiliary quantities are evaluated numerically, the starting values are set to be zero. In addition, coordinate translations are required in each iteration step to keep the origin $x_{N+1} = 0$ which then guarantees the numerical stability.

5 Longitudinal Conductivity $\sigma_{xx}(E, \omega)$

In this section we present our numerical results of the longitudinal conductivity as a function of frequency for various positions of the Fermi energy within the lowest Landau band. The real and imaginary parts of $\sigma_{xx}(E_\mathrm{F}, \omega)$ were calculated for several frequencies, but, for the sake of legibility, only four of them are plotted in Fig. 3 versus energy. While the real part exhibits a positive single Gaussian-like peak with maximum $\approx 0.5\, e^2/h$ at the critical energy, the imaginary part, which is negative almost everywhere, has a double structure and vanishes near the critical point. $\Im\, \sigma_{xx}(E_\mathrm{F}, \omega)$ almost looks like the modulus of the derivative of the real part with respect to energy. Plotting $\Im\, \sigma_{xx}(E_\mathrm{F}, \omega)$ as a function of $\Re\, \sigma_{xx}(E_\mathrm{F}, \omega)$ (see Fig. 4) we obtain for frequencies $\omega < \omega^*$ a single, approximately semi-circular curve that, up to a minus sign, closely resembles the experimental results of Hohls *et al.* [31]. However, for larger ω our data points deviate from a single curve.

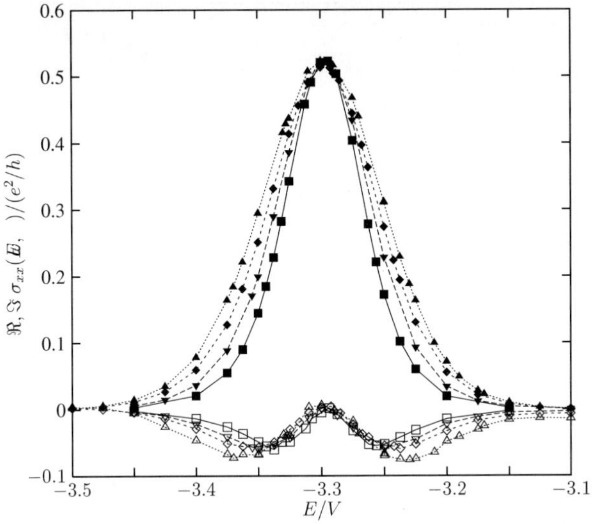

Fig. 3. The real and imaginary parts of $\sigma_{xx}(E,\omega)$ as a function of energy and frequency $\hbar\omega/V = 2 \cdot 10^{-4}$ (■), $5 \cdot 10^{-4}$ (▼), $1 \cdot 10^{-3}$ (◆), $2 \cdot 10^{-3}$ (▲). For comparison, the cyclotron frequency is $\hbar\omega_c \approx 1.57\,V$ for $\alpha_B = 1/8$ which was chosen for the magnetic flux density. The disorder strength is $W/V = 1$ and the maximal system width amounts to $L_y/a = 96$ with periodic boundary conditions applied

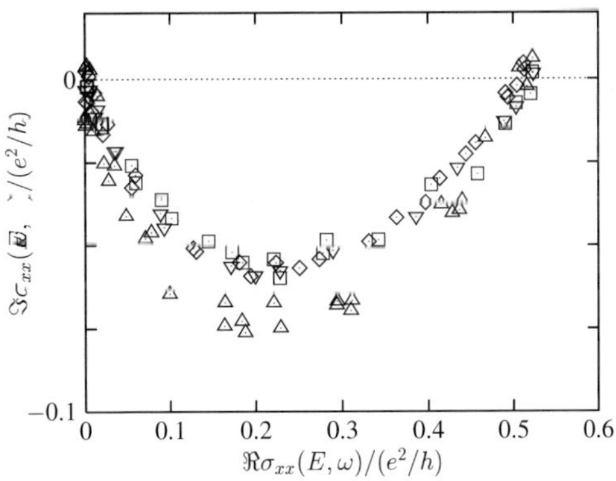

Fig. 4. The imaginary part of $\sigma_{xx}(E,\omega)$ as a function of the real part. Data are taken from Fig. 3

5.1 Frequency Dependence of Real and Imaginary Parts

The behaviour of the real and imaginary part of the longitudinal ac conductivity in the lower tail of the lowest Landau band ($E/V = -3.5$) is shown in Fig. 5. We find for the imaginary part a linear frequency dependence for small ω which is, apart from a minus sign, in accordance with (1) [27]. The real part can nicely be fitted to $\propto \omega^2 \log^2[V/(\hbar\omega)]^2$ in conformance with (1) and $b = 2$, but disagrees with the findings in [26] where $b = 1$ was proposed.

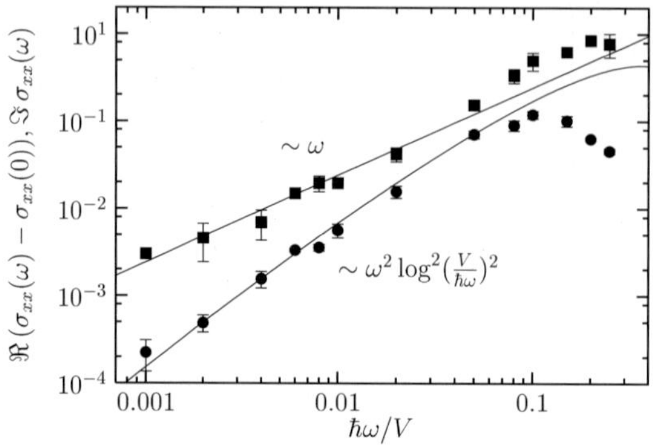

Fig. 5. The real (●) and imaginary (■) part of $\sigma_{xx}(E_F,\omega)$ in units of e^2/h at $E_F/V = -3.5$ as a function of frequency. Further parameters are the disorder strength $W/V = 1.0$, system width $L_y/a = 32$ and $\varepsilon/V = 0.0004$

5.2 Behaviour of the Maximum of $\sigma_{xx}(\omega)$

The frequency dependence of the real part of the longitudinal conductivity peak value was already investigated in [32] where for long-range correlated disorder potentials a non-Drude-like decay was observed. We obtained a similar behaviour also for spatially uncorrelated disorder potentials in a lattice model [61]. In Fig. 6 the difference $|\Re\,\sigma^{\mathrm{p}}_{xx}(\omega) - \sigma^{\mathrm{p}}_{xx}(0)|$ is plotted versus frequency in a double logarithmic plot from which a linear relation can be discerned. A linear increase with frequency was found for the imaginary part of the longitudinal conductivity at the critical point as well [61]. The standard explanation for the non-Drude behaviour in terms of long time tails in the velocity correlations, which were shown to exist in a QHE system [32,35], seems not to be adequate in our case. For the uncorrelated disorder potentials considered here, it is not clear whether

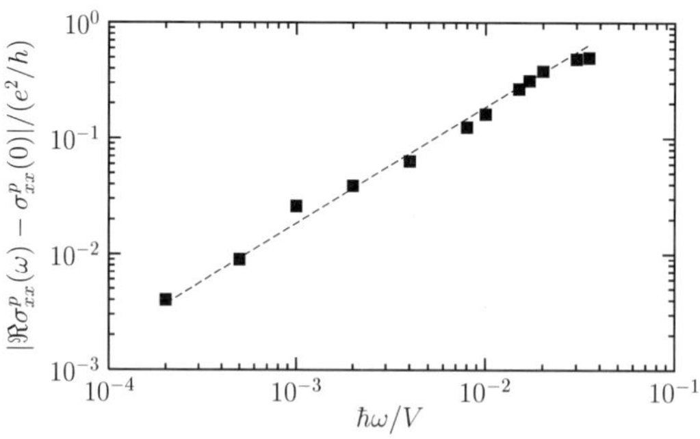

Fig. 6. The non-Drude decrease of the peak value of the longitudinal conductivity as a function of frequency. A linear behaviour of $|\Re\,\sigma^{\mathrm{p}}_{xx}(\omega) - \sigma^{\mathrm{p}}_{xx}(0)|$ is clearly observed with $\sigma^{\mathrm{p}}_{xx}(0)/(e^2/h) = 0.512$ using the following parameters $E/V = -3.29$, $W/V = 0.1$, $L_y/a = 32$, $\varepsilon/V = 0.0004$

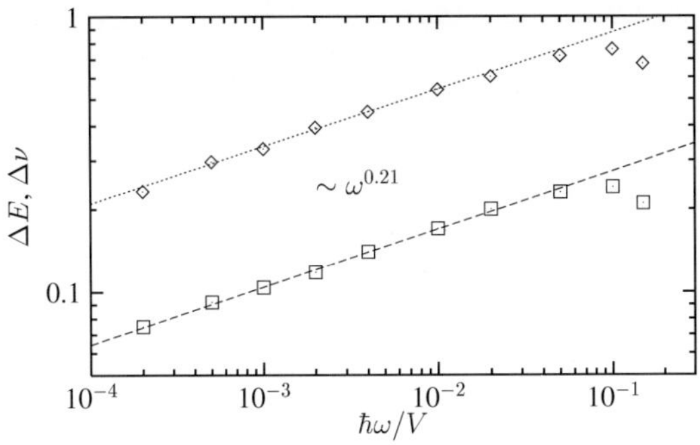

Fig. 7. Frequency scaling of the $\sigma_{xx}(\omega)$ peak width. The width in energy ΔE (□) and the width in filling factor $\Delta\nu$ (◇) show a power-law $\Delta \sim \omega^{\kappa}$ with an exponent $\kappa = (\mu z)^{-1} = 0.21$

the picture of electron motion along equipotential lines, a basic ingredient for the arguments in [32], is appropriate.

5.3 Scaling of the $\sigma_{xx}(\omega)$ Peak Width

The scaling of the width of the conductivity peaks with frequency is shown in Fig. 7 where both the $\sigma_{xx}(\omega)$ peak width expressed in energy and, due to the knowledge of the density of states, in filling factor are shown to follow a power-law $\Delta \sim \omega^{\kappa}$ with $\kappa = (\mu z)^{-1} = 0.21$ [60]. Taking $\mu = 2.35$ from [14] we get $z = 2.026$ close to what is expected for non-interacting electrons. Therefore, the result reported in [42] seems to be doubtful. However, spatial correlations in the disorder potentials as considered in [42] may influence the outcome. Alternatively, one could fix $z = 2$ and obtain $\mu = 2.38$ in close agreement with the results from numerical calculations of the scaling of the static conductivity [63] or the localisation length [14].

The experimentally observed values $\kappa \simeq 0.42$ [6] and $\kappa = 0.5 \pm 0.1$ [12] are larger by a factor of about 2. This is usually attributed to the influence of electron-electron interactions ($z = 1$) which were neglected in the numerical investigations.

6 Frequency Dependent Hall-Conductivity

The Hall-conductivity due to an external time dependent electric field as a function of filling factor ν is shown in Fig. 8 for system widths $L_y/a = 64$ and $L_y/a = 128$, respectively. While $\sigma_{xy}(\omega)$ has already converged for E_{F} lying in the upper tail of the lowest Landau band a pronounced shift can be seen in the lower tail of the next Landau band. For $\nu = 1.3$ a system width of at least $L_y/a = 192$ was necessary for $\sigma_{xy}(\omega)$ to converge. This behaviour originates in the exponential increase of the localisation length with increasing Landau band index. Due to the applied frequency $\hbar\omega/V = 0.008$ the σ_{xy} plateau is not

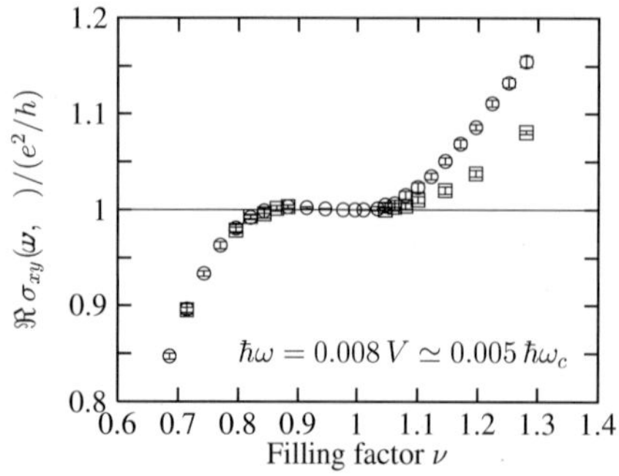

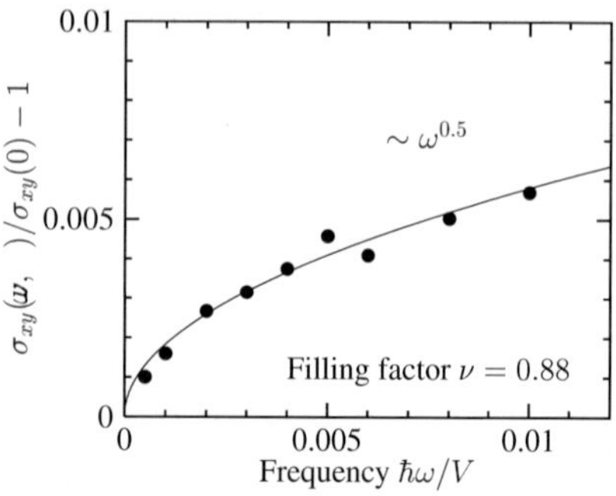

Fig. 8. Frequency dependent Hall conductivity: The figure on the l.h.s shows $\sigma_{xy}(\omega)$ versus filling factor for two system sizes $L_y/a = 64$ (○) and $L_y/a = 128$ (□). On the r.h.s., the relative deviation of the Hall-conductivity from the dc value, $\sigma_{xy}(\omega)/\sigma_{xy}(0) - 1$, is plotted as a function of frequency for $\nu = 0.88$

flat, but rather has a parabola shape near the minimum at $\nu = 1.0$, similar to what has been observed in experiment [24]. An example of the deviation of $\sigma_{xy}(\omega)$ from its quantised dc value is shown on the right hand side of Fig. 8 where $\sigma_{xy}(\omega)/\sigma_{xy}(0) - 1$ is plotted versus frequency for filling factor $\nu = 0.88$. A power-law curve $\sim \omega^{0.5}$ can be fitted to the data points. Using this empirical relation, we find a relative deviation of the order of $5 \cdot 10^{-6}$ when extrapolated down to to 1 kHz, the frequency usually applied in metrological experiments [21–24]. Therefore, there is no quantisation in the neighbourhood of integer filling even in an ideal 2d electron gas without contacts, external leads, and other experimental imperfections. Recent calculations, however, show that this deviation can be considerably reduced, even below $1 \cdot 10^{-8}$, if spatially correlated disorder potentials are considered in the model [64].

7 Conclusions

The frequency dependent electrical transport in integer quantum Hall systems has been reviewed and the various theoretical developments have been presented. Starting from a linear response expression a method has been demonstrated which is well suited for the numerical evaluation of the real and imaginary parts of both the time dependent longitudinal and the Hall-conductivity. In contrast to the analytical approaches, no further approximations or restrictions such as the position of the Fermi energy have to be considered.

We discussed recent numerical results in some detail with particular emphasis placed on the frequency scaling of the peak width of the longitudinal conductivity emerging at the quantum critical points, and on the quantisation of the ac Hall-conductivity at the plateau. As expected, the latter was found to depend on the applied frequency. The extrapolation of our calculations down to low frequencies resulted in a relative deviation of $5 \cdot 10^{-6}$ at $\sim$ 1 kHz when spatially uncorrelated disorder potentials are considered. Disorder potentials with spatial correlations, likely to exist in real samples, will probably reduce this pronounced frequency effect.

Our result for the frequency dependence of the σ_{xx} peak width showed power-law scaling, $\Delta \sim \omega^{\kappa}$, where $\kappa = (\mu z)^{-1} = 0.21$ as expected for non-interacting electrons. Therefore, electron-electron interactions have presumably to be taken into account to explain the experimentally observed $\kappa \approx 0.5$ Also, the influence of the spatial correlation of the disorder potentials may influence the value of κ.

The frequency dependences of the real and imaginary parts of the longitudinal conductivities, previously obtained analytically for Fermi energies lying deep down in the lowest tail of the lowest Landau band, have been confirmed by our numerical investigation. However, the quadratic behaviour found at low frequencies for the real part of $\sigma_{xx}(\omega)$ has to be contrasted with the linear frequency dependence that has been proposed for hopping conduction.

Finally, a non-Drude decay of the σ_{xx}-peak value with frequency, $\sigma^{p}_{xx}(0) - \sigma^{p}_{xx}(\omega) \propto \omega$, as reported earlier for correlated disorder potentials, has been observed also in the presence of uncorrelated disorder potentials. A convincing explanation for the latter behaviour is still missing.

Appendix

The iteration equations of the auxiliary quantities (required in (18)) can be written as [62] with $V_N \equiv H_{N,N+1} = H^{\dagger}_{N+1,N}$

$$\begin{aligned}
A_{N+1} &= V^{\dagger}_{N+1} R^{-}_{N+1} \left[A_N + D^1_N R^{\omega,+}_{N+1} D^2_N\right] R^{-}_{N+1} V_{N+1} && (19)\\
B_{N+1} &= V^{\dagger}_{N+1} R^{\omega,+}_{N+1} \left[B_N - D^3_N R^{+}_{N+1} D^4_N + D^2_N R^{-}_{N+1} D^1_N\right] R^{\omega,+}_{N+1} V_{N+1} && (20)\\
C_{N+1} &= V^{\dagger}_{N+1} R^{+}_{N+1} \left[C_N + D^4_N R^{\omega,+}_{N+1} D^3_N\right] R^{+}_{N+1} V_{N+1} && (21)\\
F_{N+1} &= V^{\dagger}_{N+1} R^{\omega,+}_{N+1} \left[F_N + D^3_N R^{+}_{N+1} D^4_N\right] R^{\omega,+}_{N+1} V_{N+1} && (22)\\
G_{N+1} &= V^{\dagger}_{N+1} R^{-}_{N+1} \left[G_N + D^{10}_N R^{\omega,-}_{N+1} D^{11}_N\right] R^{-}_{N+1} V_{N+1} && (23)\\
H_{N+1} &= V^{\dagger}_{N+1} R^{\omega,-}_{N+1} \left[H_N + D^{11}_N R^{-}_{N+1} D^{10}_N\right] R^{\omega,-}_{N+1} V_{N+1} && (24)\\
D^1_{N+1} &= V^{\dagger}_{N+1} R^{-}_{N+1} D^1_N R^{\omega,+}_{N+1} V_{N+1} && (25)\\
D^2_{N+1} &= V^{\dagger}_{N+1} R^{\omega,+}_{N+1} D^2_N R^{-}_{N+1} V_{N+1} && (26)\\
D^3_{N+1} &= V^{\dagger}_{N+1} R^{\omega,+}_{N+1} D^3_N R^{+}_{N+1} V_{N+1} && (27)\\
D^4_{N+1} &= V^{\dagger}_{N+1} R^{+}_{N+1} D^4_N R^{\omega,+}_{N+1} V_{N+1}. && (28)
\end{aligned}$$

Furthermore,

$$\begin{aligned}
D^5_{N+1} &= V^{\dagger}_{N+1} R^{-}_{N+1} D^5_N R^{-}_{N+1} V_{N+1} && (29)\\
D^6_{N+1} &= V^{\dagger}_{N+1} R^{+}_{N+1} D^6_N R^{+}_{N+1} V_{N+1} && (30)\\
D^{10}_{N+1} &= V^{\dagger}_{N+1} R^{-}_{N+1} D^{10}_N R^{\omega,-}_{N+1} V_{N+1} && (31)
\end{aligned}$$

$$D^{11}_{N+1} = V^{\dagger}_{N+1} R^{\omega,-}_{N+1} D^{11}_{N} R^{-}_{N+1} V_{N+1} \tag{32}$$
$$D^{14}_{N+1} = V^{\dagger}_{N+1} R^{\omega,-}_{N+1} D^{14}_{N} R^{\omega,-}_{N+1} V_{N+1} \tag{33}$$
$$D^{15}_{N+1} = V^{\dagger}_{N+1} R^{\omega,+}_{N+1} D^{15}_{N} R^{\omega,+}_{N+1} V_{N+1} \tag{34}$$
$$D^{16}_{N+1} = V^{\dagger}_{N+1} R^{-}_{N+1} D^{16}_{N} R^{-}_{N+1} V_{N+1} \tag{35}$$
$$E^{1}_{N+1} = V^{\dagger}_{N+1} R^{+}_{N+1} \Big(E^{1}_{N} + (\hbar\omega) I \Big) R^{+}_{N+1} V_{N+1} \tag{36}$$
$$E^{2}_{N+1} = V^{\dagger}_{N+1} R^{-}_{N+1} \Big(E^{2}_{N} + (\hbar\omega) I \Big) R^{-}_{N+1} V_{N+1} \tag{37}$$
$$E^{3}_{N+1} = V^{\dagger}_{N+1} R^{\omega,+}_{N+1} \Big(E^{3}_{N} + (\hbar\omega) I \Big) R^{\omega,+}_{N+1} V_{N+1} \tag{38}$$
$$E^{4}_{N+1} = V^{\dagger}_{N+1} R^{\omega,-}_{N+1} \Big(E^{4}_{N} + (\hbar\omega) I \Big) R^{\omega,-}_{N+1} V_{N+1} \; , \tag{39}$$

where the auxiliary quantities are defined by

$$A_N = V^{\dagger}_{N} \Bigg(\sum_{i,j}^{N} G^{-}_{Ni} x_i \big[(\hbar\gamma)^2 G^{\omega,+}_{ij} - 2\mathrm{i}\varepsilon \delta_{ij} \mathrm{I} \big] x_j G^{-}_{jN} \Bigg) V_N \tag{40}$$

$$B_N = V^{\dagger}_{N} \Bigg(\sum_{i,j}^{N} G^{\omega,+}_{Ni} x_i \big[(\hbar\gamma)^2 G^{-}_{ij} + 2\mathrm{i}\varepsilon \delta_{ij} \mathrm{I} + (\hbar\omega)^2 G^{+}_{ij} \big] x_j G^{\omega,+}_{jN} \Bigg) V_N \tag{41}$$

$$C_N = V^{\dagger}_{N} \Bigg(\sum_{i,j}^{N} (\hbar\omega)^2 G^{+}_{Ni} x_i G^{\omega,+}_{ij} x_j G^{+}_{jN} \Bigg) V_N \tag{42}$$

$$F_N = V^{\dagger}_{N} \Bigg(\sum_{i,j}^{N} (\hbar\omega)^2 G^{\omega,+}_{Ni} x_i G^{+}_{ij} x_j G^{\omega,+}_{jN} \Bigg) V_N \tag{43}$$

$$G_N = V^{\dagger}_{N} \Bigg(\sum_{i,j}^{N} (\hbar\omega)^2 G^{-}_{Ni} x_i G^{\omega,-}_{ij} x_j G^{-}_{jN} \Bigg) V_N \tag{44}$$

$$H_N = V^{\dagger}_{N} \Bigg(\sum_{i,j}^{N} (\hbar\omega)^2 G^{\omega,-}_{Ni} x_i G^{-}_{ij} x_j G^{\omega,-}_{jN} \Bigg) V_N , \tag{45}$$

and

$$D^{1}_{N} = V^{\dagger}_{N} \Bigg(\sum_{i}^{N} (\hbar\gamma) G^{-}_{Ni} x_i G^{\omega,+}_{iN} \Bigg) V_N \tag{46}$$

$$D^{2}_{N} = V^{\dagger}_{N} \Bigg(\sum_{i}^{N} (\hbar\gamma) G^{\omega,+}_{Ni} x_i G^{-}_{iN} \Bigg) V_N \tag{47}$$

$$D^{3}_{N} = V^{\dagger}_{N} \Bigg(\sum_{i}^{N} (\hbar\omega) G^{\omega,+}_{Ni} x_i G^{+}_{iN} \Bigg) V_N \tag{48}$$

$$D^{4}_{N} = V^{\dagger}_{N} \Bigg(\sum_{i}^{N} (\hbar\omega) G^{+}_{Ni} x_i G^{\omega,+}_{iN} \Bigg) V_N \tag{49}$$

$$D_N^5 = V_N^\dagger \left(\sum_i^N (\hbar\omega) G_{Ni}^- x_i G_{iN}^- \right) V_N \tag{50}$$

$$D_N^6 = V_N^\dagger \left(\sum_i^N (\hbar\omega) G_{Ni}^+ x_i G_{iN}^+ \right) V_N \tag{51}$$

$$D_N^{10} = V_N^\dagger \left(\sum_i^N (\hbar\omega) G_{Ni}^- x_i G_{iN}^{\omega,-} \right) V_N \tag{52}$$

$$D_N^{11} = V_N^\dagger \left(\sum_i^N (\hbar\omega) G_{Ni}^{\omega,-} x_i G_{iN}^- \right) V_N \tag{53}$$

$$D_N^{14} = V_N^\dagger \left(\sum_i^N (\hbar\omega) G_{Ni}^{\omega,-} x_i G_{iN}^{\omega,-} \right) V_N \tag{54}$$

$$D_N^{15} = V_N^\dagger \left(\sum_i^N (\hbar\omega) G_{Ni}^{\omega,+} x_i G_{iN}^{\omega,+} \right) V_N \tag{55}$$

$$D_N^{16} = V_N^\dagger \left(\sum_i^N (\hbar\omega) G_{Ni}^- x_i G_{iN}^- \right) V_N \tag{56}$$

$$E_N^1 = V_N^\dagger \left(\sum_i^N (\hbar\omega) G_{Ni}^+ G_{iN}^+ \right) V_N \tag{57}$$

$$E_N^2 = V_N^\dagger \left(\sum_i^N (\hbar\omega) G_{Ni}^- G_{iN}^- \right) V_N \tag{58}$$

$$E_N^3 = V_N^\dagger \left(\sum_i^N (\hbar\omega) G_{Ni}^{\omega,+} G_{iN}^{\omega,+} \right) V_N \tag{59}$$

$$E_N^4 = V_N^\dagger \left(\sum_i^N (\hbar\omega) G_{Ni}^{\omega,-} G_{iN}^{\omega,-} \right) V_N \, . \tag{60}$$

For the translation of the coordinates $x_i \to x_i - a$ one gets

$$A_N^{'} = A_N + H_N^1 - 2aD_N^5 + a^2 E_N^2 + aH_N^3 \tag{61}$$

$$B_N^{'} = B_N + H_N^2 - H_N^1 + a(H_N^4 - H_N^3) \tag{62}$$

$$C_N^{'} = C_N + H_N^2 - 2aD_N^6 + a^2 E_N^1 + aH_N^4 \tag{63}$$

$$F_N^{'} = F_N - aD_N^4 + 2aD_N^{15} - aD_N^3 - aH_N^4 - a^2 E_N^3 \tag{64}$$

$$G_N^{'} = G_N + a(D_N^{10} + D_N^{11}) - 2aD_N^{16} + a^2 E_N^2 + aH_N^5 \tag{65}$$

$$H_N^{'} = H_N - aD_N^{10} + 2aD_N^{14} - aD_N^{11} - aH_N^5 - a^2 E_N^4 \tag{66}$$

$$D_N^{1/2'} = D_N^{1/2} + H_N^3 \tag{67}$$

$$D_N^{3/4'} = D_N^{3/4} + H_N^4 \tag{68}$$

$$\begin{aligned}
D_N^{5'} &= D_N^5 - aE_N^2 &(69)\\
D_N^{6'} &= D_N^6 - aE_N^1 &(70)\\
D_N^{10/11'} &= D_N^{10/11} + H_N^5 &(71)\\
D_N^{14'} &= D_N^{14} - aE_N^4 &(72)\\
D_N^{15'} &= D_N^{15} - aE_N^3 &(73)\\
D_N^{16'} &= D_N^{16} - aE_N^2 \,, &(74)
\end{aligned}$$

with the following abbreviations

$$\begin{aligned}
H_N^1 &= a(D_N^1 + D_N^2) &(75)\\
H_N^2 &= a(D_N^3 + D_N^4) &(76)\\
H_N^3 &= aV_N^\dagger(R_N^{\omega,+} - R_N^-)V_N &(77)\\
H_N^4 &= aV_N^\dagger(R_N^{\omega,+} - R_N^+)V_N &(78)\\
H_N^5 &= aV_N^\dagger(R_N^{\omega,-} - R_N^-)V_N \,. &(79)
\end{aligned}$$

References

1. B. Huckestein: Rev. Mod. Phys. **67**, 357 (1995)
2. S. L. Sondhi, S. M. Girvin, J. P. Carini, D. Shahar: Rev. Mod. Phys. **69**(1), 315 (1997)
3. F. Kuchar, R. Meisels, G. Weimann, W. Schlapp: Phys. Rev. B **33**(4), 2956 (1986)
4. J.I. Lee, B.B. Goldberg, M. Heiblum, P.J. Stiles: Solid State Commun. **64**(4), 447 (1987)
5. I.E. Batov, A.V. Polisskii, M.I. Reznikov, V.I. Tal'yanskii: Solid State Commun. **76**(1), 25 (1990)
6. L.W. Engel, D. Shahar, C. Kurdak, D.C. Tsui: Phys. Rev. Lett. **71**(16), 2638 (1993)
7. N.Q. Balaban, U. Meirav, I. Bar-Joseph: Phys. Rev. Lett. **81**(22), 4967 (1998)
8. W. Belitsch, R. Meisels, F. Kuchar, G. Hein, K. Pierz: Physica B **249–251**, 119 (1998)
9. R. Meisels, F. Kuchar, W. Belitsch, G. Hein, K. Pierz: Physica B **256-258**, 74 (1998)
10. F. Kuchar, R. Meisels, B. Kramer: Advances in Solid State Physics **39**, 231 (1999)
11. D. Shahar, L. W. Engel, D. C. Tsui: In *High Magnetic Fields in the Physics of Semiconductors*, edited by D. Heiman (World Scientific Publishing Co., Singapore, 1995): pp. 256–259
12. F. Hohls, U. Zeitler, R.J. Haug, R. Meisels, K. Dybko, F. Kuchar: cond-mat/0207426 (2002)
13. B. Huckestein B. Kramer: Phys. Rev. Lett. **64**(12), 1437 (1990)
14. B. Huckestein: Europhysics Letters **20**(5), 451 (1992)
15. S. Koch, R.J. Haug, K. v. Klitzing, K. Ploog: Phys. Rev. Lett. **67**, 883 (1991)
16. D.-H. Lee Z. Wang: Phys. Rev. Lett. **76**(21), 4014 (1996)
17. S.-R.E. Yang A.H. MacDonald: Phys. Rev. Lett. **70**(26), 4110 (1993)

18. B. Huckestein M. Backhaus: Phys. Rev. Lett. **82**(25), 5100 (1999)
19. Z. Wang, M.P.A. Fisher, S.M. Girvin, J.T. Chalker: Phys. Rev. B **61**(12), 8326 (2000)
20. Z. Wang S. Xiong: Phys. Rev. B **65**, 195316 (2002)
21. F. Delahaye: J. Appl. Phys. **73**, 7914 (1993)
22. F. Delahaye: Metrologia **31**, 367 (1994/95)
23. S.W. Chua, A. Hartland, B.P. Kibble: IEEE Trans. Instrum. Meas. **48**, 309 (1999)
24. J. Schurr, J. Melcher, A. von Campenhausen, G. Hein, F.-J. Ahlers, K. Pierz: Metrologia **39**, 3 (2002)
25. R. Joynt: J. Phys. C: Solid State Phys. **18**, L331 (1985)
26. W. Apel: J. Phys.: Condens. Matter **1**, 9387 (1989)
27. O. Viehweger K. B. Efetov: J. Phys.: Condens. Matter **2**(33), 7049 (1990)
28. O. Viehweger K. B. Efetov: J. Phys.: Condens. Matter **3**(11), 1675 (1991)
29. O. Viehweger K.B. Efetov: Phys. Rev. B **44**(3), 1168 (1991)
30. D.G. Polyakov B.I. Shklovskii: Phys. Rev. B **48**(15), 11167 (1993)
31. F. Hohls, U. Zeitler, R.J. Haug: Phys. Rev. Lett. **86**(22), 5124 (2001)
32. B.M. Gammel W. Brenig: Phys. Rev. B **53**(20), R13279 (1996)
33. A. MacKinnon: Z. Phys. B **59**, 385 (1985)
34. F. Evers W. Brenig: Z. Phys. B **94**, 155 (1994)
35. W. Brenig, B.M. Gammel, P. Kratzer: Z. Phys. B **103**, 417 (1997)
36. B. Huckestein L. Schweitzer: In *High Magnetic Fields in Semiconductor Physics III: Proceedings of the International Conference, Würzburg 1990*, edited by G. Landwehr (Springer Series in Solid-State Sciences 101, Springer, Berlin, 1992): pp. 84–88
37. W. Pook M. Janßen: Z. Phys. B **82**, 295 (1991)
38. B. Huckestein, B. Kramer, L. Schweitzer: Surface Science **263**, 125 (1992)
39. M. Janßen: International Journal of Modern Physics B **8**(8), 943 (1994)
40. J.T. Chalker G.J. Daniell: Phys. Rev. Lett. **61**(5), 593 (1988)
41. B. Huckestein L. Schweitzer: Phys. Rev. Lett. **72**(5), 713 (1994)
42. Y. Avishay J.M. Luck: preprint, cond-mat/9609265 (1996)
43. T. Ando: J. Phys. Soc. Jpn. **53**(9), 3101 (1984)
44. G. Jug K. Ziegler: Phys. Rev. B **59**(8), 5738 (1999)
45. A.W.W. Ludwig, M.P.A. Fisher, R. Shankar, G. Grinstein: Phys. Rev. B **50**, 7526 (1994)
46. K. Ziegler: Phys. Rev. B **55**(16), 10661 (1997)
47. A. MacKinnon: J. Phys. C **13**, L1031 (1980)
48. A. MacKinnon B. Kramer: Z. Phys. B **53**, 1 (1983)
49. A. MacKinnon B. Kramer: Phys. Rev. Lett. **47**(21), 1546 (1981)
50. A. MacKinnon B. Kramer: In *Lecture Notes in Physics*, edited by G. Landwehr (Springer-Verlag, Berlin, Heidelberg, New York, Tokyo, 1982): pp. 74–86
51. L. Schweitzer, B. Kramer, A. MacKinnon: J. Phys. C **17**, 4111 (1984)
52. T. Koschny, H. Potempa, L. Schweitzer: Phys. Rev. Lett. **86**(17), 3863 (2001)
53. B. Kramer, L. Schweitzer, A. MacKinnon: Z. Phys. B - Condensed Matter **56**, 297 (1984)
54. L. Schweitzer, B. Kramer, A. MacKinnon: Z. Phys. B **59**, 379 (1985)
55. C. Villagonzalo, R.A. Römer, M. Schreiber, A. MacKinnon: Phys. Rev. B **62**, 16446 (2000)
56. T. Saso, C.I. Kim, T. Kasua: J. Phys. Soc. Jap. **52**, 1888 (1983)
57. T. Saso: J. Phys. C **17**, 2905 (1984)
58. B.M. Gammel: Ph. D. Thesis (Technical-University Munich) (1994)
59. B.M. Gammel F. Evers: unpublished report, 10 pages (1999)

60. A. Bäker L. Schweitzer: Ann. Phys. (Leipzig) **8**, SI-21 (1999)
61. A. Bäker L. Schweitzer: In *Proc. 25th Int. Conf. Phys. Semicond., Osaka 2000*, edited by N. Miura T. Ando (Springer, Berlin, 2001): vol. 78 of *Springer Proceedings in Physics*: pp. 975–976
62. A. Bäker L. Schweitzer: PTB-report, unpublished, 94 pages (2002)
63. B.M. Gammel W. Brenig: Phys. Rev. Lett. **73**(24), 3286 (1994)
64. L. Schweitzer: to be published (2003)

Part II

Localization and Phase Coherence in Disordered Media

What Do Phase Space Methods Tell Us about Disordered Quantum Systems?

Gert-Ludwig Ingold, André Wobst, Christian Aulbach, and Peter Hänggi

Institut für Physik, Universität Augsburg, D-86135 Augsburg, Germany

1 Introduction

At a summer school held in fall 1991 at the PTB in Braunschweig, Bernhard Kramer proposed a book project that should encompass novel solid state research topics ranging from quantum transport to quantum chaos. This very book finally appeared at the end of 1997 under the title "Quantum transport and dissipation" [1]. Compiled between the book covers is a series of topics, some of which are usually not connected with each other in present day's research. The main subject in the first chapter is coherent transport in disordered systems. The last chapter, on the other hand, dwells on concepts such as phase space, Wigner and Husimi functions, and alike. In any case, in the description of disordered systems we do not find many works that explicitly make use of phase space concepts. This is so, although the existence of a mapping between the Anderson model and the kicked rotor [2] indeed suggests that methods employed in quantum chaos may advantageously be used to elucidate the physics at work in disordered quantum systems.

There exists, however, another, almost obvious physical motivation to utilize the powerful phase space concepts to study disordered quantum systems. as a function of the disorder strength, the nature of the eigenstates changes from ballistic to localized behavior, containing possibly in-between a diffusive regime [3]. In the case of ballistic transport it is appropriate to think in terms of plane waves which are occasionally scattered by the weak disorder potential. Then a momentum space description clearly imposes itself. For strong disorder, however, the natural physical space is real space. A unified description, being valid at arbitrary disorder strength, can thus be achieved by focusing on suitable (quantum) phase space concepts.

2 Phase Space Methods in Quantum Mechanics

When an attempt is made to represent a quantum state in phase space, there are two observations which deserve our attention. First, phase space combines conjugate quantum observables which brings into play a Heisenberg uncertainty relation. While in classical mechanics, the state of a point particle can be described in phase space with infinite precision, this no longer holds true in quantum mechanics. Therefore, we expect some difficulties if for quantum systems one attempts to obtain perfect resolution in phase space. Second, the full information about a quantum state is contained already in the corresponding wave function which can either be expressed in position or momentum space representation. Thus, the phase space representation of a quantum state cannot provide more information than the wave function itself. This information, however, may be presented in a more advantageous form, as we shall demonstrate next.

2.1 The Wigner Function

As a first possibility to define a phase space representation of a quantum state possessing the wave function $\psi(x)$ we introduce the Wigner function

$$W(x,k) = \int \mathrm{d}y \mathrm{e}^{\mathrm{i}ky} \psi^*(x+\frac{y}{2})\psi(x-\frac{y}{2})\,. \tag{1}$$

Here, and in the following we use the wave number k instead of the momentum $p = \hbar k$. The definition (1) is useful and appropriate as can be seen by considering the moments of position and momentum observables. Upon evaluation of the momentum integral, one readily finds that

$$\int \frac{\mathrm{d}x\mathrm{d}k}{2\pi} x^n W(x,k) = \int \mathrm{d}x\, x^n |\psi(x)|^2 . \tag{2}$$

With a little more effort one further can establish that

$$\begin{aligned}\int \frac{\mathrm{d}x\mathrm{d}k}{2\pi} k^n W(x,k) &= \int \mathrm{d}x \int \mathrm{d}y \int \frac{\mathrm{d}k}{2\pi} \left[(-\mathrm{i})^n \frac{\mathrm{d}^n}{\mathrm{d}y^n} \mathrm{e}^{\mathrm{i}ky}\right] \psi^*(x)\psi(x-y) \\ &= \int \mathrm{d}x\, \psi^*(x) \left(\frac{1}{\mathrm{i}}\frac{\mathrm{d}}{\mathrm{d}x}\right)^n \psi(x)\,.\end{aligned} \tag{3}$$

A generalization to moments containing both position and momentum operators is possible if a Weyl ordering for the operators is respected (for further details see [4]).

To start, it is instructive to consider a few special quantum states of simple structure which also play a key role in the discussion of disordered systems. Let us begin with a plane wave, i.e. $\psi(x) = \exp(\mathrm{i}k_0 x)/\sqrt{2\pi}$. Inserting this wave function into (1) one finds

$$W(x,k) = \int \frac{\mathrm{d}y}{2\pi} \mathrm{e}^{\mathrm{i}(k-k_0)y} = \delta(k-k_0)\,. \tag{4}$$

The intermediate result emphasizes the fact that the off-diagonal contributions in (1), which are parameterized by the coordinate y, contain the information about the momentum. For a localized state, $\psi(x) = \delta(x-x_0)$, the Wigner function emerges as

$$W(x,k) = \mathrm{e}^{2\mathrm{i}k(x-x_0)}\delta(x-x_0) = \delta(x-x_0)\,. \tag{5}$$

Next we consider a quantum state localized at two positions, i.e. $\psi(x) = [\delta(x + a) + \delta(x - a)]/\sqrt{2}$. Proceeding as before, one obtains the Wigner function

$$W(x, k) = \frac{1}{2}\left[\delta(x + a) + \delta(x - a)\right] + \cos(2ka)\delta(x)\,. \tag{6}$$

The first two terms describe the localization of the particle at $x = \pm a$. In addition, there occurs an oscillatory term in the middle between the two localization centers which accounts for the coherent superposition of two localized states. This simple example demonstrates that the Wigner function generally is not positive even though it could be treated as a phase space density in (2) and (3). This characteristic feature is a direct consequence of the fact that by virtue of the definition (1) one attempts to define a phase space representation which allows for perfect localization in position and momentum as indicated by the results (4) and (5). This formal attempt to circumvent the Heisenberg uncertainty relation is generally paid for by negative parts of the Wigner function.

Thus far, we have considered a particle on a continuous and infinitely extended one-dimensional state space. The situation changes when we try to apply phase space concepts to the Anderson model of disordered systems. For such a lattice model the integral in (1) has to be replaced by a sum. Moreover, in numerical calculations, the system size has to be taken finite. This results in a phase space that is twisted to a torus so that additionally periodic boundary conditions appear in momentum space apart from those usually imposed in real space. Therefore, interference terms in momentum space analogous to those appearing in real space as in (6) occur also across the boundaries of the Brillouin zone. Such artifacts can also be understood as arising from the discretization of the integral in (1) due to the lattice structure of the model. In particular for the states located at the boundaries of the Brillouin zone, the wave number k is so large that a discretized version of the Fourier integral no longer represents a good approximation. The usage of Wigner functions for lattice models therefore becomes problematic.

In Fig. 1a we present an example for the Wigner function of an eigenstate of the Anderson model at relatively low disorder. While a certain spatial localization has already set in, two dominant wave numbers $\pm k$ can still be clearly recognized. At $k = 0$, interference effects analogous to those found in (6) are visible while the two low ridges at larger wave numbers represent the artifacts due to the lattice structure of the Anderson model discussed in the previous paragraph.

2.2 The Husimi Function

The above discussed shortcomings of the Wigner function can be circumvented by use of the so-called Husimi function. The latter one is obtained from the Wigner function by a Gaussian smearing according to

$$\rho(x, k) = \frac{1}{\pi}\int \mathrm{d}x'\mathrm{d}k' \exp\left[-\frac{(x - x')^2}{2\sigma^2} - 2\sigma^2(k - k')^2\right] W(x', k')\,. \tag{7}$$

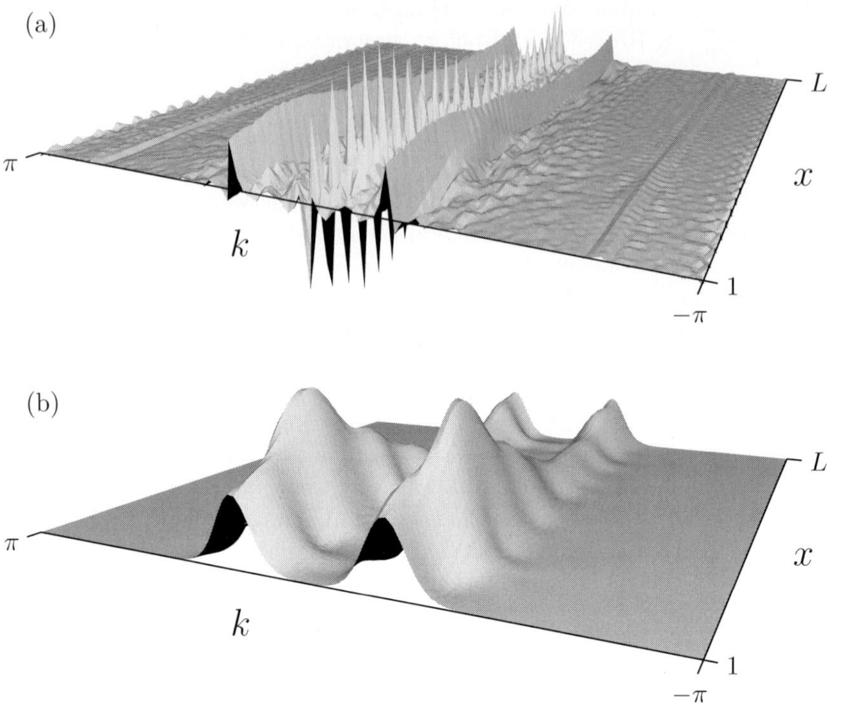

Fig. 1. a Wigner function and **b** Husimi function for an eigenstate of the Anderson model

Even though one might assume that one discards information by transforming the Wigner function $W(x, k)$ into the Husimi function $\rho(x, k)$, this is actually not the case as one can show by inverting (7): The Wigner function can then be recast as

$$W(x,k) = \int \frac{\mathrm{d}u\mathrm{d}v}{2\pi}\frac{\mathrm{d}x'\mathrm{d}k'}{2\pi}\exp\left[\frac{\sigma^2u^2}{2}+\frac{v^2}{8\sigma^2}+\mathrm{i}(x'-x)u+\mathrm{i}(k'-k)v\right]\rho(x',k')\,. \tag{8}$$

The existence of the integrals is guaranteed by the inherent asymptotic decay of the Husimi function provided that the integrals over x' and k' are evaluated first.

From a physical point of view, the most important aspect of the Gaussian smearing introduced in (7) is that the Heisenberg uncertainty relation is now accounted for in a natural way. The Husimi function will never localize a quantum state in phase space beyond the limits set by the requirement $\Delta x \Delta k \geq 1/2$. In return, the Husimi function yields a nonnegative-valued phase space distribution. Indeed, by inserting (1) into (7) one can express the Husimi function in terms of the wave function $\psi(x)$ as the squared quantity

$$\rho(x,k) = \left|\int \mathrm{d}x' \psi(x') \frac{1}{(2\pi\sigma^2)^{1/4}} \exp\left[-\frac{(x'-x)^2}{4\sigma^2} - \mathrm{i}kx'\right]\right|^2 , \tag{9}$$

which manifestly is nonnegative. According to (9), the Husimi function may be viewed as the projection of the wave function onto a minimal uncertainty state or, using the language of quantum optics, onto a coherent state. At this point, the width σ is still a parameter which can be chosen freely. We will fix its value later on when applying these phase space concepts to the Anderson model.

Yet another aspect of the Gaussian smearing pertains to the artifacts mentioned in the previous section for the Wigner function on a lattice. While in (1) all values of y contribute, this is no longer the case for the Husimi function where the convolution with a Gaussian restricts the difference in the arguments of the two wave functions. In fact, even the interference term appearing in the Wigner function (6) for a continuum model is strongly suppressed in the corresponding Husimi function provided the distance a is sufficiently larger than the width σ; this is evident from

$$\begin{aligned}\rho(x,k) = \frac{1}{2\sqrt{2\pi\sigma^2}}\Bigg[&\exp\left(-\frac{(x-a)^2}{2\sigma^2}\right) + \exp\left(-\frac{(x+a)^2}{2\sigma^2}\right)\\ &+ 2\exp\left(-\frac{x^2}{2\sigma^2} - \frac{a^2}{2\sigma^2}\right)\cos(2ka)\Bigg] .\end{aligned} \tag{10}$$

A comparison of the Wigner function in Fig. 1a and the Husimi function in Fig. 1b demonstrates the effect of the Gaussian smearing. The spatial localization and the two mainly contributing wave numbers are the dominant features of the Husimi function. The interferences leading to negative contributions to the Wigner function have almost disappeared except for some wiggles which, however, do not render the Husimi function negative. Finally, the features stemming from the lattice have now disappeared.

2.3 Inverse Participation Ratio

The Husimi function, in particular for quantum states at intermediate disorder strengths, does contain a rich structure. In general, it is desirable for various reasons to reduce this wealth of information and to characterize the state by a single number which is derived from the Husimi function. First, in clear contrast to the one-dimensional case depicted in Fig. 1b it is generally not possible to visualize the Husimi function for systems in two and higher space dimensions. Second, the details of the Husimi functions depend sensitively on the chosen disorder realization. In order to perform averages, it is necessary to reduce the characterization of the states to a number.

Since the Husimi function is nonnegative, in principle all quantities familiar from classical phase space analysis can be employed in the quantum case as well.

One possibility is given by the so-called Wehrl entropy [5]

$$S = -\int \frac{\mathrm{d}x\mathrm{d}k}{2\pi} \rho(x,k) \ln[\rho(x,k)] \,, \tag{11}$$

which has been used, e.g., in the discussion of the driven rotor [6]. The above-mentioned existence of a mapping between the kicked rotor and the Anderson model has motivated a study of the latter by means of the Wehrl entropy [7].

From a numerical point of view, it is advantageous to linearize the Wehrl entropy (11). Replacing $\ln(x)$ by its linear approximation $x - 1$, one obtains $-x\ln(x) \approx x - x^2$. The approximation shows qualitatively the same behavior as the original function and, in particular, yields the same values at the boundaries at $x = 0$ and 1. Performing this linearization in (11), the first order term gives rise to a constant due to normalization and we are left with the second order term, the inverse participation ratio (IPR) in phase space

$$P = \int \frac{\mathrm{d}x\mathrm{d}k}{2\pi} [\rho(x,k)]^2 \,. \tag{12}$$

While both the Wehrl entropy S and the phase space IPR P are entirely determined in terms of the wave function $\psi(x)$, it is necessary for the evaluation of the Wehrl entropy to first calculate the Husimi function explicitly. This requires a significant numerical effort which can be reduced by resorting to the phase space IPR [8]. In turn, this makes possible the study of the Anderson model in two and even in three dimensions [9]. Inserting (9) into (12) one obtains the phase space IPR expressed directly in terms of the wave function as

$$P = \frac{1}{8\sqrt{\pi}\sigma} \int \mathrm{d}u \left| \int \mathrm{d}v \, \psi\left(\frac{u-v}{2}\right) \psi\left(\frac{u+v}{2}\right) \exp\left(-\frac{v^2}{8\sigma^2}\right) \right|^2 \,. \tag{13}$$

Apart from the numerical aspects, another advantage relates to the fact that inverse participation ratios may be defined as well in real and momentum space, i.e.,

$$P_x = \int \mathrm{d}x \, |\psi(x)|^4 \qquad \text{and} \qquad P_k = \int \mathrm{d}k \, |\widetilde{\psi}(k)|^4 \,, \tag{14}$$

where $\widetilde{\psi}(k)$ is the wave function in momentum representation. The availability of the inverse participation ratio in different spaces allows for instructive physical comparisons. The real space IPR P_x is a well studied quantity as it is related to the return probability of a diffusing particle [10] while the phase space IPR has recently been used to describe the complexity of quantum states [11,12]. We add that a quantity containing information similar to the phase space IPR can be defined on the basis of marginal distributions in real and momentum space [13].

In order to demonstrate that it is justified to refer to the above-mentioned quantities as inverse participation ratios, we consider a few special cases for quantum states on a one-dimensional lattice with L lattice sites. Let us first assume that the state is localized on a single site. Then we find $P_x = 1$ and

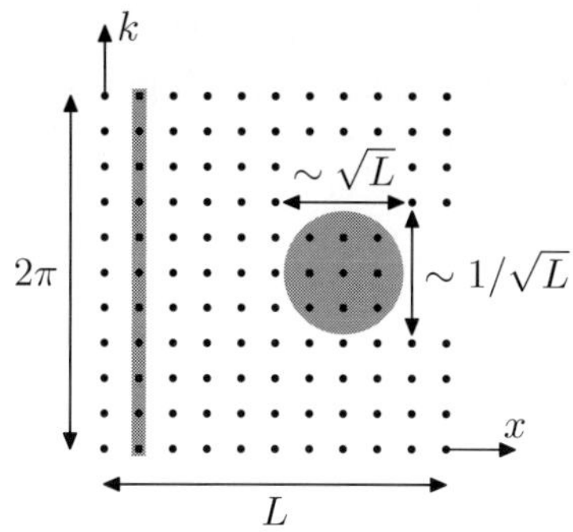

Fig. 2. The different phase space resolution provided by real space wave function and Husimi function is visualized by the *gray shaded areas* on the left and right, respectively

$P_k = 1/L$. The inverse of these results indeed indicates that in real space the state occupies only one site while in momentum space L sites are occupied. In phase space, one obtains $P = 1/2\sqrt{\pi}\sigma$ which primarily reflects the Gaussian width in real space.

As a second example, we consider a real-valued ballistic wave function at momentum $\bar{k}$, i.e. $\psi(x) = \sqrt{2/L}\cos(\bar{k}x)$. In real and momentum space, the inverse participation ratios become $P_x = 3/2L$ and $P_k = 1/2$, respectively. The first result accounts for the nonuniform extension in real space while the latter indicates the contribution of the two momenta $\pm\bar{k}$. For $\bar{k} \neq 0, \pm\pi$, the phase space IPR $P = (\sigma\sqrt{\pi}/L)[1 + 2\exp(-4\sigma^2\bar{k}^2)]$ depends on the distance between the two contributing momenta.

So far, we have not specified the width σ entering both, the definition (9) of the Husimi function and the results for the phase space IPR P mentioned in the previous two paragraphs. Although one is rather free in choosing σ, the most impartial choice consists in selecting an equal relative resolution in real and momentum space, i.e. $\Delta x/L = \Delta k/2\pi$. Together with the requirement of minimal uncertainty, $\Delta x\Delta k = 1/2$, one finds $\Delta x = \sigma = \sqrt{L/4\pi}$. For our two examples mentioned above (which will appear later in the limits of vanishing and very strong disorder) this implies, that the IPR P will scale like $1/\sqrt{L}$ as a function of system size. This result for one-dimensional systems is found to generalize to $P \sim L^{-d/2}$ in d dimensions.

An intuition for the differences between inverse participation ratios in real and momentum space, on the one hand, and the IPR in phase space, on the other hand, can be obtained by considering the resolution provided by the wave function and the Husimi function in real and momentum space. The vertical gray stripe on the left in Fig. 2 corresponds to the real space wave function. Independent of system size, this wave function always leads to perfect resolution in real space. The price to be paid, however, consists in the absence of *any* resolution in momentum space. The converse holds true for the momentum space wave function. The situation is quite different for the Husimi function symbolized by the gray disk in Fig. 2. Although the discussion can be generalized [14], we will consider the case where $\sigma \sim \sqrt{L}$. Then, structures occurring on $\sqrt{L}$ lattice points, or less, cannot be resolved. While this resolution becomes worse as the system size is increased, the resolution relative to the system size improves with $1/\sqrt{L}$. For large system sizes, the *phase space approach* therefore allows for a detailed description of a quantum state both in real and momentum space, while a specific wave function representation will always neglect one of the two possibilities.

3 Anderson Model in Phase Space

In the preceding section we have demonstrated that a phase space approach indeed presents advantages when both, real and momentum space properties are of interest as it is the case for disordered systems when the disorder strength is varied from zero to infinity. Although, in principle, the Husimi function is equivalent to the wave function, it will make the relevant information more readily accessible. We next want to illustrate this very fact by considering the Anderson model for disordered systems. Its Hamiltonian [15]

$$H = -t \sum_{<\mathbf{x},\mathbf{x}'>} (|\mathbf{x}'\rangle\langle\mathbf{x}| + |\mathbf{x}\rangle\langle\mathbf{x}'|) + W \sum_{n} v_n |\mathbf{x}\rangle\langle\mathbf{x}|, \tag{15}$$

is defined on a d dimensional square lattice with L sites in each direction. In order to avoid boundary effects, periodic boundary conditions are imposed. The first term describes the kinetic energy which allows for hopping between nearest neighbor sites $< \mathbf{x}, \mathbf{x}' >$. In the following, the hopping matrix element will set the energy scale, i.e. $t = 1$. The second term on the right-hand side of (15) represents the disordered on-site potential where the energies v_n are independently drawn from a box distribution on the interval $[-1/2; 1/2]$. The respective disorder strength is then determined by W.

3.1 Husimi Functions

Before we investigate the inverse participation ratio for the Anderson model, it is instructive to first take a look at the underlying Husimi functions. Since a visualization is readily possible only for one-dimensional models, we compare in Fig. 3 the one-dimensional Anderson model and the Aubry-André model [16]. The latter is based on a periodic potential $\lambda \sum_n \cos(2\pi\beta n)|n\rangle\langle n|$ replacing the disorder potential in (15). Choosing β as the golden mean, $(\sqrt{5}-1)/2$, the potential is incommensurate with the underlying lattice and displays a phase transition from delocalized to localized states [16]. We have chosen this model for comparison since its phase space IPR is comparable to that obtained for the two- and three-dimensional Anderson model [14]. In Fig. 3, the Aubry-André model therefore serves as a substitute for the higher-dimensional Anderson models.

Before we discuss in further detail the Husimi functions depicted in Fig. 3, we mention that for real wave functions the Husimi function, according to its

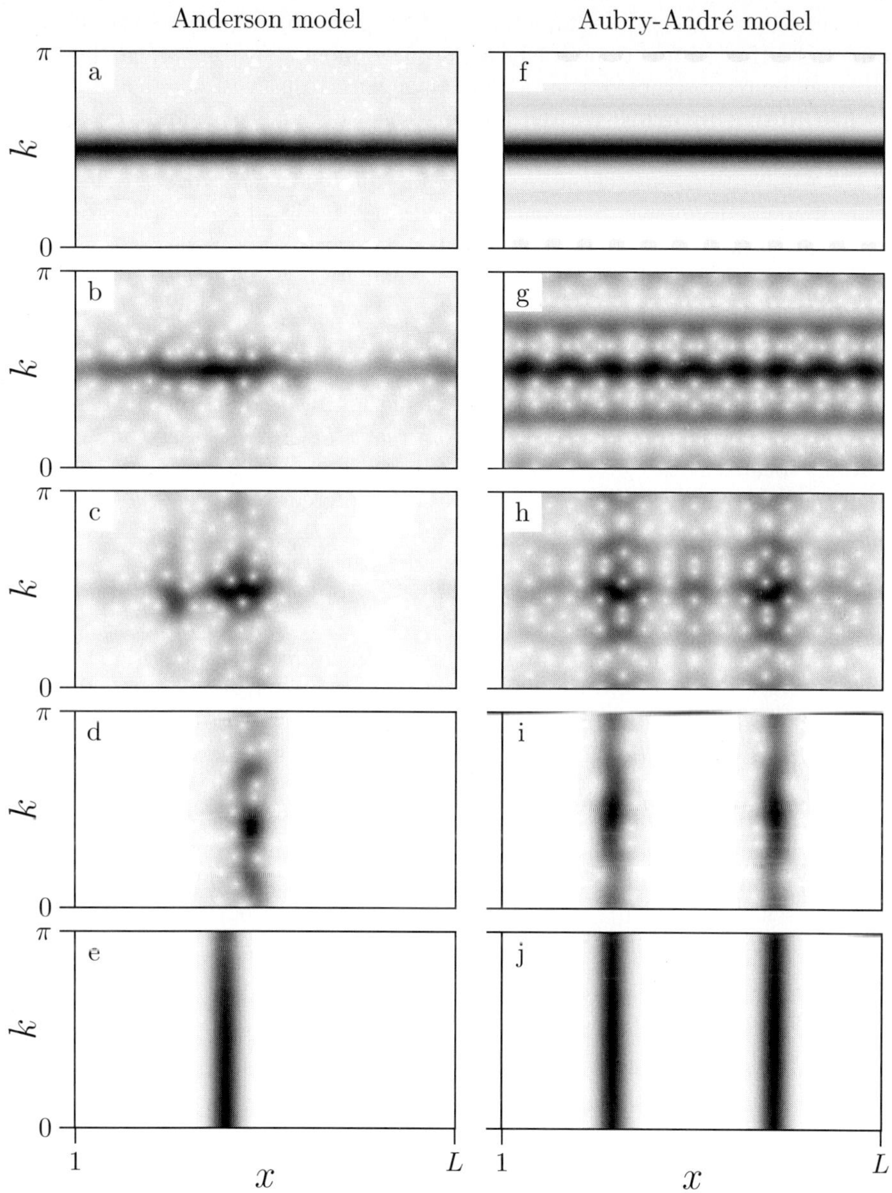

Fig. 3. Husimi functions for **a–e** the Anderson model with disorder strength $W = 0.1, 1, 2, 4$, and 40 and **f–j** the Aubry-André model with potential strength $\lambda = 0.1, 1.5, 2, 2.5$, and 10 for a state close to the band center and a system size $L = 377$. In view of the symmetry $\rho(x, -k) = \rho(x, k)$ only the upper half of the phase space is shown

definition (9), respects the symmetry $\rho(x,k) = \rho(x,-k)$. It is therefore sufficient to plot the Husimi functions solely for positive momenta from which the lower part may be reconstructed as a mirror image with respect to $k = 0$.

The Husimi functions for the limiting cases of very weak and very strong potential depicted in Figs. 3a and 3e for the Anderson model and Figs. 3f and 3j for the Aubry-André model can readily be interpreted in terms of states localized either in momentum or in real space. In addition, in Fig. 3f the coupling between plane waves due to the periodic potential becomes visible while the localization at two sites seen in Fig. 3j arises because the calculation was restricted to the class of antisymmetric states [17]. Furthermore, we keep the state number fixed while passing through avoided crossings so that states may be localized at different positions as a function of potential strength (cf. e.g. Figs. 3d and 3e).

More interesting than the interpretation of the Husimi function in the limiting cases is a discussion of the transition from plane waves for a weak potential to spatially localized states for a strong potential. The one-dimensional Anderson model and the Aubry-André model display two different scenarios with the latter being also characteristic for higher-dimensional Anderson models (cf. Sect. 3.2).

For the one-dimensional Anderson model, the Husimi function for a plane wave (Fig. 3a) contracts in real space (Fig. 3b) and displays a very well localized core at intermediate disorder strength W (Fig. 3c). A further increase of the potential strength leads to a spreading in momentum direction (Fig. 3d) and finally a state localized in real space (Fig. 3e) is approached.

For the Aubry-André model the scenario is quite different. As mentioned above, already a weak potential strength λ can lead to a coupling between plane waves of very different momenta (Fig. 3f). This type of coupling is accompanied by an increased filling of phase space (Fig. 3g) up to potential strengths λ located just below the critical value of $\lambda = 2$ where the localization transition takes place (Fig. 3h). There, for increasing system size, a more and more abrupt contraction in phase space occurs, cf. Fig. 3i. Finally, for very strong potentials, the Husimi function of a localized antisymmetric state depicted in Fig. 3j forms.

3.2 Inverse Participation Ratios

While the Husimi functions depicted in Fig. 3 nicely elucidate the phase space behavior of a state as a function of potential strength, a more quantitative analysis is desirable. On the basis of inverse participation ratios, a study of the higher-dimensional Anderson model becomes feasible and averages over disorder realizations can be performed. This in turn will allow us to draw general conclusions about the physics that rules the localization transition in the Anderson model.

In Fig. 4 we present distributions of the inverse participation ratios P_x, P, and P_k in real space, phase space, and momentum space, respectively, for the Anderson model in one, two, and three dimensions. The distributions have been obtained by diagonalizing the Anderson model for 50 disorder realizations in $d = 1$ and 2 and taking half of the eigenstates around the bandcenter for each realization. In $d = 3$ it was sufficient to consider only 20 disorder realizations.

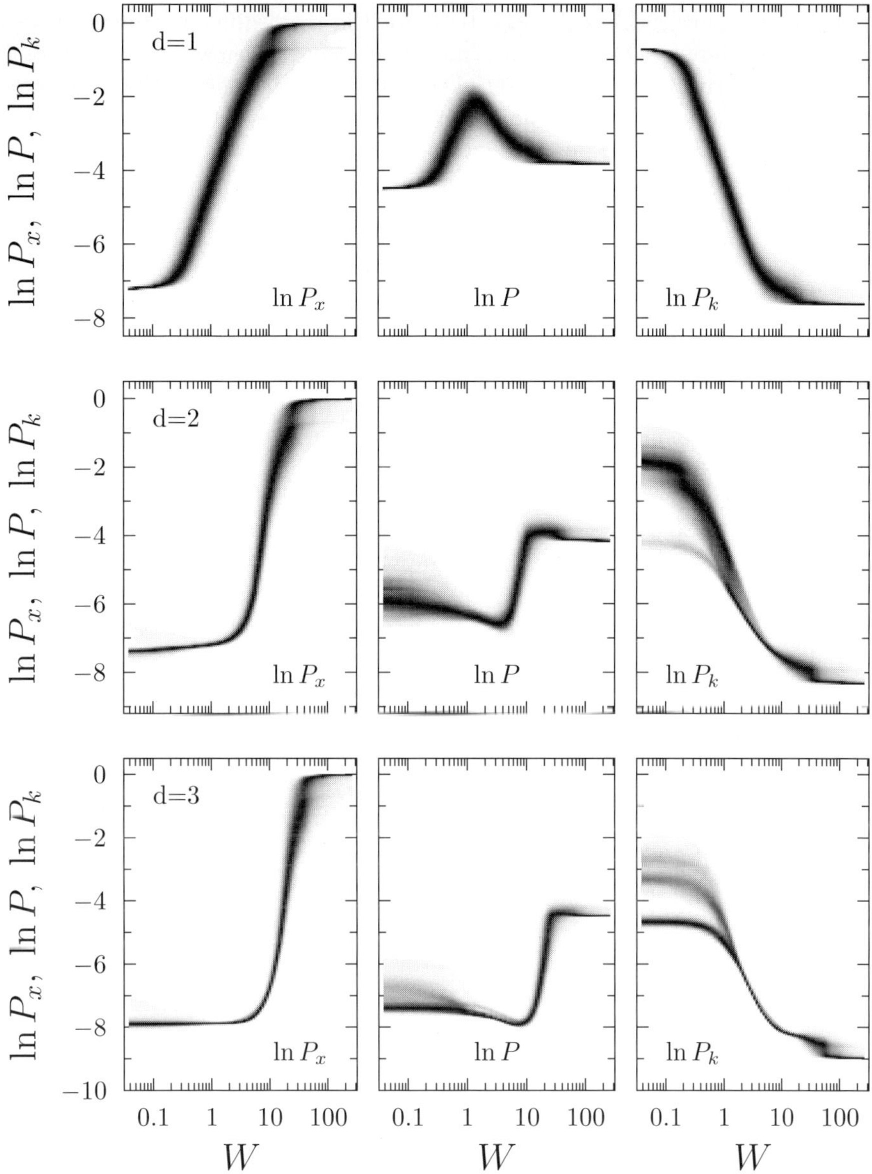

Fig. 4. Distribution of inverse participation ratios in real space (left), phase space (*middle*), and momentum space (*right*) for the Anderson model in $d = 1$ (*upper row*, $L = 2048$), $d = 2$ (*middle row*, $L = 64$), and $d = 3$ (*lower row*, $L = 20$)

Upon comparing the real space IPRs for the three different dimensions, no qualitative differences can be seen. The overall behavior shows an increase of P_x (left column) with increasing potential strength reflecting the spatial localization of the wave functions. Correspondingly, the momentum space IPR (right column) decreases with increasing potential strength.

In clear contrast, the phase space IPR behaves qualitatively different for the Anderson model in $d = 1$ and $d \geq 2$. For $d = 1$, the peak in P is consistent with our observations for the Husimi function in Figs. 3a–e, namely the contraction of the Husimi function at intermediate disorder strength.

For $d = 2$ and 3, the behavior of the phase space IPR rather corresponds to what we have observed for the Husimi function of the Aubry-André model in Figs. 3f–j. With increasing potential strength, the Husimi function spreads in phase space implying a decrease of P. Then, at the transition, a sudden contraction occurs, implying a jump to relatively large values of P. This scenario can be understood in physical terms as follows. The large spread in phase space is associated with the presence of a diffusive regime which is known to exist in $d = 2$ and larger. This interpretation is corroborated by a comparison with results from energy level statistics [9]. The jump in P, on the other hand, indicates the Anderson transition [18,19]. Here, one might object that the case $d = 2$ is the marginal case where strictly speaking no phase transition occurs. One finds indeed, that both, the minimum and the maximum of P, shift towards vanishing disorder as the system size is increased [20]. In contrast, in $d = 3$, the minimum of P shifts to larger disorder strength W, while the maximum shifts in the opposite direction, in agreement with the existence of the Anderson transition.

A comparison of the Husimi functions for the one-dimensional Anderson model and the Aubry-André model reveals the differences between the Anderson model in $d = 1$ and in higher dimensions. In the one-dimensional Anderson model, a weak disorder potential predominantly couples a plane wave to other, energetically almost degenerate plane waves. Therefore, the coupling occurs between plane waves of almost the same momentum. Due to the finite resolution in phase space, this coupling cannot be observed in momentum space. In real space, however, the coupling results in large scale variations of the Husimi function which in the end cause a contraction in phase space as discussed above. In clear contrast, for the Aubry-André model the situation is quite different: Here, the coupling predominantly occurs to distant momentum values as can be seen in Fig. 3f. This finally leads to a spreading of the Husimi function in phase space [14]. The very same scenario is valid for the Anderson model in two and higher dimensions. Again, there exist energetically almost degenerate momenta which are very different from those present in the original plane wave. The coupling due to a weak disorder potential thus again leads to a spreading in phase space. On the other hand, perturbation theory for strong disorder shows that the limiting value of the phase space IPR for large W is approached from above [20]. Therefore, a jump from small to large values of P arises and thus a phase transition (in $d = 3$) is to be expected at an intermediate disorder strength W.

In conclusion, these considerations in phase space in terms of concepts such as the Husimi function and the corresponding inverse participation ratio prove indeed very valuable in order to explore in greater detail the physics for a class of quantum systems which is so dear to Bernhard Kramer, namely disordered quantum systems [3].

Acknowledgment

The authors have enjoyed constructive and insightful discussions with S. Kohler, I. Varga, and D. Weinmann. This work was supported by the Sonderforschungsbereich 484 of the Deutsche Forschungsgemeinschaft. The numerical calculations were partly carried out at the Leibniz-Rechenzentrum München. Moreover, two of us (P.H., G-L.I.) are eagerly looking forward to see many more insightful and provoking achievements by Bernhard Kramer; he is still young and vivacious enough to contribute to great science.

References

1. T. Dittrich, P. Hänggi, G.-L. Ingold, B. Kramer, G. Schön, W. Zwerger: *Quantum transport and dissipation* (Wiley-VCH 1998)
2. S. Fishman, D.R. Grempel, R.E. Prange: Phys. Rev. Lett. **49**, 509 (1982)
3. B. Kramer, A. MacKinnon: Rep. Prog. Phys. **56**, 1469 (1993)
4. M. Hillery, R.F. O'Connell, M.O. Scully, E.P. Wigner: Phys. Rep. **106**, 121 (1984)
5. A. Wehrl: Rep. Math. Phys. **16**, 353 (1979)
6. T. Gorin, H.J. Korsch, B. Mirbach: Chem. Phys. **217**, 145 (1997)
7. D. Weinmann, S. Kohler, G.-L. Ingold, P. Hänggi: Ann. Phys. (Leipzig), **8**, SI-277 (1999)
8. G. Manfredi, M.R. Feix: Phys. Rev. E **62**, 4665 (2000)
9. A. Wobst, G.-L. Ingold, P. Hänggi, D. Weinmann: Eur. Phys. J. B **27**, 11 (2002)
10. D.J. Thouless: Phys. Rep. **13**, 93 (1974)
11. A. Sugita, H. Aiba: Phys. Rev. E **65**, 036205 (2002)
12. A. Sugita: arXiv:nlin.CD/0112042
13. I. Varga, J. Pipek: arXiv:cond-mat/0204041
14. G.-L. Ingold, A. Wobst, C. Aulbach, P. Hänggi: Eur. Phys. J. B **30**, 175 (2002)
15. P.W. Anderson: Phys. Rev. **109**, 1492 (1958)
16. S. Aubry, G. André: Ann. Israel Phys. Soc. **3**, 133 (1980)
17. D.J. Thouless: Phys. Rev. B **28**, 4272 (1983)
18. E. Abrahams, P.W. Anderson, D.C. Licciardello, T. V. Ramakrishnan: Phys. Rev. Lett. **42**, 673 (1979)
19. P.A. Lee, T.V. Ramakrishnan: Rev. Mod. Phys. **57**, 287 (1985)
20. A. Wobst, G.-L. Ingold, P. Hänggi, D. Weinmann: in preparation

Transfer Matrix Studies of Left-Handed Materials

C.M. Soukoulis[1] and P. Markoš[2]

[1] Ames Laboratory and Dept. of Physics Ames, Iowa, 50011, USA
[2] Institute of Physics, Slovak Academy of Sciences, Dubravska cesta 9, 842 28 Bratislava, Slovakia

1 Introduction

Photonic band gap (PBG) materials are by definition composites, whose properties are not determined by the fundamental physical properties of their constituents but by the shape and distribution of specific patterns included in them. They were originally introduced to control the electromagnetic wave propagation [1,2]. Not only dielectric [1,2] but also metallic structures [3–5] were proposed for applications in the microwave and infrared regions.

Very recently, a new area of research, called left-handed materials (LHMs), has been introduced by Pendry *et al.* [6,7] and Smith *et al.* [8,9]. Left-handed materials, as were constructed until now, represent a new class of PBG structures, which possess, due to certain patterns and distributions of thin metallic components, in a given frequency region, both negative effective permittivity ϵ_{eff} and permeability μ_{eff}. No such materials are available in nature.

The first step towards the construction of the left-handed materials was done by Pendry *et al.* [7] who did the theoretical analysis [7] of the magnetic response of the split ring resonators (SRRs) (for the shape of a SRR see Fig. 1) to time dependent external magnetic fields. For the periodic array of SSRs, resonant frequency dependent effective magnetic permeability was predicted [7]. Then, for frequencies close to the resonance frequency f_m, the real part of the magnetic permeability μ_{eff} is negative. It is also well known [6,3] that an array of thin metallic wires behaves like a high-pass filter. This means that the effective dielectric constant is negative at low frequencies. By combining a two-dimensional (2D) array of SRRs with a 2D array of wires, Smith *et al.* [8] demonstrated for the first time the existence of left-handed materials.

For electromagnetic waves, propagating through materials which have both $\epsilon_{\text{eff}} < 0$ and $\mu_{\text{eff}} < 0$, the phase and group velocity have opposite sign. This inspired Veselago [10] to introduce a term "left-handedness", to emphasize the fact that the electric intensity $\boldsymbol{E}$, magnetic intensity $\boldsymbol{H}$ and the vector $\boldsymbol{k}$ are related by a left-handed rule.

Left-handedness gives rise to a number of novel physical properties, not observable in the conventional right-handed world. Some of them were predicted already by Veselago over 30 years ago [10], others are described in [9]. Shelby *et al.* [11] demonstrated experimentally one of the most challenging properties of LHMs, namely the negative refraction of the electromagnetic waves. This con-

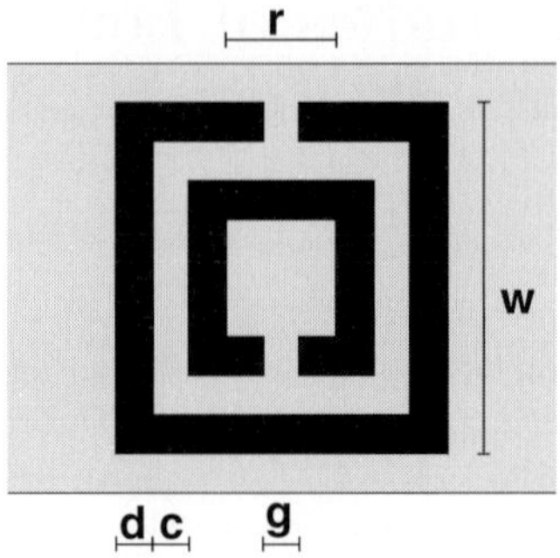

Fig. 1. The shape of the split ring resonator (SRR) and definition of its parameters. The SRR consists of thin metallic rectangles with small gaps. For a given configuration, the SRR exhibits a resonant response to an external magnetic field. In the resonance interval, an array of SRRs behaves as medium with negative effective permeability [7]. Resonance frequency f_m is of order of 10 GHz for SRRs of the size of 3 mm

firmed that the index or refraction n is negative in a LHM. Also, Pendry [12] has suggested that a LHM with negative n can make a perfect lens. These unusual results have raised objections both to the theoretical analysis and to the interpretation of the experimental data [15,16]. Nevertheless, measurements of the transmission properties of new designs of LHM have been performed [17,18], which support the predictions of the theory.

All the experiments that showed left-handed behavior were performed in the microwave regime. It is of interest to examine if it will be possible to observe left-handed behavior at the optical or the far-infrared regime. Using the metallic structures needed in the microwave experiments might not be possible to fabricate them in such small length scales. It was therefore proposed that photonic crystals might have some frequency region, which will show left-handed behavior [19–24,24]. Notomi [19] studied light propagation in strongly modulated two dimensional (2D) photonic crystals. Such photonic crystal (PC) behaves as a material having an effective refractive index n_{eff} controllable by the band structure. In these PC structures the permittivity is periodically modulated in space and is positive. The permeability is equal to one. Negative n_{eff} for a frequency range was found. The existence of negative n_{eff} was demonstrated [20,23] by a finite difference time domain (FDTD) simulation. Negative refraction on the interface of a 3D PC structure has been experimentally observed by Kosada *et al.* [24]. Very recently negative refraction and superlensing has been experimentally observed [24] in 2D photonic crystals. Similar unusual light propagation was observed [25] in 1D and 2D diffraction gratings.

Specific properties of LHMs make them interesting for physical and technological applications. While experimental preparation of the LHM structures is rather difficult, especially when isotropic structures are required, numerical simulations could predict how the transmission properties depend on various structural parameters of the system. It would be extremely difficult, if not impossible, to predict the transmission properties of such materials analytically. Numerical simulations of various configurations of SRRs and of LHMs could be therefore very useful in searching the direction of the technological development.

Here we review recent progress in the transfer matrix numerical simulations of left-handed structures. We have employed the transfer matrix technique to calculate the transmission and the reflection properties of the LHM structures [13] and proved that the data for the transmission and reflection can be fitted by length independent and frequency dependent effective permittivity and permeability [14]. Our results confirm that in a resonance frequency region both ϵ_{eff} and μ_{eff} are negative. The index of refraction n_{eff} was found to be unambiguously negative [14,28].

2 Transfer Matrix Method

Transfer-matrix calculations are based on the scattering formalism. The sample is considered as the scatterer of an incoming wave. The incident wave (say, from the left) is either reflected or transmitted through the sample. Then, the transfer matrix is defined by relating the incident fields on one side of the structure with the outgoing fields

$$\mathcal{T} = \begin{pmatrix} t_-^{-1} & -t_-^{-1} r_- \\ r_+ t_-^{-1} & t_+ - r_+ t_-^{-1} r_- \end{pmatrix} . \tag{1}$$

Here t_- (t_+) is the transmission matrix for the wave incident from the right (left) of the slab, respectively, and r_- (r_+) is the corresponding reflection matrix.

The transfer matrix method (TMM) was used by MacKinnon and Kramer [29] to confirm numerically the scaling theory of localization. Since then, TMM became the standard tool for the numerical analysis of transport properties of disordered electronic structures. Later, Pendry and MacKinnon [30] generalized the TMM to the numerical simulation of the transmission of electromagnetic waves through non-homogeneous media [31,32]. TMM was also used in numerical simulations of photonic band-gap materials (for references see ref. [1]). In all these examples, agreement between the theoretical predictions and the experimental measurements was very good [1].

As was discussed above, LHM is a periodic structure. Therefore, when the TMM is applied to the studies of LHMs, we can concentrate on single unit cell and use periodic boundary conditions along the lateral directions. A typical unit cell is shown in Fig. 2. In numerical calculations, the total volume of the unit cell is divided into small cells and fields in each cell coupled to those in the neighboring cell following discretized Maxwell's equations [31,33] with periodic boundary conditions in the x and y directions. If the number of mesh points in the plane perpendicular to the propagation is N, then the transfer matrix is of the size $4N \times 4N$. Corresponding transmission and reflection matrices t and r are then of the size $2N \times 2N$. The factor 2 is due to two polarizations of the EM wave. N varies from 10^2 to 20^2 in the numerical simulations.

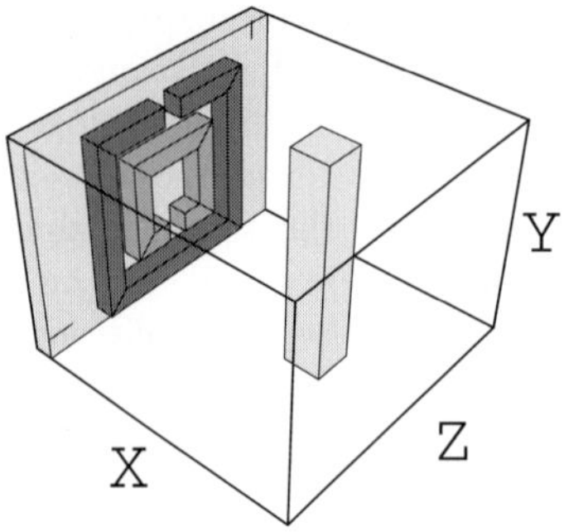

Fig. 2. The structure of the unit cell of the left-handed metamaterials as was used in the present simulations. The split ring resonator is deposited on the planar dielectric board, the wire is located opposite to SRR. The structure acts as the left-handed metamaterial if the electromagnetic wave propagates along the z direction and is polarized with electric field $\boldsymbol{E}$ parallel to the wire and magnetic field $\boldsymbol{H}$ parallel to the axis of the SRR. Periodic boundary conditions along transversal directions (x and y) are used in the simulations to mimic a structure of infinite size. The size of the unit cell is typically 3-5 mm, which is much less than the wavelength of the electromagnetic wave used in the experiments. Thus, in studies of the electromagnetic response, the LHM can be considered as a homogeneous material, in spite of rather complicated structure of the unit cell

In the leads, attached to the sample, both $\epsilon = 1$ and $\mu = 1$ and the transfer matrix can be diagonalized analytically [32]. One half of the $4N$ eigenvectors correspond to the propagation to left, and the second half to propagation to the right. Using the general form of the transfer matrix as given by (1) we can express the transmission and reflection matrices explicitly as

$$t_-^{-1} = \mathcal{L}_1 T \mathcal{R} \quad \text{and} \qquad r_+ t_-^{-1} = \mathcal{L}_2 \mathcal{T} \mathcal{R}. \tag{2}$$

In (2), $\mathcal{L}_1$ and $\mathcal{R}$ are $4N \times 2N$ ($2N \times 4N$) matrices, that contain the left (right) eigenvectors of the left-going wave, and $\mathcal{L}_2$ is a $2N \times 4N$ matrix containing the left eigenvectors of the right-going wave.

For the frequency range of interest, the transfer matrix has only one propagating mode for each polarization. Hence, the transmission coefficient is $T_{ij} = \sum_{k=1}^{2N} t_{ik} t_{jk}^*$ with $i, j = 1$ or 2 for the p or s polarized wave.

The most important technical problem of the application of the TMM is the numerical stability of the simulated data. This problem is crucial in the present simulations, since the values of the permittivity of the components of the LHM structure differ from each other in many orders of magnitude: while for air $\epsilon_a = 1$, the typical permittivity of metal in our simulation is $\epsilon_m = (-3 + 6\,i) \times 10^5$.

Following the algorithm described by Pendry *et al.* [33], we have developed our version of the transfer matrix code. The main change from the standard algorithm commonly used in the PBG studies [31,32] is the faster normalization of the transmitted waves in the calculation of the transmission coefficient.

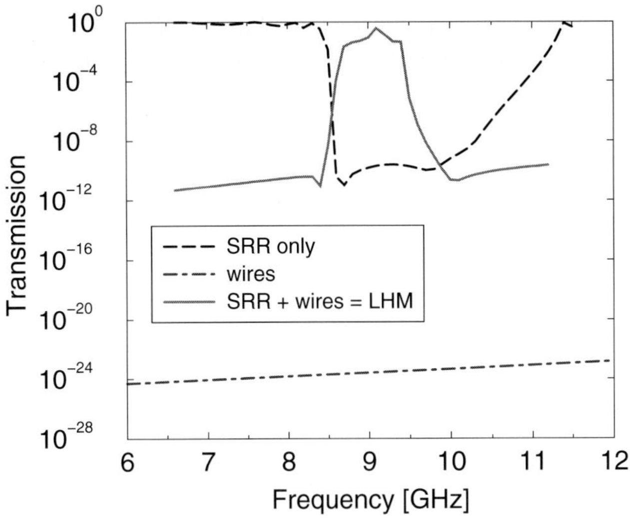

Fig. 3. Frequency dependence of the transmission, as is typically obtained from the numerical analysis. Three composites are considered: (1) a periodic array of wires, which is supposed to have $\mu_{\text{eff}} > 0$ and $\epsilon_{\text{eff}} < 0$. The transmission is very small (*dot-dashed line*). (2) a periodic array of split ring resonators that has $\epsilon_{\text{eff}} > 0$, while μ_{eff} is negative within the resonance interval (8.5-10 GHz in this particular case). Consequently, transmission is of the order of one for frequencies outside the resonance interval, and drops to very small values inside the resonance interval. (3) Quite opposite behavior is expected for a combination of both wires and split ring resonators. This structure possesses $\epsilon_{\text{eff}} < 0$ and $\mu_{\text{eff}} > 0$ outside the resonance interval. Inside the resonance interval, both ϵ_{eff} and μ_{eff} are negative, which enables transmission through the structure (*solid line*). The same scenario was observed experimentally [8]

The TMM has been used to simulate the reflection, transmission and the absorption from arrays of wires, square SRRs and LHMs. Figure 3 describes typical results of the transmission versus frequency for three different structures: an array of metallic wires, an array of split ring resonators, and finally of the left-handed structure which consists of both the SRR and of metallic wires placed uniformly between the SRRs.

The square array of metal wires alone behaves as a high pass filter with a cutoff frequency f_p, which is approximately given [3,4] by the relation $f_p = c_{\text{clight}}/2a\sqrt{\epsilon}$ (a is the distance between the wires, c_{light} is the velocity of light in the air, and ϵ is the dielectric constant of the background). The cutoff frequency is almost independent of the value of Im ϵ_{m}. In the case shown in Fig. 3, $a = 5$mm, $\epsilon = 1$ and $f_p \approx 19$ GHz. The dashed curve, shown in Fig. 3 shows transmission through the SRR only. By adding wires uniformly between the SRRs, as is shown in Fig. 2 a pass band occurs where it is believed that both ϵ_{eff} and μ_{eff} are negative (solid line).

We have studied the dependence of the resonance frequency f_m of an array of SRRs on the ring thickness d, inner radius r, radial (c) and azimuthal (g) gaps (for understanding of the parameters, see Fig. 1), as well as on the electrical permittivity of the embedding medium, where the SRR and wire reside. We found also that the resonance frequency, as well as the width of the resonance interval strongly depends on the mutual position of the SRR and wire, on the orientation of the azimuthal gaps of the SRR, thickness of wire and, last but not least, on the size and the shape of the unit cell [13]. The transmission losses of the left-handed structures and the role of absorption of the metallic components of SRRs and LHMs were investigated [28].

We believe that the TMM can be reliably used to calculate the transmission and reflection properties of left-handed materials and SRRs. However, one disadvantage of the present transfer matrix method should be noted, namely that it cannot treat structures with smaller length scales than our discretization mesh. For example, the thickness of the SRR is an order of magnitude smaller in experiments than in our simulations. Also the structural parameters of SRRs can be changed only discontinuously as multiples of the unit mesh length. This difficulty could be partially overcome by generalizing the present code to a nonuniform mesh discretization. Nevertheless, already uniform discretization enables us to obtain credible results. A comparison of our results with that obtained by the commercial software MAFIA and Microwave studio [34] confirmed that the three methods find the same position of the resonant gap provided that they use the same mesh discretization.

3 Determination of Effective Parameters of LHMs

It is assumed that LHMs respond to electromagnetic radiation as continuous materials, at least in the long wavelength limit. In recent experiments and simulations [6–8,13,14], it has been demonstrated that certain LHMs exhibit scattering behavior consistent with the assumption of an approximate frequency dependent form for $\epsilon_{\rm eff}$ and $\mu_{\rm eff}$. However, the techniques applied in those studies probed the materials indirectly, and did not provide an explicit measurement that would assign values for permittivity and permeability.

We have used in [14] the reflection and transmission coefficients calculated from the transfer matrix simulations on finite length of the LHM to determine $\epsilon_{\rm eff}$ and $\mu_{\rm eff}$. The wavelength of EM wave of a frequency of 10 GHz is an order of magnitude larger than the period of the left-handed structure. We therefore consider the LHM as a homogeneous slab of thickness d (given as the number of unit cells along the propagation direction). We then can invert the well-known relations for the transmission and reflection as a function of refraction index n and the impedance z

$$\frac{1}{t} = \cos(nkd) - \frac{i}{2}\left(z + \frac{1}{z}\right)\sin(nkd) \quad \text{and} \quad \frac{r}{t} = -\frac{i}{2}\left(z - \frac{1}{z}\right)\sin(nkd), \tag{3}$$

(k is a wave vector of the EM wave in vacuum). For passive materials, Re z and Im n must be greater than zero. These conditions enabled us to choose the proper solution of (3).

Using the numerical data for systems of several lengths d, we obtain from (3) the effective $n_{\rm eff}$ and z which determines the effective permittivity and permeability as $\epsilon_{\rm eff} = n_{\rm eff}/z$ and $\mu_{\rm eff} = z n_{\rm eff}$. In [14] we have estimated $\epsilon_{\rm eff}$ and $\mu_{\rm eff}$ for periodic arrangements of wires, SRRs and both wires and SRRs (i.e. LHMs) Our numerical analysis clearly demonstrated that the frequency dependence of the effective permittivity for wires and permeability for SRRs are given by

$$\epsilon_{\rm eff} = 1 - \frac{f_p^2}{f^2 + i\gamma_e f}, \tag{4}$$

and

$$\mu_{\rm eff} = 1 - \frac{F f^2}{f^2 - f_m^2 + i\gamma_m f}, \tag{5}$$

respectively. In (4) and (5), f_p is the electronic plasma frequency, f_m is the magnetic resonance frequency, and γ_e (γ_m) represent the losses of the system.

Figure 4 presents the recovered frequency dependence of real and imaginary part of $n_{\rm eff}$, z, $\epsilon_{\rm eff}$ and $\mu_{\rm eff}$ for a LHM. Notice that indeed the Re $n_{\rm eff}$ is negative in the resonance interval of the LHM. In addition, the Im $n_{\rm eff}$ in this resonance regime is only of the order of 10^{-2}. This guarantees that transmission losses of a LHM are relatively small. As expected Re $\epsilon_{\rm eff}$ and Re $\mu_{\rm eff}$ are negative in the resonance interval. Surprisingly, the imaginary part of the effective permittivity $\epsilon_{\rm eff}$ is also found to be negative. This seems to contradict our physical intuition, which requires that in a passive material the electromagnetic losses

$$Q = \frac{1}{2\pi}\int {\rm d}\omega\ \omega \left[{\rm Im}\ \epsilon_{\rm eff}|\boldsymbol{E}|^2 + {\rm Im}\ \mu_{\rm eff}|\boldsymbol{H}|^2\right] \tag{6}$$

[35] must be positive. This is trivially fulfilled if both Im $\epsilon_{\rm eff}$ and Re $\mu_{\rm eff}$ are positive. Nevertheless, it can be shown that, at least for our particular case of normally incident plane wave, Q could be expressed in terms of Im $n_{\rm eff}$ and Re z instead of Im $\epsilon_{\rm eff}$ and Im $\mu_{\rm eff}$:

$$Q = \frac{1}{2\pi}\int {\rm d}\omega\ \omega|\boldsymbol{H}|^2\ {\rm Im}\ n_{\rm eff}\ {\rm Re}\ z \tag{7}$$

so that Q is always positive as Re z and Im $n_{\rm eff}$ are always positive. There is therefore no physical constrain to the sign of the imaginary part of the permittivity and permeability.

4 Conclusion

We have used the transfer-matrix method to calculate the transmission properties of left-handed materials and arrays of split ring resonators. We have also demonstrated that the traditional procedure of obtaining material parameters from transmission/reflection data can be successfully applied to LHMs

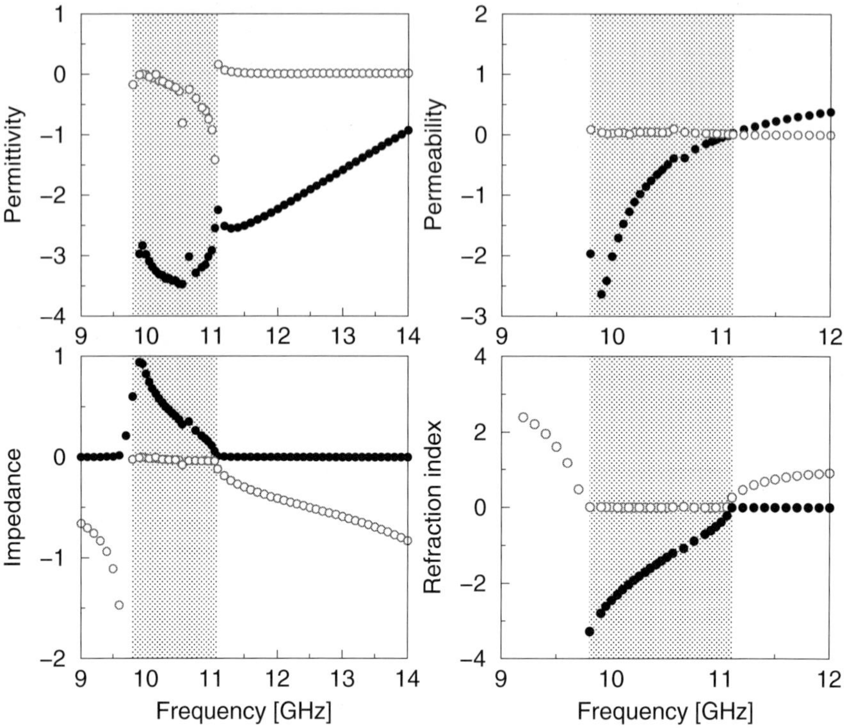

Fig. 4. Frequency dependence of the effective index of refraction n_{eff}, impedance z, permittivity ϵ_{eff} and permeability μ_{eff} in the resonance interval (highlited by *dashed area*). Real (imaginary) part is represented by *solid (open) circles*, respectively. The main features of the effective parameters are visible: Real parts of ϵ_{eff}, μ_{eff} and n_{eff} are negative, in agreement with theoretical expectations. The effective permeability μ_{eff} exhibits resonant behavior as predicted by (5). The frequency dependence of the effective permittivity ϵ_{eff} is more complicated. For an array of wires, we found that ϵ_{eff} agree with the predicted relation (4). However, the electrical response of SRRs superposes to (4) an "anti-resonant" behavior. This is a reason why Re ϵ_{eff} possesses a minimum within the resonance interval [14]. The imaginary part of ϵ_{eff} is also negative, which is nevertheless consistent with theory. For frequencies below the pass band, we were not able to estimate the real part of the refraction index. The spatial oscillations of the transmission are so large that the wavelength of the transmitted wave is comparable with the size of the unit length. In this limit, relations (3) are not applicable any more

and SRRs. We successfully obtained the frequency dependence of effective permittivity, permeability and of the index of refraction for LHMs and SRRs. The technique we have used, in obtaining ϵ_{eff} and μ_{eff} from the scattering data, will be readily applicable to the experimental characterization of metamaterial samples whenever the scattering parameter (their amplitudes as well as phases) are known.

Our results demonstrate that the transfer-matrix method can be reliably used to calculate the transmission and reflection properties of left-handed materials. Thus, numerical simulations could answer some practical questions about different proposed structures, which might be too complicated to be treated by analytical studies. The transfer-matrix method can be used in the future for detailed studies of two dimensional or even three dimensional structures, which are expected to exhibit left-handed behavior for the EM wave propagating in any direction. These structures should contain more SRR and wires per unit cell, which makes their analytical analysis difficult. It is important to find the best design and test the transmission properties of proposed metamaterial even before their fabrication and experimental measurements start.

Ames Laboratory is operated for the U.S. Department of Energy by Iowa State University under Contract No. W-7405-Eng-82. This work was supported by the Director of Energy Research, Office of Basic Science, DARPA (MDA972-01-2-0016), NATO Grant No. PST.CLG.978088, APVT (51-021602) and EU project DALHM.

References

1. Photonic Band Gap Materials, ed. by C.M. Soukoulis (Kluwer, Dordrecht, 1996); Photonic crystals and Light Localization in the 21st Century, ed. by C.M. Soukoulis (Kluwer, Dodrecht, 2001)
2. J.D. Joannopoulos, R.D. Mead, and J.N. Winn, Photonic Crystal (Princeton University Press, Princeton, 1995); J. D. Joannopoulos, P.R. Villeneuve, and S. Fan, Nature (London) **386** 143 (1997)
3. D.R. Smith, S. Schultz , N. Kroll, M. Sigalas, K.M. Ho, and C.M. Soukoulis, Appl. Phys. Lett. **65** 645 (1994)
4. E. Ozbay, B. Temelkuran, M. Sigalas, G. Tuttle, C.M. Soukoulis and K.M. Ho, Appl. Phys. Lett. **69** 3397 (1996); M. Sigalas, C.T. Chan, K.M. Ho, and C.M. Soukoulis, and C.M. Soukoulis, Phys. Rev. B **52** 11 744 (1995)
5. D.F. Sievenpiper, E.H. Yablonovitch, J.N. Winn, S. Fan, P.R. Villeneuve, and J.D. Joannopoulos, Phys. Rev. Lett. **80** 2829 (1998)
6. J.B. Pendry, A.J. Holden, W.J. Stewart, and I. Youngs, Phys. Rev. Lett. **76** 4773 (1996); J.B. Pendry, A.J. Holden, D.J. Robbins, and W.J. Stewart, J. Phys.: Condes. Matter **10** 4785 (1998)
7. J.B. Pendry, A.J. Holden, D.J. Robbins and W.J. Stewart, IEEE Trans. Microwave Theory Techn. **47** 2075 (1999)
8. D.R. Smith, W.J. Padilla, D.C. Vier, S.C. Nemat-Nasser, and S. Schultz, Phys, Rev. Lett. **84** 4184 (2000)
9. R.A. Shelby, D.R. Smith and S. Shultz, Appl. Phys. Lett. **78** 489 (2001)
10. V.G. Veselago, Usp. Fiz. Nauk **92** 517 (1968) [Sov. Phys. Usp. **10** 509 (1968)
11. J.B. Pendry, Phys. World **13** 27 (2000)
12. R.A. Shelby, D.R. Smith, and S. Schultz, Science **292** 77 (2001)
13. J.B. Pendry, Phys, Rev. Lett. **85** 3966 (2000)
14. N. Garcia and M. Nieto-Vesperinas, Phys. Rev. Lett. **88** 207403 (2002)
15. P.M. Valanju, R.M. Walser and A.O. Valanju, Phys. Rev. Lett. **88** 187401 (2002)
16. M. Bayindir, K. Aydin, E. Ozbay, P. Markoš, and C.M. Soukoulis, Appl. Phys. Lett. **81** 120 (2002)

17. K. Li, S.J. Mc Lean, R.B. Greegor, C.G. Parazolli and M.H. Tanielian, Appl. Phys. Lett. (to be published)
18. M. Notomi, Phys. Rev. B **62** 10696 (2002)
19. B. Gralak, S. Enoch and G. Tayed, J. Opt. Soc. Amer A **17** 1012 (2002)
20. C. Luo et al., Phys. Rev. B **65** 201104 (R) (2002)
21. S. Foteinopoulou and C.M. Soukoulis, unpublished
22. S. Foteinopoulou, E.N. Economou and C.M. Soukoulis submitted to Phys. Rev. Lett. (cond-mat/0210346)
23. H. Kosaka, T. Kanashima, A. Tomita, M. Notomi, T. Tamamura, T. Sato, and S. Kawakami, Phys. Rev. B **58** R10096 (1998)
24. E. Cubukcu, K. Aydin, E. Ozbay, S. Foteinopoulou and C.M. Soukoulis, submitted to Nature
25. P. St. J. Russel, Phys. Rev. A **33** 3232 (1986)
26. P. Markoš and C.M. Soukoulis, Phys. Rev. B **65** 033401 (2001); P. Markoš and C.M. Soukoulis, Phys. Rev. E **65** 03622 (2002)
27. D.R. Smith, S. Schultz, P. Markoš and C.M. Soukoulis, Phys. Rev. B **65** 195104 (2002)
28. P. Markoš, I. Rousochatzakis and C.M. Soukoulis, Phys. Rev. E **66**, 045601(R) (2002)
29. A. MacKinnon and B. Kramer, Phys. Rev. Lett. **47** 1546 (1981)
30. J.B. Pendry and A. MacKinnon Phys. Rev. Lett. **69** 2772 (1992)
31. J.B. Pendry, J. Mod. Opt. **41** 209 (1994)
32. J.B. Pendry and P.M. Bell, in Photonic Band Gap Materials, Ref. 1 p. 203
33. J.B. Pendry, A. MacKinnon, and P.J. Roberts, Proc. R. Soc. London, Ser. A **437** 67 (1992)
34. D. Vier, private communication
35. L.D. Landau, E.M. Lifshitz, and C.P. Pitaevskii Electrodynamics of Continuous Media, Pergamon Press, 1984

Emission Spectrum of Random Lasers

Gregor Hackenbroich and Fritz Haake

Fachbereich Physik, Universität Essen, D-45117 Essen, Germany

1 Introduction

In recent years advances in microstructuring techniques have made it possible to manufacture novel mirrorless cavities known as random lasers. Examples for this new type of laser include strongly disordered semiconductors [1,2], polymer systems [3], solutions of TiO_2 nanoparticles [4,5], as well as synthetic opals infiltrated with laser dyes [6]. The feedback in random lasers is provided by the random scattering of light in a dielectric with a spatially fluctuating refractive index. Light amplification results from the interaction with a medium of active atoms. The interest in random lasers is motivated both by possible applications and by their novel emission properties. Due to their intrinsic randomness such lasers often emit light in a large solid angle (up to 4π). Moreover, random lasers are cheap and can easily be designed with different shape and size. These properties may prove useful for applications in computer displays and photonic devices. In this paper we discuss our present theoretical understanding [7,8] of random lasers.

Random lasers differ from conventional lasers in two fundamental respects. First, while standard laser cavities are defined by mirrors no such mirrors are required for random lasers. As a result, light in random lasers is only weakly confined so that such lasers typically support a large number of overlapping modes. Second, the modes of standard lasers often have a simple spatial structure while random laser modes generically form complicated patterns that reflect the chaotic nature of disordered or wave-chaotic media. Random laser theory therefore must address lasing in wave-chaotic weakly confining laser resonators. In Fig. 1 we reproduce the emission spectrum of a pumped disordered ZnO powder film measured by Cao et al. [1]. From bottom to top the intensity of pumping was increased. At low pumping strength one observes the smooth and broad emission spectrum characteristic for amplified spontaneous emission. However, above a critical pump value sharp peaks emerge that signal the onset of laser action. With increasing pump strength, more and more such peaks become visible and a rather complex emission spectrum results.

In this paper we address the emission spectrum of random lasers both below and above the lasing threshold. Above threshold, we focus on the regime of single-mode lasing where the laser supports a single self-sustained laser oscillation. Weak confinement and random scattering of light manifest themselves in two ways. First, the spectral width of the laser line above threshold may be greatly enhanced compared with the line width of standard lasers. The enhancement

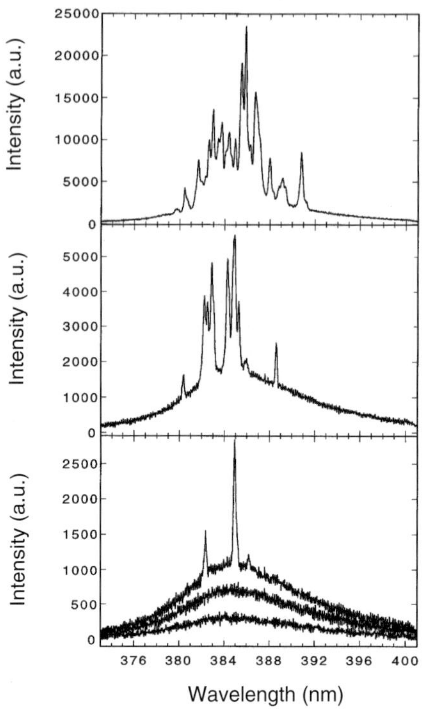

Fig. 1. Emission spectra from ZnO powder as measured by Cao et al. [1]. The pump strength increases from bottom to top. Clearly visible is the broad spontaneous emission peak at low pump intensity and sharp laser peaks above the threshold value of pump. From [1]

factor is known as the Petermann factor, and may be expressed in terms of left and right eigenstates of a non-Hermitian operator characterizing the open laser resonator. Second, the line shape of the laser line may significantly deviate from the Lorentzian found in conventional lasers. The deviation results from the interplay of the lasing mode with a large number of non-lasing oscillations below the laser threshold. In random media the line shape may be analyzed in a statistical fashion using random-matrix theory.

2 Linear Optical Media

A theory of random lasers must be based on a quantum description of the electromagnetic field in disordered dielectrics. The field quantization must account both for the internal field within the dielectric and for the external field which carries the laser input and output radiation. To quantize the field we recently employed [7,8] a system-and-bath approach. This approach allows the quantization of both scalar [7] and vector fields [8]. The dielectric resonator with position dependent dielectric function $\epsilon(\mathbf{r})$ is the system while the free space outside comprises the bath. We assume that $\epsilon(\mathbf{r})$ is real and independent of frequency. The shape of the resonator is arbitrary. Resonators defined by mirrors are a special case with appropriate boundary conditions on the mirrors. Openings between the mirrors and/or the finite transmission of the dielectric material allow for a coupling between the resonator and the external radiation field. We note that we do not restrict ourselves to good cavities with very large cavity quality factors Q. The quantization method, in fact, works for arbitrary values of Q.

The basic idea of the quantization method is to separate the electromagnetic field into two contributions [9]. These contributions account, respectively, for the field within the resonator and in the channel region outside. In both the resonator and the channels one may expand the field in terms of a complete and orthonormal set of modes. These modes are found upon solving the wave equation subject to a prescribed boundary condition at the surface of the resonator such that the eigenvalue problems both in the resonator and in the channel region are self-adjoint. This procedure yields a set of resonator and channel modes that would be eigenmodes of the respective subsystems if they were *completely isolated* from each other. In terms of the creation and destruction operators associated with these modes the Hamiltonian of the total system takes the form

$$\begin{aligned} H = & \sum_\lambda \hbar\omega_\lambda \, a_\lambda^\dagger a_\lambda + \sum_m \int d\omega \; \hbar\omega \, b_m^\dagger(\omega) b_m(\omega) \\ & + \hbar \sum_\lambda \sum_m \int d\omega \Big[\mathcal{W}_{\lambda m}(\omega) \, a_\lambda^\dagger b_m(\omega) + \mathcal{V}_{\lambda m}(\omega) \, a_\lambda b_m(\omega) + \text{h.c.} \Big] . \end{aligned} \tag{1}$$

Here the discrete index λ labels the resonator modes while the channel modes are labeled by the continuous frequency ω and a discrete channel index m. The Hamiltonian is a bilinear form of the field operators as expected for a free electromagnetic field. We note that (1) is the exact Hamiltonian of the problem. The continuity condition of the field at the boundary of the resonator gives rise to terms that couple resonator and channel operators. There are both resonant ($a^\dagger b$, $ab^\dagger$) and non-resonant (ab, $a^\dagger b^\dagger$) terms. The non-resonant terms can be discarded here since we are not interested in overdamping (where mode widths would be larger than or at least comparable to the optical frequencies [10]). We rather focus on the case of overlapping modes which is fully compatible with the rotating-wave approximation, where only the resonant terms are kept.

To investigate the effects of the resonator-channel coupling we study the field *dynamics*. The Heisenberg equations of motion for the resonator operators a_λ and the channel operators $b_m(\omega)$ are given by

$$\dot{a}_\lambda = -\mathrm{i}\omega_\lambda a_\lambda - \mathrm{i} \sum_m \int d\omega \; \mathcal{W}_{\lambda m}(\omega) b_m(\omega), \tag{2}$$

$$\dot{b}_m(\omega) = -\mathrm{i}\omega b_m(\omega) - \mathrm{i} \sum_\lambda \mathcal{W}_{\lambda m}^*(\omega) a_\lambda . \tag{3}$$

Integration of (3) starting from some initial time $t_0 < t$ yields

$$b_m(\omega, t) = e^{-i\omega(t-t_0)} b_m(\omega, t_0) - i \sum_\lambda \mathcal{W}^*_{\lambda m}(\omega) \int_{t_0}^{t} dt' \, e^{-i\omega(t-t')} a_\lambda(t'), \quad (4)$$

where $b_m(\omega, t_0)$ denotes the channel operator $b_m(\omega)$ at time t_0. In an analogous fashion, one can express $b_m(\omega, t)$ in terms of the channel operators at the final time $t_1 > t$,

$$b_m(\omega, t) = e^{-i\omega(t-t_1)} b_m(\omega, t_1) + i \sum_\lambda \mathcal{W}^*_{\lambda m}(\omega) \int_{t}^{t_1} dt' \, e^{-i\omega(t-t')} a_\lambda(t'). \quad (5)$$

Subtracting (4) from (5) and performing the asymptotic limits $t_0 \longrightarrow -\infty$ and $t_1 \longrightarrow \infty$ we arrive at the relation

$$b^{\text{out}}_m(\omega) - b^{\text{in}}_m(\omega) = -i \sum_\lambda \mathcal{W}^*_{\lambda m}(\omega) a_\lambda(\omega), \quad (6)$$

which connects the outgoing field $b^{\text{out}}(\omega) \equiv e^{i\omega t_1} b(\omega, t_1)$ with the incoming field $b^{\text{in}}(\omega) \equiv e^{i\omega t_0} b(\omega, t_0)$ and the resonator field a_λ. Equation (6) defines the input-output relation [11] of the open resonator. The input-output theory of quantum optics [11] has its origin in the LSZ scattering formulation of quantum field theory [12]. The connection with scattering theory becomes explicit when one eliminates the internal operators from the input-output relation. We substitute (4) into (2) and solve the resulting equation by Fourier transformation. Substitution into the input-output relation (6) yields a linear relation between the input and output field

$$b^{\text{out}}(\omega) = S(\omega) b^{\text{in}}(\omega), \quad (7)$$

where S is the scattering matrix. Here, we have combined the input and output operators to M-component vectors, where M is the number of open channels at frequency ω. The scattering matrix S is an $M \times M$ matrix. Likewise, the coupling amplitudes $\mathcal{W}_{\lambda m}$ form an $N \times M$ coupling matrix where N denotes the number of cavity modes. The finite number of cavity modes is artificial, and may eventually be taken to infinity. The scattering matrix takes a form well-known in nuclear and condensed-matter physics [13,14]

$$S(\omega) = 1 - 2\pi i \mathcal{W}^\dagger(\omega) D^{-1}(\omega) \mathcal{W}(\omega), \quad (8)$$

where D is an $N \times N$ matrix with the matrix elements

$$D_{\lambda\lambda'}(\omega) = (\omega - \omega_\lambda)\delta_{\lambda\lambda'} + \Delta_{\lambda\lambda'}(\omega) + i\pi \left(\mathcal{W}(\omega)\mathcal{W}^\dagger(\omega)\right)_{\lambda\lambda'}, \quad (9)$$

and $\Delta_{\lambda\lambda'}$ is the principal value integral

$$\Delta_{\lambda\lambda'}(\omega) = \sum_m \mathcal{P} \int d\omega' \frac{\mathcal{W}_{\lambda m}(\omega')\mathcal{W}^*_{m\lambda}(\omega')}{\omega' - \omega}. \quad (10)$$

Using the commutation relations $[b_n(\omega), b_m^\dagger(\omega')] = \delta_{nm}\,\delta(\omega-\omega')$ for both $b = b^{\text{in}}$ and $b = b^{\text{out}}$, one can easily show that S is unitary, $SS^\dagger = S^\dagger S = 1$. Equation (8) and its generalization to absorbing and amplifying media has recently [15,16] been used by Beenakker and co-workers to compute the emission from *linear* optical media in the presence of chaotic scattering.

According to (7) and (8), the output field dynamics is determined by the input field and the scattering matrix. The complex poles of S are the scattering *resonances*. At the same time the resonances are the solutions of the equation $\det D(\omega) = 0$. Thus they represent the complex eigenvalues of the internal field dynamics in the presence of damping inflicted by the coupling to the external radiation field. We note that the resonances determine the field dynamics of the open resonator even though the underlying field quantization is formulated in terms of closed-cavity eigenmodes. To estimate the photon occupation numbers associated with a typical input field we assume a thermal input at a characteristic temperature of $T \sim 3000$ K corresponding to $k_B T \sim 0.3$ eV. This is much smaller than the characteristic energy $\hbar\omega \sim 3$eV of the photon emission in the experiment of Cao et al. [1]. Hence, the thermal photon occupation numbers are rather small and cannot account for the output observed in that experiment. Rather, the bulk of the output is due to amplified emission from the active medium within the dielectric. To discuss this output theoretically we must couple the electromagnetic field to a medium of active atoms.

3 Laser Equations

To study the amplification of light we employ the laser model of Haken [17] and generalize it to open resonators: We couple the electromagnetic field of the open resonator to an ensemble of pumped two-level atoms. The analogy between a two-level atom and a spin-1/2 system may then be employed to represent the atoms in terms of Pauli spin operators. For each atom we introduce a set of operators $S_\pm$, S_z with the commutation relations $[S_+, S_-] = 2S_z$ and $[S_z, S_\pm] = \pm S_\pm$. Physically, the ladder operators $S_\pm$ are connected with the atomic dipole moment while $2S_z$ represents the atomic inversion. The free-atom Hamiltonian takes the form $H_0 = \hbar\nu S_z$ where ν denotes the transition frequency between the two energy levels. To model pumping and losses (e.g. due to non-radiative transitions out of the excited level) we couple the atoms to external pumping and loss baths. The interaction of the radiation field with the two-level atoms is described within the electric-dipole and the rotating-wave approximation. This model leads to the following set of Heisenberg–Langevin equations for the coupled field-atom dynamics

$$\dot{a}_\lambda = -\mathrm{i}\omega_\lambda a_\lambda - \pi \sum_{\lambda'} (\mathcal{W}\mathcal{W}^\dagger)_{\lambda\lambda'} a_{\lambda'} + \sum_p g_{\lambda p} S_{-p} + F_\lambda, \tag{11}$$

$$\dot{S}_{-p} = (-\mathrm{i}\nu - \gamma_\perp) S_{-p} + 2\sum_\lambda g^*_{\lambda p} S_{zp} a_\lambda + F_{-p}, \tag{12}$$

$$\dot{S}_{zp} = \gamma_\parallel (S_0 - S_{zp}) - \sum_\lambda (g_{\lambda p} a^\dagger_\lambda S_{-p} + g^*_{\lambda p} S^\dagger_{-p} a_\lambda) + F_{zp}, \tag{13}$$

where λ labels the resonator modes and p the atoms. The dynamics of the resonator field is damped by the coupling $\mathcal{W}$ to the external radiation field. Likewise, the atomic polarization and inversion are damped with damping amplitudes $\gamma_\perp$, $\gamma_\parallel$. The operators F_λ, F_{-p}, and F_{zp} represent quantum noise. The existence of such noise terms is a consequence of the coupling to external baths and the resulting irreversible system dynamics. The joint appearance of damping and noise terms may be understood as manifestation of the fluctuation-dissipation theorem. The derivation shows that F_λ is simply connected with the input field introduced before, $F_\lambda = -2\pi\mathrm{i}\sum_m \mathcal{W}_{\lambda m} b_m^{in}$. The coupling $g_{\lambda p}$ between the resonator mode λ and atom p is given by

$$g_{\lambda p} = -\mathrm{i}\frac{\nu d_\lambda}{\sqrt{2\hbar\omega_\lambda\varepsilon_0}}\phi_\lambda(\mathbf{r}_p), \tag{14}$$

where d_λ is the projection of the atomic dipole matrix element on the mode polarization vector, ω_λ the frequency of the mode λ, and $\phi_\lambda(\mathbf{r}_p)$ the value of the mode function at the position of the p-th atom. Finally, S_0 is a parameter for the pump strength. To derive (11)–(13) we assumed that the amplitudes $W_\lambda(\omega)$ are frequency independent over a sufficiently large frequency band around the atomic transition frequency ν. This approximation is known as the Markov approximation.

Random scattering of light and wave chaos enter our laser dynamics in several ways. First, the eigenfunctions of wave chaotic resonators vary quasi-randomly on the scale of the optical wavelength $2\pi/\omega_\lambda$ [18,19]. The coupling constants $g_{\lambda p}$ associated with atoms separated by more than a wavelength therefore behave as independent random numbers. For the same reason the outcoupling amplitudes $\mathcal{W}_{\lambda m}$ fluctuate statistically from mode to mode. Second, the frequencies ω_λ of wave-chaotic resonator display level repulsion and universal statistical properties. Random lasers therefore must be analyzed in a statistical fashion performing averages over appropriate *ensembles of random media*. Using ensembles known from random matrix theory this has recently lead to predictions for the photocount statistics of random lasers below [15] and above [20] the laser threshold.

Our Heisenberg equations differ from the independent-oscillator equations of standard laser theory [17] in two aspects: First, the modes are coupled by the damping matrix $\mathcal{W}\mathcal{W}^\dagger$; and second, the noise operators F_λ are correlated, $\langle F_\lambda^\dagger F_{\lambda'}\rangle \neq \delta_{\lambda\lambda'}$, as different modes couple to the same external channels (the expectation value is defined with respect to the channel oscillators at time t_0). The origin of these deviations from the independent oscillator dynamics may be understood in the limiting case of weak damping. This is the regime where all matrix elements of $\mathcal{W}\mathcal{W}^\dagger$ are much smaller than the resonator mode spacing $\Delta\omega$. This regime is realized in dielectrics that strongly confine light due to a large mismatch in the refractive index. To leading order in $\mathcal{W}\mathcal{W}^\dagger/\Delta\omega$ only diagonal elements contribute to the damping matrix, and (11) reduces to the standard equation of motion for independent oscillators. The deviations from the independent-oscillator dynamics become important when the damping rates are of the order of or exceed the mean frequency spacing of the resonator modes.

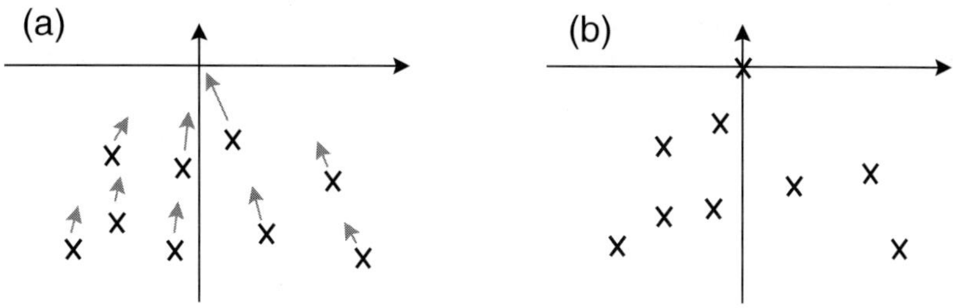

Fig. 2. a Schematic illustration of the resonances (crosses) of the linear amplifier. The resonances are the complex eigenvalues of the internal dynamics of the resonator field below threshold. With increasing pump the resonances move towards the real axis. **b** Eigenvalues of the matrix $\mathcal{D}$ above the laser threshold. One eigenvalue is zero and gives rise to phase diffusion of the laser field

The laser equations are non-linear due to the interaction terms in (12) and (13). The significance of the interactions depends crucially on the pumping strength: For weak pumping (small values of S_0) one is below the threshold of lasing. Then the field amplitudes are so small that we may neglect the interaction terms in (13). The inversion may be approximated by its steady-state value, $S_{zp} \equiv S_0$. Substitution into (12) replaces the non-linear laser equations by a set of linear equations that describe a linear amplifier. The solution of the below-threshold problem is easily found by Fourier transformation and yields

$$b^{\text{out}}(\omega) = S(\omega)b^{\text{in}}(\omega) + U(\omega)F_-(\omega), \tag{15}$$

where b^{in} is the channel input field and F_- the atomic polarization noise. The polarization noise forms a L-component vector where L denotes the number of atoms in the resonator. According to (15) there are now two sources of resonator output: The scattering of incoming radiation and the decay of excited atoms. The first process is described by the S-matrix, and the latter by the $N \times L$ matrix

$$U(\omega) = \frac{1}{\mathrm{i}(\nu - \omega) + \gamma_\perp} \mathcal{W}^\dagger D^{-1}(\omega)\mathcal{G}, \tag{16}$$

where $\mathcal{G}$ is a matrix comprising the matrix elements $g_{\lambda p}$, and

$$D_{\lambda\lambda'}(\omega) = (\omega - \omega_\lambda)\delta_{\lambda\lambda'} + \mathrm{i}\pi\left(\mathcal{W}\mathcal{W}^\dagger\right)_{\lambda\lambda'} - 2\mathrm{i}S_0 \sum_p \frac{g_{\lambda p} g^*_{\lambda' p}}{\mathrm{i}(\nu - \omega) + \gamma_\perp}. \tag{17}$$

The internal field dynamics described by D now includes a damping as well as a gain term. The latter is proportional to the pumping strength S_0 and amplifies modes in a frequency range of size $\gamma_\perp$ around the transition frequency ν. The presence of the gain shifts the complex resonances upwards towards the real axis. We illustrate this shift schematically in Fig. 2a. The emission spectrum of the resonator shows a broad peak of width $\sim \gamma_\perp$ around the frequency ν due to amplified spontaneous emission. This enhancement is easily visible in the experiment of Cao et al. (see Fig. 1).

4 Laser Linewidth

Clearly, the model of a linearly amplifying medium is only meaningful when all resonances stay below the real axis as resonances above the real axis would correspond to field amplitudes that increase exponentially in time. The amplifier model breaks down when one of the resonances touches the real axis. This happens at the threshold of lasing. The non-linear terms in our dynamical equations must then be included to stabilize the field intensities. We [7] employed the method of quasi-linearization [17] to study the laser dynamics sufficiently far above threshold. We focus on the case of a single self-sustained oscillation above threshold and compute the laser linewidth. To simplify the calculation, we assume (i) a large number of atoms $L \gg 1$ that provide a spatially uniform gain, (ii) atomic decay rates that much exceed the field decay rates, and (iii) exact resonance between the laser frequency $\bar{\omega}$ and the atomic transition frequency. We decompose the field into a classical steady-state value and quantum fluctuations, $a_\lambda = (\bar{a}_\lambda + \delta a_\lambda)\exp(-\mathrm{i}\bar{\omega}t)$. The steady-state conditions take the form $0 = \mathcal{D} \cdot \bar{a}$ where $\bar{a}$ comprises the N steady-state field amplitudes $\bar{a}_\lambda$. The non-Hermitian matrix

$$\mathcal{D}_{\lambda\lambda'} = (\bar{\omega} - \omega_\lambda)\delta_{\lambda\lambda'} + \mathrm{i}\pi(\mathcal{W}\mathcal{W}^\dagger)_{\lambda\lambda'} - \mathrm{i}G\delta_{\lambda\lambda'}, \tag{18}$$

depends on the laser field intensity $I = \sum_\lambda |\bar{a}_\lambda|^2$ via the gain G. For spatially uniform gain one finds $G = (2S_0Lg^2/\gamma_\perp)(1 + 4g^2I/\gamma_\parallel\gamma_\perp)^{-1}$ where $g^2 = (L\nu d^2)/(2\hbar\epsilon_0 V)$ is the atom-field coupling and V the volume of the resonator (the coupling no longer depends on the atom index p as we assumed spatially uniform gain). The equations of motion for the quantum fluctuations follow upon linearization around the steady-state solution

$$\begin{pmatrix} \delta\dot{a} \\ \delta\dot{a}^\dagger \end{pmatrix} = \mathcal{L}\begin{pmatrix} \delta a \\ \delta a^\dagger \end{pmatrix} + \begin{pmatrix} \mathcal{F} \\ \mathcal{F}^\dagger \end{pmatrix}. \tag{19}$$

The noise operators $\mathcal{F}, \mathcal{F}^\dagger$ incorporate both field noise and noise from the atomic reservoirs. The dynamics of δa and $\delta a^\dagger$ is coupled by the $2N \times 2N$ matrix

$$\mathcal{L} = \begin{pmatrix} \mathrm{i}\mathcal{D} & 0 \\ 0 & -\mathrm{i}\mathcal{D}^* \end{pmatrix} + \frac{\partial G}{\partial I}\begin{pmatrix} \bar{a}\cdot\bar{a}^\dagger & \bar{a}\cdot\bar{a}^T \\ \bar{a}^*\cdot\bar{a}^\dagger & \bar{a}^*\cdot\bar{a}^T \end{pmatrix}, \tag{20}$$

which depends explicitly on the steady-state solution $\bar{a}$. Equations (19) and (20) reduce the computation of the field fluctuations to the spectral decomposition of the non-Hermitian matrix $\mathcal{L}$. One easily shows that $\mathcal{L}$ has a zero eigenvalue connected with the well-known process of phase diffusion [17,21]. The corresponding right eigenvector has the form $(r, -r^*)$, where $r \propto \bar{a}$ is a right eigenvector to eigenvalue 0 of the $N \times N$ matrix $\mathcal{D}$; the existence of r and the corresponding left eigenvector l is guaranteed by the steady-state equations, $0 = \mathcal{D} \cdot \bar{a}$. The phase-diffusion coefficient and the laser linewidth $\delta\omega$ can now be computed along standard lines [17,21]: We solve the equations of motion (19) and calculate the Fourier transform of the stationary correlator $\langle \delta a^\dagger(t) \delta a(0) \rangle$. Keeping only the zero-eigenvalue contribution in the spectral decomposition of $\mathcal{L}$, we obtain the linewidth

$$\delta\omega = K \delta\omega_{\mathrm{ST}}, \tag{21}$$

which is larger than the fundamental (Schawlow–Townes) linewidth $\delta\omega_{\mathrm{ST}}$ by the Petermann factor [22,23]

$$K = \langle l | l \rangle \langle r | r \rangle. \tag{22}$$

The Petermann factor was previously known from studies of the linear subthreshold dynamics; our derivations hold in the laser regime above threshold. The non-zero eigenvalues of $\mathcal{L}$ will generally modify the Lorentzian spectrum and the laser lineshape [24]. The deviations from Lorentzian shape are most pronounced in the tails of the laser line. Their origin are slowly decaying oscillations of the resonator field that are close to, but below, the laser threshold. In the presence of such oscillations one also expects large Petermann factors suggesting a possible connection between large K-values and non-Lorentzian laser lineshapes. A detailed investigation of this connection, the pertinent photon statistics, and the lineshape statistics in random media is in preparation.

References

1. H. Cao et al.: Phys. Rev. Lett. **82**, 2278 (1999)
2. H. Cao et al.: Phys. Rev. Lett. **84**, 5584 (2000); H. Cao *et al.*: Phys. Rev. Lett. **86**, 4524 (2001)
3. S.V. Frolov, Z.V. Vardeny, K. Yoshino: Phys. Rev. B **57**, 9141 (1998)
4. H. Cao, J.Y. Xu, S.H. Chang, S.T. Ho: Phys. Rev. E **61**, 1985 (2000)
5. Y. Ling, H. Cao, A. Burin, M.A. Ratner, X. Liu, R.P.H. Chang: Phys. Rev. A **64**, 063808 (2001)
6. D.S. Wiersma, S. Cavalieri: Nature (London) **414**, 708 (2001)
7. G. Hackenbroich, C. Viviescas, F. Haake: Phys. Rev. Lett. **89**, 083902 (2002)
8. C. Viviescas, G. Hackenbroich: quant-ph/0203122, Phys. Rev. A (in press)
9. The separation is formally achieved using the Feshbach projector method known from nuclear theory, see e.g. H. Feshbach, Ann. Phys. (N.Y.) **19**, 287 (1962)
10. F. Haake and R. Reibold, Phys. Rev. A **32**, 2462 (1985)
11. C.W. Gardiner, P. Zoller: *Quantum Noise*, 2nd edn. (Springer, Berlin, 2000)
12. see e.g. C. Itzykson, J.-B. Zuber: *Quantum Field Theory* (Mc Graw-Hill, New York, 1980)
13. C. Mahaux, H.A. Weidenmüller: *Shell-Model Approach to Nuclear Reactions* (North-Holland, Amsterdam, 1969)
14. C.W.J. Beenakker: Rev. Mod. Phys. **69**, 731 (1997)
15. C.W.J. Beenakker: Phys. Rev. Lett. **81**, 1829 (1998)

16. E.G. Mishchenko, M. Patra, C.W.J. Beenakker: Eur. Phys. J. D **13**, 289 (2001)
17. H. Haken: *Laser theory*, 2nd edn. (Springer, Berlin, 1984); *Light* (North-Holland, Amsterdam, 1985)
18. M.V. Berry: J. Phys. A **10**, 2083 (1977)
19. F. Haake: *Quantum Signatures of Chaos*, 2nd edn. (Springer, Berlin, 2000)
20. G. Hackenbroich, C. Viviescas, B. Elattari, and F. Haake, Phys. Rev. Lett. **86**, 5262 (2001)
21. M. Sargent III, M.O. Scully, and W.E. Lamb, *Laser Physics* (Addison Wesley, Reading, MA, 1974)
22. K. Petermann, IEEE J. Quantum Electron. **15**, 566 (1979)
23. A.E. Siegman, Phys. Rev. A **39**, 1253 (1989); Phys. Rev. A **39**, 1264 (1989)
24. S.M. Dutra et al., Phys. Rev. A **59**, 4699 (1999)

Coherent Random Lasing and 'Almost Localized' Photon Modes

V.M. Apalkov[1], M.E. Raikh[1], and B. Shapiro[2]

[1] Department of Physics, University of Utah, Salt Lake City, UT 84112, USA
[2] Department of Physics, Technion-Israel Institute of Technology, Haifa 32000, Israel

1 Introduction

There exists a formal analogy between the Schrödinger equation describing electron motion in a random potential, and the scalar wave equation for light propagation in a medium with fluctuating dielectric constant. Consider, for concreteness, a two-dimensional geometry (quantum well for an electron and a thin film for light). Then both equations can be presented in the form

$$\Delta_{\boldsymbol{\rho}}\psi + \left[k^2 - U(\boldsymbol{\rho})\right]\psi = 0, \tag{1}$$

where $\boldsymbol{\rho}$ is the in-plane coordinate. For an electron with mass m and energy E, moving in a random potential $V(\boldsymbol{\rho})$, the parameters k^2 and $U(\boldsymbol{\rho})$ are

$$k^2 = 2mE\ , \quad U(\boldsymbol{\rho}) = 2mV(\boldsymbol{\rho}). \tag{2}$$

For a light wave with frequency ω, traveling in a medium with dielectric constant ϵ, corresponding expressions for k^2 and $U(\boldsymbol{\rho})$ take the form

$$k^2 = \epsilon\left(\frac{\omega}{c}\right)^2\ , \quad U(\boldsymbol{\rho}) = -\delta\epsilon(\boldsymbol{\rho})\left(\frac{\omega}{c}\right)^2\ , \tag{3}$$

where $\delta\epsilon(\boldsymbol{\rho})$ is the fluctuating part of the dielectric constant.

The analogy between electrons and light is incomplete. For one thing, the "scattering potential" for light depends on frequency and it vanishes in the limit of zero frequency. Moreover, in a dielectric medium (positive ϵ) k^2 must be positive which implies that there can be no bound states for light, as long as $\delta\epsilon(\boldsymbol{\rho})$ is assumed to vanish outside some finite region in space. Therefore the simple notion of a binding potential well does not exist for light: for instance, a dielectric sphere embedded in a uniform medium cannot bind photons, regardless of whether its dielectric constant is larger or smaller than that of the surrounding medium. Thus, unlike electrons, photons will always escape from a dielectric medium into the surrounding air. However, under appropriate conditions, they can be trapped within the sample for a long time. The main purpose of this paper is to discuss trapping of light in a weakly disordered dielectric film.

The random term in (1) is zero on average; its statistical properties are described by the r.m.s. value, U_0, and the correlator $K(\boldsymbol{\rho_1} - \boldsymbol{\rho_2})$

$$\langle U(\boldsymbol{\rho})\rangle = 0\ , \quad \langle U(\boldsymbol{\rho_1})U(\boldsymbol{\rho_2})\rangle = U_0^2 K(\boldsymbol{\rho_1} - \boldsymbol{\rho_2})\ , \quad K(0) = 1. \tag{4}$$

It is known [4] that in two dimensions the nature of solutions of (1) is governed by the dimensionless conductance kl, where l is the transport mean free path. Using the golden rule, the product kl can be expressed through the correlator, K, as follows

$$(kl)^{-1} = \frac{U_0^2}{4\pi k^4} \int dq\, d\phi\, q^3 \delta(q^2 + 2kq\cos\phi) \int d^2\rho\, K(\rho) \exp(iq\rho) = \\ = \frac{2U_0^2}{k^2} \int_0^{\pi/2} d\alpha\, \sin^2\alpha \int_0^\infty d\rho\, \rho\, K(\rho)\, J_0(2k\rho\sin\alpha), \tag{5}$$

where J_0 is the Bessel function of zero order. The above integral can be evaluated analytically if we choose a Gaussian form for the correlator $K(\rho) = \exp(-\rho^2/R_c^2)$. Substituting this form into (5), we obtain

$$(kl)^{-1} = \pi \left(\frac{U_0 R_c}{2k}\right)^2 F\left(\frac{k^2 R_c^2}{2}\right), \tag{6}$$

where the dimensionless function F is defined as

$$F(x) = e^{-x}\Big[I_0(x) - I_1(x)\Big]. \tag{7}$$

Here I_0 and I_1 are the modified Bessel functions of zero and first order, respectively.

As a simple example of a realistic two-dimensional disorder, consider a system of disks with random positions of the centers and with fluctuating radii described by a normalized distribution function, $\Phi(R)$. Within a disk, we have $U(\boldsymbol{\rho}) = U_d$, whereas outside the disk $U(\boldsymbol{\rho}) = 0$. If the filling fraction, f, is low, $f \ll 1$, the positions of the centers of the disks are uncorrelated. Then $K(\rho)$ is determined by the overlap between two circles of the same radius with centers shifted by ρ. A straightforward calculation yields

$$\langle U(\boldsymbol{\rho_1}) U(\boldsymbol{\rho_1} + \boldsymbol{\rho})\rangle = \frac{2U_d^2 f}{\pi} \int dR\, \theta\Big(R - \frac{\rho}{2}\Big)\, \Phi(R) \\ \times \left\{ \arcsin\sqrt{1 - \Big(\frac{\rho}{2R}\Big)^2} - \frac{\rho}{2R}\sqrt{1 - \Big(\frac{\rho}{2R}\Big)^2} \right\}, \tag{8}$$

where $\theta(x)$ is the step-function. The actual distribution function, $\Phi(R)$, is governed by various technological factors. However, it is reasonable to assume that small values of R are strongly unlikely, and that $\Phi(R)$ falls off abruptly at large R. Consider as an example the distribution $\Phi(x) = c^3 x^4 \exp(-cx^2)$, where $x = R/\overline{R}$ ($\overline{R}$ is the average radius) and $c = 64/9\pi$. The distribution is designed so as to yield the value 50% for the relative spread in R, i.e., $\langle |R - \overline{R}|\rangle/\overline{R} = 0.5$. The result of calculation of the integral (8) using this distribution is shown in Fig. 1 together with its Gaussian fit, which yields $U_0^2 = 0.86 U_d^2 f$ and $K(\rho) = \exp\left(-3.4\rho^2/\overline{R}^2\right)$. From this fit we conclude that the Gaussian form of the correlator, $K(\rho)$, corresponds to a quite generic distribution of R.

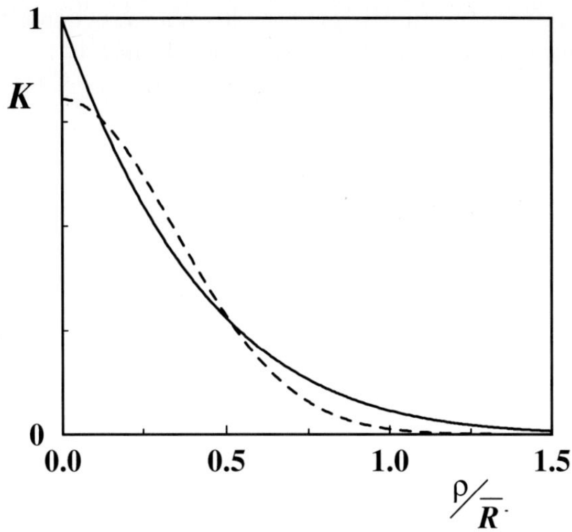

Fig. 1. Correlation function (solid line) for the model of randomly distributed disks $U(\boldsymbol{\rho}) = U_d\theta(R-\rho)$ with fluctuating radii is shown together with its Gaussian fit (*dashed line*)

Equation (6) provides the explanation why the criterion for strong localization, $kl < 1$, can be easily satisfied for electrons, but very hard to achieve in the case of photons. Indeed, for electrons (6) can be presented as

$$(kl)^{-1} = \left(\frac{\pi m}{2E}\right) V_0^2 R_c^2\ F\left(mER_c^2\right), \tag{9}$$

while for photons (6) takes the form

$$(kl)^{-1} = \left(\frac{\pi}{4\epsilon}\right)\left(\frac{\omega}{c}\right)^2 \Delta^2 R_c^2\ F\left(\frac{\epsilon\omega^2 R_c^2}{2c^2}\right), \tag{10}$$

where V_0 and Δ are the r.m.s. fluctuations of the potential and dielectric constant, respectively. It is seen from (9) that, since the function F is always smaller than one, the electrons are strongly localized in the *low-energy* domain $E < E_c$. The value of E_c is different for the short-range, $R_c \ll (mV_0)^{-1/2}$, and for the smooth, $R_c \gg (mV_0)^{-1/2}$, potentials. For short-range potential we can use the asymptotics $F(x)\big|_{x\ll 1} \approx 1$, which leads us to the estimate $E_c \sim mV_0^2R_c^2$. Hence, $mE_cR_c^2 \sim \left(mV_0R_c^2\right)^2 \ll 1$, so that the argument of F is indeed small at $E = E_c$. In the case of the smooth potential, the condition $mV_0R_c^2 \gg 1$ suggests that the semiclassical description applies, so that $E_c \sim V_0$. Calculation of the mean free path based on the golden rule is inadequate in this case. To see this, note that at $E \sim V_0$, the argument of F in (9) is large, so that we can use the asymptotics $F(x)\big|_{x\gg 1} \propto x^{-3/2}$. Then (9) yields $l \sim R_c$ for $E \sim V_0$. Thus, for smaller E, namely $E < V_0$, we have $l < R_c$. The latter relation indicates the failure of the perturbation theory for $E < V_0$.

Now let us perform the similar analysis for photons. Using the small-x and large-x asymptotics of the function F, we obtain from (10)

$$kl|_{k_0 R_c \ll 1} = \frac{4\epsilon}{\pi (k_0 R_c \Delta)^2}\ , \ kl|_{k_0 R_c \gg 1} = \frac{4\epsilon^{5/2} k_0 R_c}{\pi^{1/2} \Delta^2}, \tag{11}$$

where $k_0 = \omega/c$. Equation (11) suggests that when the r.m.s. fluctuation Δ is weak, $\Delta \ll \epsilon$, we have $kl \gg 1$ *both* in the low-frequency ($k_0 R_c \ll 1$) and in the high-frequency ($k_0 R_c \gg 1$) domains. This peculiar result is due to the already mentioned frequency dependence of the "optical potential".

The above analysis, however, does not rule out the possibility of light localization in the *strongly scattering* media. In fact, first experimental indications of localization effects for microwaves were reported more than a decade ago [2,3]. In these experiments localization was inferred from the measurements of various transmission characteristics of the microwaves through the tube filled with a random mixture of aluminum and Teflon spheres. For optical frequencies [4–6], the strongly scattering medium used in first experiments, aimed at light localization, was a semiconductor (*GaAs*) powder. Measurements of the transmission vs. the sample thickness were complimented with measurements of the coherent backscattering (CBS) cone [7,8]. In the latter measurements localization manifests itself through the rounding of the CBS cone by limiting the maximal length of coherent path [9].

The early reports of the observation of wave localization both for microwaves and for light [2–4] were inconclusive because of the possibility that the results were affected by absorption. To get rid of this ambiguity, in the later experiments the CBS measurements [10] were performed on the macroporous *GaP* networks, which scatter light stronger than a powder [4]. This allowed the authors to rule out the absorption or the finite sample size as a source of rounding of the CBS cone. For microwaves [11–13], the recent progress in detecting localization is due to a novel approach to the analysis of the transmission data based on analysis of the relative size of the transmission fluctuations. This approach permits one to detect localization even in the presence of absorption.

In the weakly scattering *active* medium the propagation of light remains diffusive. However, the interplay of the diffusion and the gain-induced amplification can be very nontrivial. Namely, this interplay can give rise to the *incoherent* random lasing, predicted by V. I. Letokhov [14]. As it was pointed out in Ref. [14], there is a close analogy between multiplication of neutrons in course of the chain reaction and photons in the amplifying disordered medium (photonic bomb). Upon first experimental observation of incoherent random lasing [15], it was subsequently reproduced for various realizations of the gain media and different types of disorder. Comparison of theoretical [16–21] and experimental results [15,22–42] has confirmed that the diffusion theory, which neglects the interference effects, is quite sufficient for the description of *incoherent* lasing. Except for studies on powder grains of laser crystal materials [37–40] and π-conjugated polymer films [41,42], the majority of experiments [15,22–36] have used dye solutions as amplifying media. Colloidal particles suspended in a solution served

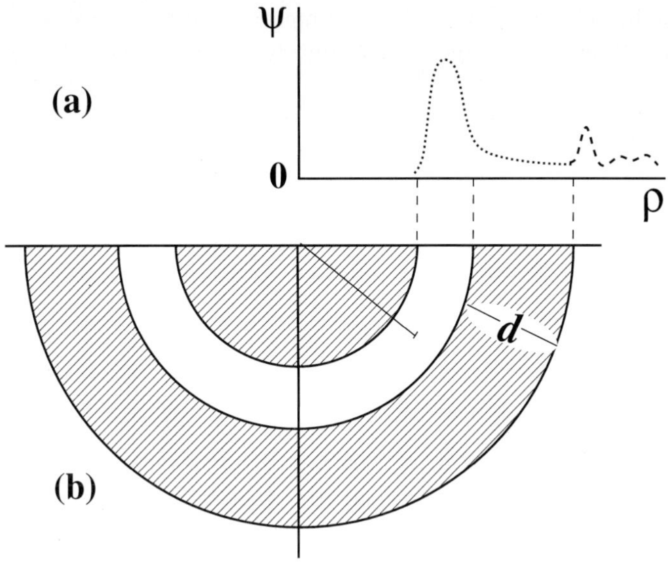

Fig. 2. a Spatial distribution of the wave function of the anomalously localized solution, $\psi(\rho)$, of (1) is shown schematically. **b** Spatial distribution of the dielectric constant, $\epsilon(\rho)$, corresponding to the trap, responsible for the solution $\psi(\rho)$; $\epsilon(\rho) = \epsilon$ outside the blank region. Only the lower half of the trap is shown

as random scatterers. These scatterers are responsible for *nonresonant feedback* required for incoherent lasing. The essence of this feedback is that the light amplification length, $\tilde{l}_a$ (in the absence of disorder), is significantly shortened [to $\sim (l\tilde{l}_a)^{1/2}$] in the disordered medium, when the light propagation is diffusive. The threshold condition for incoherent lasing corresponds to the gain magnitude at which $(l\tilde{l}_a)^{1/2}$ becomes of the order of the sample size.

In contrast to incoherent lasing, the recent discovery of *coherent* random lasing [43–46] adds a new dimension to the physics of light propagation in disordered media. Coherent random lasing emerges as the degree of disorder increases, so that the mean free path, l, becomes progressively smaller. The fact that the light, emitted from a disordered sample, is truly coherent, which does not necessarily follow [47,48] from drastic narrowing of the emission spectrum, observed in Refs. [43–46], was later convincingly demonstrated in photon statistics experiments [49,50]. In the absence of mirrors, it is evident that, in order to support the coherent lasing, the disordered medium *itself* should assume their role. The latter is feasible only due to the *interference effects*, that are not captured within the diffusion picture. More quantitatively, in order for the random medium to play the role of a Fabry-Perot resonator, it is necessary that certain eigenfunctions of (1) were either completely localized or *almost* localized. Almost localized solution can be, roughly, envisaged as a very high *local* maximum of the extended eigenfunction $\psi(\boldsymbol{\rho})$ of (1). If this maximum is viewed as a *core* of $\psi(\boldsymbol{\rho})$, then the delocalized *tail* (see Fig. 2) can be viewed as a source of *leakage*. In other words, the core itself, being not an exact eigenfunction, can be viewed as a solution of (1) corresponding to a *complex* eigenvalue $\mathrm{Im}\, k^2 \neq 0$. Then the weak leakage translates into a small value of the imaginary part of k^2. The inverse of this imaginary part determines the *lifetime* of the core. The higher the local maximum of $\psi(\boldsymbol{\rho})$ is, the longer is the lifetime. Making link to the Fabry-Perot resonator, the ratio $k^2/\mathrm{Im}\, k^2$ can be identified with a quality factor, Q. Thus, from the perspective of the coherent random lasing, the right question to be asked is: how high are the attainable quality factors of the almost localized solutions of (1) at a given parameter of disorder, i.e., magnitude, Δ, and correlation radius, R_c.

We address this question in the present paper. However, prior to discussing the almost localized solutions of (1), a subtle point must be clarified. Namely, (1) with the "potential" being real, describes propagation of light in a *passive* medium. It might be argued that, in the presence of gain, which is required for lasing, the spatial structure of the solutions of (1) undergoes a drastic change, so that the almost localized modes of a passive random medium and actual *lasing* modes have little in common. In fact, it is well known both from theory [51] and from CBS experiments [52] that the *diffusive* trajectories of light within a random medium get *elongated* in the presence of the gain. With regard to coherent random lasing, it was initially claimed [53] that the gain *facilitates localization* of light. This claim was even supported by the numerical simulations [53]. However, in the later theoretical [54–56] and experimental [57] papers it was explicitly stated that, similarly to the conventional lasers with Fabry-Perot resonators, the gain only *reveals* high-Q solutions of (1) existing in the passive medium. Thus, in the present paper, we will focus exclusively on the passive disordered media. The question about the likelihood of formation of the disorder-induced resonators is central to the understanding of the coherent random lasing. This question is at the core of the ongoing in-depth experimental studies [57–65]. Except for Ref. [66], theoretical papers on random lasing [56,67–71] do not address this question.

Let us finally mention that a similar question of trapping electrons, for a long time, in a weakly disordered conductor was addressed long ago in the context of electronic transport [72–75]. However, the "almost localized" states discussed in [66,76,77] and, further, in this paper are quite different from the "prelocalized" states studied in [72–75]. Wave functions of the "almost localized" states are confined primarily to a small ring (a high-Q resonator). Moreover, these states are extremely sensitive to the value of the correlation radius, R_c, of the disordered potential. We shall see below that, for a given parameter $kl > 1$, the likelihood of high-Q "almost localized" modes is very low for the short-range disorder but sharply increases with R_c. Such features are absent for the "prelocalized" states [72–75], with their comparatively large spatial extent.

2 Likelihood of Random Resonators in a Disordered Film

2.1 Intuitive Scenario of Random Cavities Based on Recurrent Scattering

The first experimental work on coherent random lasing by Cao *et. al.* [43] was carried out on thin (with a width $\sim 2\pi/k_0$) zinc oxide (ZnO) polycrystalline films. Laser action manifested itself through a drastic narrowing of the emission spectrum when the optical excitation power exceeded a certain threshold. The authors of Ref. [43] realized at the time that coherent lasing requires a resonator (cavity). They conjectured that, in the absence of a traditional, well-defined resonator (as in a semiconductor laser), the cavities in polycrystalline films are "self-formed" due to strong optical scattering. For a microscopic scenario of the formation of such cavities, they alluded to the remark made in Ref. [18] that closed-loop paths of light can serve as "random ring cavities". The importance of these closed-loop paths (recurrent scattering events) was pointed out earlier [78], when they were invoked for the explanation of the magnitude of the CBS albedo. With regard to CBS, the effect of recurrent scattering events, e.g., the events in which the *first* scattering (of the incident wave) and the *last* scattering (of the outgoing wave) are provided by the *same* scatterer, is that these events *do not* contribute to the CBS, thus reducing the albedo in the backscattering direction.

The fact that recurrent events show up in CBS certainly does not allow to automatically identify closed-loop paths with resonator cavities. Therefore, the feasibility of the scenario of random cavities, adopted in Refs. [43,44,58,59] for interpretation of experimental results, was later put in question [79]. The arguments against this scenario were the following. Since in each scattering act most of the energy gets scattered out of the loop, an unrealistically high gain would be required to achieve the lasing threshold condition for such a loop. Also the loops of scatterers are likely to generate a broad frequency spectrum rather than isolated resonances.

Certainly the picture of random cavities representing a certain spatial arrangement of *isolated* scatterers is too naive. This, however, does not rule out the entire concept of disorder-induced resonators. Although sparse, the disorder configurations that trap the light for long enough time can be found in a sample of a large enough size, and a single such configuration is already sufficient for lasing to occur. Therefore, under the condition $kl \gg 1$, which implies that *overall* scattering is weak, the conclusion about the relevance of random cavities can be drawn only upon the *calculation* of their likelihood. This calculation is decribed in the rest of this section first on qualitative, and then on quantitative levels.

Making link to the discussion in the Introduction: by pursuing the random cavity scenario Cao et. al. have intuitively arrived at the concept of prelocalized states, that was introduced in transport more than a decade ago. The fundamental difference between the prelocalized states and recurrent events is that recurrent events emerge in the higher order [in parameter kl^{-1}] of the *perturbation theory* (they correspond to the specific type of diagrams called Hikami-boxes

[80]), whereas the formation of prelocalized states is a genuinely *nonperturbative* effect. Already in the first analytical approach to the problem of prelocalized states [81], it was demonstrated that, in order to capture an anomalously slow tail of the conductance relaxation (which is due to trapping), *all* the orders of the perturbation theory must be taken into account.

2.2 Optimal Fluctuation Approach to the Problem

Qualitative Discussion. Similarly to the treatment in [43,53,59] we restrict our consideration to the two-dimensional case (a disordered film). Regarding the geometry of a random resonator, we adopt the idea proposed by Karpov for trapping acoustic waves [76] and electrons [77] in three dimensions.

Suppose that within a certain stripe the effective in-plane dielectric constant of a film is enhanced by some small value $\epsilon_1 \ll \epsilon$. Then such a stripe can play a role of a *waveguide*, i.e., it can capture a transverse mode, as it is illustrated in Fig. 3. There is no threshold for such a waveguiding, which means that the transverse mode will be captured even if the width of the stripe is small. Now, in order to form a resonator, one has to roll the stripe into a ring. Upon this procedure, the mode propagating along the waveguide transforms into a whispering-gallery mode of a ring. An immediate consequence of the curving of the waveguide is emergence of the evanescent leakage – the optical analog of the under-the-barrier tunneling in quantum mechanics (see Fig. 2). This leakage is responsible for a finite lifetime of the whispering-gallery mode. Thus we have specified the structure of the weakly decaying solutions of (1), discussed in the Introduction. Namely, the mode of the waveguide plays the role of the *core*, while delocalized tail (see Fig. 2) reflects the evanescent leakage. Due to the azimuthal symmetry, the modes of the resonator are characterized by the angular momentum, m. Denote by $\mathbf{N}_m(kl, Q)$ the areal density of resonators with quality factor Q in the film with a transport mean free path l. Obviously, in the diffusive regime, $kl > 1$, the density $\mathbf{N}_m(kl, Q)$ is exponentially small for $Q \gg 1$. In this

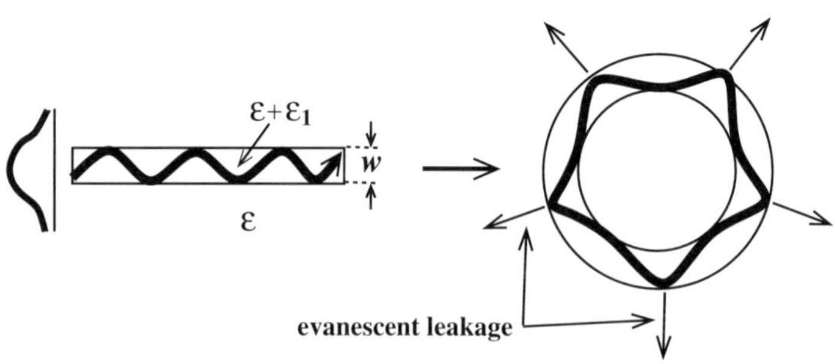

Fig. 3. Rationale for the structure of the resonator. Upon wrapping a stripe with enhanced dielectric constant into a ring, a waveguided mode transforms into a whispering-gallery mode

domain $\mathbf{N}_m(kl, Q)$ can be presented as

$$\mathbf{N}_m(kl, Q) \propto e^{-S_m(kl,Q)} . \quad (12)$$

Let us first give a qualitative estimate for S_m, which reveals its sensitivity to the strength and the range of the disorder (Δ^2 and R_c). Since m is the number of wave lengths along the ring, its radius is $\rho_0 = m/k$. The ring waveguide can support a weakly decaying mode only if its width w satisfies the condition $w(\epsilon_1/\epsilon)^{1/2} > k^{-1}$. A straightforward estimate for the decay time due to evanescent leakage, i.e., the quality factor Q of the waveguide results in $\ln Q \sim k\rho_0(\epsilon_1/\epsilon)^{3/2}$. Since the number $k\rho_0 = m$ is large, a relatively small fluctuation of the dielectric constant within the area $2\pi\rho_0 w$ of the ring can produce a large value of Q.

The probability $\mathcal{W}$ for creating the required fluctuation strongly depends on R_c. For a short range disorder ($k_0 R_c \ll 1$) fluctuations of order ϵ_1 should occur *independently* in a large number, $N \sim \rho_0 w/R_c^2$, of spots within the ring, so that the probability $\mathcal{W} \sim \exp\left(-N\epsilon_1^2/\Delta^2\right)$. In the other extreme of strongly correlated disorder, when $R_c \gg w$, the number of independent spots is much smaller, $N' \sim \rho_0/R_c$ (the number of squares with a size R_c needed to cover the ring). Correspondingly, the probability $\mathcal{W} \sim \exp\left(-N'\epsilon_1^2/\Delta^2\right)$ is much larger than for the short range case. Finally, using the relation between ϵ_1 and Q, the probability $\mathcal{W} = \exp(-S_m)$ can be rewritten in terms of Q, thus, yielding an estimate for S_m. For the short range case ($k_0 R_c \ll 1$) we obtain $S_m \sim kl \ln Q$, where the mean free path l is proportional to $(R_c\Delta)^{-2}$ [see (11)]. In the opposite limit of a smooth disorder (11) yields for the transport mean free path $l \sim R_c/\Delta^2$. Then we have $S_m \sim N'\epsilon_1^2/\Delta^2 \sim l(\ln Q)^{4/3}/(kR_c^2 m^{1/3})$. Thus, for given kl and Q, the density of resonators is the higher the smoother is the disorder. This conclusion is central to our study and will be addressed below in more detail.

Quantitative Results from the Optimal Fluctuation Approach. The above program can be carried out analytically [66] with the use of the optimal fluctuation approach [82,83]. This approach is based on the idea that, when the exponent, S_m, in $\mathbf{N}_m(kl, Q)$ is large, then the major contribution to $\mathbf{N}_m(kl, Q)$ comes from a certain specific disorder realization. In application to random resonators, the optimal fluctuation procedure reduces to finding the *most probable* fluctuation of the dielectric constant which is able to trap the light for a long time $\sim \omega^{-1}Q$. Assuming that the fluctuation is azimuthally symmetric (see Fig. 2), the shape of the optimal fluctuation can be found explicitly [66] yielding the following expression for the exponent S_m

$$S_m = 2^4 3^{-3/2}\pi^{1/2} m \left(\frac{\epsilon_1^3}{\epsilon}\right)^{1/2} \frac{\Phi(\epsilon_1^{1/2} k_0 R_c)}{(\Delta k_0 R_c)^2} , \quad (13)$$

where $\epsilon_1 = \epsilon(3\ln Q/2m)^{2/3}$. The analytical expression for the function $\Phi(u)$ is the following

$$\Phi(u) = \frac{3^{3/2}}{2^6} \frac{(5+\sqrt{9+16u^2})^{5/2}}{(3+\sqrt{9+16u^2})^{3/2}}. \tag{14}$$

Recall now, that we are interested in the density of random resonators at a *given* value of kl. The transition from Δ to l is accomplished by using (11). For the short range case, when $R_c \to 0$ and $\Phi(u) \to 1$, we obtain

$$S_m(k_0 R_c \ll 1) = 2\left(\frac{\pi^3}{3}\right)^{1/2} kl \ln Q\ . \tag{15}$$

To trace the change of S_m with increasing R_c it is convenient, after using (11), to present (13) in the form

$$\frac{S_m(k_0 R_c > 1)}{S_m(k_0 R_c \ll 1)} = \frac{\Phi(\epsilon_1^{1/2} k_0 R_c)}{\pi^{1/2}(\epsilon^{1/2} k_0 R_c)^3}\ . \tag{16}$$

It is seen from (16) that S_m falls off rapidly with increasing R_c. In the domain $k_0 R_c > 1$, but $\epsilon_1^{1/2} k_0 R_c < 1$ we have $\Phi \approx 1$, so that $S_m \propto (k_0 R_c)^{-3}$. For larger R_c we have $\Phi(u) \propto u$. In this domain S_m decreases more slowly with R_c: $S_m \propto (k_0 R_c)^{-2}$,

$$S_m(k_0 R_c \gg 1) = \frac{3^{4/3}\pi}{4^{5/3}} \frac{kl \ln^{4/3} Q}{m^{1/3}(kR_c)^2}. \tag{17}$$

Asymptotic expressions (15) and (17) agree with the results of the qualitative derivation, with all numerical factors now being determined. We emphasize that (15) and (17) apply *for a given* kl value, so that the decrease of S_m with R_c *leaves the backscattering cone unchanged.*

Estimates. Equation (15) quantifies the effectiveness of trapping of light in a random medium with point-like scatterers. It follows from (15) that the likelihood of high-Q cavity is really small. Indeed, even for rather strong disorder, $kl = 5$, the exponent, S_m, in the probability of having a cavity with a quality factor $Q = 50$ is close to $S_m = 120$. We emphasize that in the two dimensional case under consideration, this exponent does not depend on m and, thus, on the cavity radius $\rho_0 = m/\epsilon^{1/2} k_0$. A more accurate calculation [84], taking into account the corrections to (15), indicates that S_m as a function of m has a minimum at $m \sim (kl \ln Q)^{1/2}$.

To estimate the degree to which finite size of scatterers ($\sim R_c$) improves the situation, we choose $k_0 R_c \approx 2$, which already corresponds to the limit $k_0 R_c \gg 1$ in (11), but still allows to set $\Phi = 1$. Then for $Q = 50$, $kl = 5$ we obtain $S_m \approx 1.1$, suggesting that the resonators with this Q are quite frequent. In the latter estimate we have set $\epsilon = 4$.

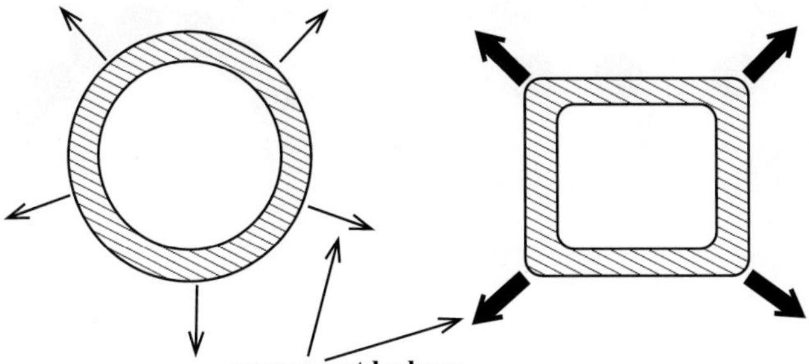

Fig. 4. This drawing illustrates the optimal character of the ring-shape resonator

2.3 Frequently Asked Questions

Why Rings? The answer to this question is illustrated in Fig. 4. The distinguishing property of a ring is that the local curvature radius is the same at each point. Upon any deviation from the ring geometry, the curvature in a certain region of the fluctuation would be higher than in all other regions. Since the evanescent losses are governed by this curvature, the quality factor of the resonator would be determined exclusively by this region (see Fig. 4), so that the remaining low-curvature part would be "unnecessary", in the sense, that a ring with a radius corresponding to the maximal curvature would have the same quality factor as a square in Fig. 4 but significantly higher probability of formation. It is also quite obvious that, for the purpose of supporting a wave-guided mode of the whispering-gallery type, a ring is much superior to a disk of the same radius: indeed, the internal area of the disk remains unused in the guiding process, whereas a heavy penalty in terms of probability is paid in creating this area.

Why Smooth Disorder Facilitates Trapping? At the qualitative level, the enhancement of the probability of formation of the cavity with increasing R_c can be understood for a toy model of the disorder, illustrated in Fig. 5. Suppose that all the disks that model the scatterers, are identical. Then R_c scales with the radius of the disk, R. Since the disks cannot interpenetrate, the ring-shaped cavity corresponds to their arrangement in the form of a necklace. The probability of formation of such a cavity can be estimated as follows. Suppose that a sector, $\delta\phi$, is "allocated" for a single disk. The probability to find a disk within this sector, at the distance ρ_0 from the center , is $\sim n(\rho_0\delta\phi)^2$, where n is the concentration of the disks. Thus, the probability of formation of the necklace is $\exp\left[-\frac{2\pi}{\delta\phi}\ln\left(\frac{1}{n\rho_0^2(\delta\phi)^2}\right)\right]$, where $\frac{2\pi}{\delta\phi}$ is the number of sectors. It is obvious that if a necklace is "loose", the quality factor of the corresponding cavity would be low. In order for Q to be high, neighboring disks must almost touch each other. This implies that $\delta\phi \approx 2R/\rho_0$. Then the above estimate for probability takes the form $\exp\left[-\frac{\pi\rho_0}{R}\ln\left(\frac{1}{f}\right)\right]$, where $f = n\pi R^2$ is the filling fraction. This probability increases exponentially with R, i.e., with R_c, reflecting the fact that, for

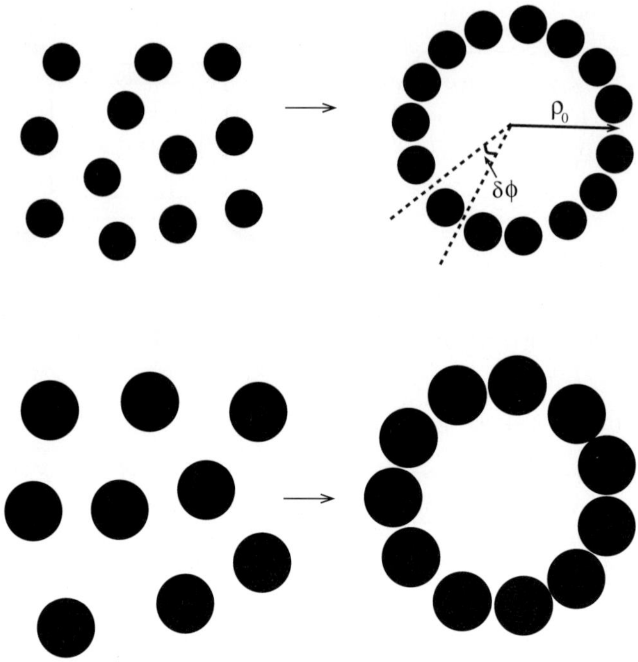

Fig. 5. A schematic illustration explaining why larger correlation radius for a fixed filling fraction facilitates trapping. A sector, shown with dashed lines, illustrates the tolerance in the arrangement of disks into a necklace

a given ρ_0, the number of disks to be arranged is smaller when R is larger. The above estimate was based on the assumption that the positions of the disks are uncorrelated, i.e., $f \ll 1$ (in contrast to [54] where $f = 0.4$). We have used the model of *hard* disks as an easiest illustration of the role of R_c. Obviously, (13) does not apply to this model, since it was derived under the assumption that the statistics of the fluctuations is Gaussian.

"Vulnerability" of the Ring-Shaped Cavities. The value S_m given by (13), which was derived within the optimal fluctuation approach, is the exponent in the probability of formation of an *ideal* ring. Obviously, any actual disorder realization is not ideal, in the sense, that the actual distribution of the dielectric constant differs from the optimal. For the same reason, the probability of formation of ideal necklace of the type shown in Fig. 5 is zero. In order for the probability to be finite, we should allow a certain *tolerance* in the positions of the centers of the disks, as illustrated in Fig. 5. In the conventional applications of the optimal fluctuation approach [85], deviations from the optimal distribution do not affect the value of the exponent, S_m. However, in application to the random cavities, we have searched for the fluctuation which is optimal *for a given trapping time*, $\omega^{-1}Q$. In this particular application, a "normal" Gaussian deviation from the optimal geometry can have a catastrophic effect on trapping by scattering the light wave out of the whispering-gallery trajectory. This scattering is discussed below.

Scattering within the Plane. The two-dimensional picture adopted throughout this paper, implies that the electromagnetic field is confined within a thin film in the z-direction. This confinement results from the fact that the average dielectric constant of the film is higher than in the adjacent regions. Then the field distribution, $\mathcal{E}_0(z)$, along z corresponds to a transverse waveguide mode. For a given frequency, ω, the almost localized state on a ring can be destroyed due to the scattering into states with the *same* distribution of the field in the z-direction, which propagate freely along the film. More precisely, the almost localized state with a given angular momentum m, which is protected from the outside world by the centrifugal barrier, can be scattered out to the continua of states with smaller m's, for which there is no barrier. It is essential to estimate the lifetime, τ, with respect to these scattering processes and to verify that it is feasible to have τ larger than the prescribed trapping time, $\omega^{-1}Q$, so that the almost localized state is not destroyed. A rigorous treatment of this "scattering out" effect is quite involved and is done in Sect. 3, for a Gaussian potential.

The effect can be also illustrated with the model of randomly positioned hard disks (Fig. 5), although the disorder in this model is non-Gaussian. It is seen from Fig. 5, that spacings between the rings, which are due to tolerance, open a channel for the light escape, that is different from evanescent leakage. A *typical* lifetime with respect to such an escape is quite short, i.e., even a small tolerance, which affects weakly the exponent in the probability of the cavity formation, seems to be detrimental for trapping. At this point we emphasize that, in calculation of the scattering rate out of the whispering-gallery trajectory, the disks constituting the necklace must be considered as a *single entity.* As a result, for a *given* configuration of the disks, the rate of scattering out caused by a *single* disk must be multiplied by the following *form-factor*

$$\mathcal{F} = \int \frac{d\varphi_{\boldsymbol{k}}}{2\pi} \left| \sum_i \exp\left(i\boldsymbol{k}\boldsymbol{\rho}_i\right) \right|^2 = \sum_{i,j} J_0\left(k|\boldsymbol{\rho}_i - \boldsymbol{\rho}_j|\right), \tag{18}$$

where $\boldsymbol{\rho}_i$ is the position of the center of the i-th disk in the necklace. The form-factor, $\mathcal{F}$, is the sum of $\left(\frac{\pi\rho_0}{R}\right)^2 \gg 1$ terms. Out of this number, $(kR)^{-1}\left(\frac{\pi\rho_0}{R}\right)$ terms (for $kR < 1$) and $\frac{\pi\rho_0}{R}$ terms (for $kR > 1$), for which $k|\boldsymbol{\rho}_i - \boldsymbol{\rho}_j| < 1$, are close to unity. The portion of these terms is small. Other terms have random sign. This leads us to the important conclusion that, for certain realizations of the necklace in Fig. 5, the form-factor can take anomalously small values. For these realizations the quality factor will be still determined by the evanescent leakage. The "phase volume" of these realizations is exponentially small and depends strongly on the model of the disorder.

Scattering out of the Plane. Compared to the previous case, two modifications are in order. Firstly, since the final state of the scattered cavity mode is a plane wave with the wave vector pointing in a certain direction within the solid angle

4π, the expression (18) for the form-factor should be replaced by

$$\tilde{\mathcal{F}} = \sum_{i,j} \frac{\sin\left(k_0|\boldsymbol{\rho}_i - \boldsymbol{\rho}_j|\right)}{k_0|\boldsymbol{\rho}_i - \boldsymbol{\rho}_j|}. \tag{19}$$

Secondly, for $kR > 1$, i.e., when the disorder is smooth, scattering out of the plane that is caused by a single disk, requires a large wave vector transfer, $\sim k$. Thus, the corresponding rate is suppressed as compared to the in-plane scattering.

3 Prefactor

3.1 Qualitative Discussion

In Sect. 2 we calculated the *probability* of the formation of a trap that captures light for a long time $\tau = \omega^{-1}Q$. However, the relevant characteristics of traps is their *areal density.* Within the optimal fluctuation approach, the relation between the probability and the areal density emerges in course of calculation of the prefactor [85,86]. Namely, the combination with dimensionality of the inverse area comes from the so called "zero modes", which reflect the fact that the fluctuation can be shifted *as a whole* in both x and y directions. A typical shift, making two fluctuations independent, is of the order of the extent of the fluctuation in each direction. This suggests that the proportionality coefficient between the density of traps and the probability of the formation of the trap is roughly the inverse area of the fluctuation [85]. In our particular case, when the fluctuation is ring-shaped, the estimate for the *dimensional* prefactor is $\sim w^{-2}$, where w is the width of the ring (see Sect. 2). The dimensionless part of the prefactor within the optimal fluctuation approach reflects the "phase volume" of the fluctuations, which perturb the shape of the optimal fluctuation leaving the "energy" k^2 unchanged.

For the almost localized modes, considered above, the situation with prefactor is qualitatively different from the case of the truly localized states, for which the optimal fluctuation approach was devised [82]. The specifics of the almost localized states is that their "energy", k^2, [see (1)] is degenerate with the continuum of the propagating modes. As a result, a typical small perturbation of the shape of the ring will not only shift k^2, but also cause the coupling of the trapped mode to the continuum, or in other words, the *additional leakage* will emerge due to the fluctuations. In order to incorporate this effect into the theory, the density of traps should be multiplied by the probability, $P(\tau)$, that the lifetime with respect to this additional leakage is longer than τ [84]. To estimate this probability, we consider the additonal leakage for a *given* disorder realization

$$\frac{1}{\tau_U} = 2\pi \left(\frac{c}{k}\right) \sum_{\mu} \left| \int d\boldsymbol{\rho}\, \psi_0^*(\boldsymbol{\rho})\, U(\boldsymbol{\rho})\, \psi_\mu(\boldsymbol{\rho}) \right|^2 \delta(k_\mu^2 - k^2), \tag{20}$$

where ψ_0 is the "localized" solution of (1) (without evanescent leakage), ψ_μ and k_μ are the propagating eigenfunctions and eigenvalues of (1), respectively. It is seen from (20) that small additional leakage requires small matrix elements $\langle\psi_0|U|\psi_\mu\rangle$. Since the functions ψ_μ belong to the continuous spectrum, it might seem that this requirement is impossible to meet. This is, however, not the case, since the normalization factor of $|\psi_\mu|^2$ is the inverse system size. To demonstrate that additional leakage can be small, it is convenient to use the explicit form of ψ_μ, namely, the plane waves, and rewrite (20) in the form

$$\frac{1}{\tau_U} = \iint d\boldsymbol{\rho}_1 d\boldsymbol{\rho}_2 U(\boldsymbol{\rho}_1) S(\boldsymbol{\rho}_1,\boldsymbol{\rho}_2) U(\boldsymbol{\rho}_2), \tag{21}$$

where the kernel $S(\boldsymbol{\rho}_1,\boldsymbol{\rho}_2)$ is defined as

$$\begin{aligned} S(\boldsymbol{\rho}_1,\boldsymbol{\rho}_2) &= \psi_0^*(\boldsymbol{\rho}_1)\psi_0(\boldsymbol{\rho}_2)\left(\frac{c}{k}\right)\int \frac{d\boldsymbol{q}}{2\pi} e^{i\boldsymbol{q}(\boldsymbol{\rho}_1-\boldsymbol{\rho}_2)}\delta(q^2-k^2) \\ &= \frac{c}{2k}\psi_0^*(\boldsymbol{\rho}_1)\psi_0(\boldsymbol{\rho}_2) J_0\left(k|\boldsymbol{\rho}_1-\boldsymbol{\rho}_2|\right), \end{aligned} \tag{22}$$

It is seen from (22) that the kernel $S(\boldsymbol{\rho}_1,\boldsymbol{\rho}_2)$ is exponentially small if one of the points, $\boldsymbol{\rho}_1$ or $\boldsymbol{\rho}_2$, is located outside the region, occupied by the "body" of $\psi_0(\boldsymbol{\rho})$. When both $\boldsymbol{\rho}_1$ and $\boldsymbol{\rho}_2$ are located inside this region, then the characteristic spatial scale of the kernel, S, is $|\boldsymbol{\rho}_1-\boldsymbol{\rho}_2| \sim k^{-1}$. Thus, for the sake of our qualitative discussion, we can replace $S(\boldsymbol{\rho}_1,\boldsymbol{\rho}_2)$ by $A^{-1}\theta\left(k^{-1}-|\boldsymbol{\rho}_1-\boldsymbol{\rho}_2|\right)$, where $A \sim |\psi_0(0)|^{-2}$ is the area of the fluctuation, and restrict integration in (21) to the region of the area A.

Averaging over disorder configurations in (21) yields the mean value of the additional leakage

$$\frac{1}{\tau_e} = \left(\frac{c}{k}\right)\int \frac{d\boldsymbol{q}}{2\pi}\left\langle\left|\int d\boldsymbol{\rho}\ \psi_0^*(\boldsymbol{\rho})\, U(\boldsymbol{\rho})\, e^{i\boldsymbol{q}\boldsymbol{\rho}}\right|^2\right\rangle \delta(q^2-k^2) = \frac{c}{l_e}, \tag{23}$$

which is determined by *typical* values of $U(\boldsymbol{\rho})$. We note that l_e is of the order of the mean free path, l. The fact that $l_e \sim l$ can be seen from (21)–(23), taking into account that the area A is always bigger than R_c^2. Moreover, for short-range disorder, (21)–(23) suggest that $l_e \approx l$.

In order to have τ_U^{-1} small, the actual value of $U(\boldsymbol{\rho})$ should be suppressed with respect to typical in *each* box of a size k^{-1}. Then the condition $\tau_U > \tau$ requires the suppression factor to exceed τ/τ_e. The corresponding probability can be estimated as

$$P(\tau) = (\tau_e/\tau)^N = (Q/\omega\tau_e)^{-N}, \tag{24}$$

where $N \sim k^2 A$ is the number of boxes that "cover" the area of localization of ψ_0. In particular, for the ring-shaped fluctuations we have $N \sim w\rho_0 k^2$. We conclude that, due to additional leakage, the prefactor in the density of traps is *exponentially* small, i.e., $P(\tau) \sim \exp[-k^2 A \ln(Q/kl)]$. The exponent

$k^2 A \ln Q$ should be compared with the main exponent, found in Sect. 2. Since the above estimate assumed a short-range disorder, $kR_c \ll 1$, the principal exponent is $2\left(\pi^3/3\right)^{1/2} kl \ln Q$ [(15)]. Recall that, for short-range disorder we have $w \sim k^{-1}(k\rho_0/\ln Q)^{1/3}$, $\rho_0 \sim m/k$, so that the product $k^2 A$ is $\sim m^{4/3} \ln^{-1/3}(Q/kl)$. The final estimate for the exponent, originating from the prefactor, is $\sim m^{4/3} \ln^{2/3} Q$. This suggests that traps are the more frequent the smaller is the angular momentum, m. On the other hand, in derivation of the main exponent we have assumed that $m > \ln Q$. So that the minimal value of $(m^2 \ln Q)^{2/3}$ is $\sim \ln^2 Q$. This leads us to the conclusion that the main exponent *dominates* over the exponent, originating from the prefactor. This is because for the main exponent we have $kl \ln Q > \ln^2 Q$, since the relation $kl > \ln Q$ was assumed in the derivation of the main exponent.

The above qualitative analysis leading to (24) was restricted, for simplicity, to the case of the short-range disorder. In the next subsection we present a rigorous calculation of $P(\tau)$ for the more realistic case of a smooth disorder.

3.2 Derivation of $P(\tau)$

Denote by $p(\tau)$ the probability *density* that the lifetime with respect to additional leakage is equal to τ, so that $p(\tau) = dP(\tau)/d\tau$. The rigorous definition of this density reads

$$p(\tau) = \mathcal{N} \int \mathcal{D}\{U\}\, e^{-\mathcal{P}\{U\}}\, \delta(\tau - \tau_U)\,, \tag{25}$$

where the normalization constant is defined as

$$\mathcal{N} = \left[\int \mathcal{D}\{U\}\, e^{-\mathcal{P}\{U\}}\right]^{-1}, \tag{26}$$

and $\mathcal{P}\{U\}$ is given by

$$\mathcal{P}\{U\} = \frac{1}{2U_0^2} \iint d\boldsymbol{\rho}_1 d\boldsymbol{\rho}_2 U(\boldsymbol{\rho}_1) \mathcal{K}(\boldsymbol{\rho}_1, \boldsymbol{\rho}_2) U(\boldsymbol{\rho}_2), \tag{27}$$

where $\mathcal{K}(\boldsymbol{\rho}_1, \boldsymbol{\rho}_2)$ is related to the correlator $K(\boldsymbol{\rho_1}, \boldsymbol{\rho_2})$, defined by (4), as

$$\int d\boldsymbol{\rho}' \mathcal{K}(\boldsymbol{\rho}_1, \boldsymbol{\rho}') K(\boldsymbol{\rho}', \boldsymbol{\rho}_2) = \delta\left(\boldsymbol{\rho}_1 - \boldsymbol{\rho}_2\right). \tag{28}$$

Since we are dealing with photons, the value U_0 can be expressed through the r.m.s. fluctuation of the dielectric constant as $U_0 = k_0^2 \Delta$ [see (3)]. Using the integral representation of the δ-function, (25) can be rewritten in the form

$$p(\tau) = \frac{\mathcal{N}}{2\pi} \int_{-\infty}^{\infty} dt\, e^{it/\tau} \int \mathcal{D}\{U\}\, e^{-\mathcal{P}_t\{U\}}, \tag{29}$$

where the auxilary functional, $\mathcal{P}_t\{U\}$, is defined as

$$\mathcal{P}_t\{U\} = \mathcal{P}\{U\} + i\frac{t}{\tau_U} = \iint d\boldsymbol{\rho}_1 d\boldsymbol{\rho}_2 U(\boldsymbol{\rho}_1)\mathcal{K}_t(\boldsymbol{\rho}_1,\boldsymbol{\rho}_2)U(\boldsymbol{\rho}_2). \tag{30}$$

The kernel, $\mathcal{K}_t(\boldsymbol{\rho}_1,\boldsymbol{\rho}_2)$, of the functional $\mathcal{P}_t\{U\}$ has the form

$$\mathcal{K}_t(\boldsymbol{\rho}_1,\boldsymbol{\rho}_2) = \frac{1}{2}k_0^{-4}\Delta^{-2}\kappa(\boldsymbol{\rho}_1,\boldsymbol{\rho}_2) + itS(\boldsymbol{\rho}_1,\boldsymbol{\rho}_2). \tag{31}$$

Following the standard procedure of functional integration, we present the fluctuation $U(\boldsymbol{\rho})$ as a linear combination

$$U(\boldsymbol{\rho}) = \sum_\mu C_{t,\mu}\phi_{t,\mu}(\boldsymbol{\rho}), \tag{32}$$

where $\phi_{t,\mu}(\boldsymbol{\rho})$ are the eigenfunctions of the operator $\hat{\mathcal{K}}_t$ with the kernel $\mathcal{K}_t$

$$\hat{\mathcal{K}}_t\phi_{t,\mu}(\boldsymbol{\rho}) = \int d\boldsymbol{\rho}_1\mathcal{K}_t(\boldsymbol{\rho},\boldsymbol{\rho}_1)\phi_{t,\mu}(\boldsymbol{\rho}_1) = \Lambda_{t,\mu}\phi_{t,\mu}(\boldsymbol{\rho}), \tag{33}$$

and $\Lambda_{t,\mu}$ are the corresponding eigenvalues.

At this point, we note that the operator $\hat{\mathcal{K}}_t$ is non-hermitian. As a consequence, the functions $\phi_{t,\mu}$ form an orthogonal basis if the definition of the scalar product is modified to $\langle\phi_1|\phi_2\rangle = \int d\boldsymbol{\rho}\,\phi_1(\boldsymbol{\rho})\,\phi_2(\boldsymbol{\rho})$. To see this, note that the kernel $\mathcal{K}_t(\boldsymbol{\rho}_1,\boldsymbol{\rho}_2)$ is symmetric with respect to interchange $\boldsymbol{\rho}_1 \leftrightarrow \boldsymbol{\rho}_2$, $\mathcal{K}_t(\boldsymbol{\rho}_1,\boldsymbol{\rho}_2) = \mathcal{K}_t(\boldsymbol{\rho}_2,\boldsymbol{\rho}_1)$. Then it follows from (33)

$$\begin{aligned}\iint d\boldsymbol{\rho}_1 d\boldsymbol{\rho}_2\,\phi_{t,\mu_1}(\boldsymbol{\rho}_1)\,\mathcal{K}_t(\boldsymbol{\rho}_1,\boldsymbol{\rho}_2)\,\phi_{t,\mu_2}(\boldsymbol{\rho}_2) &= \\ &= \int d\boldsymbol{\rho}_1\phi_{t,\mu_1}(\boldsymbol{\rho}_1)\int d\boldsymbol{\rho}_2\,\mathcal{K}_t(\boldsymbol{\rho}_1,\boldsymbol{\rho}_2)\,\phi_{t,\mu_2}(\boldsymbol{\rho}_2) \\ &= \Lambda_{t,\mu_2}\int d\boldsymbol{\rho}_1\phi_{t,\mu_1}(\boldsymbol{\rho}_1)\,\phi_{t,\mu_2}(\boldsymbol{\rho}_1).\end{aligned} \tag{34}$$

On the other hand, it also follows from (33)

$$\begin{aligned}\iint d\boldsymbol{\rho}_1 d\boldsymbol{\rho}_2\,\phi_{t,\mu_1}(\boldsymbol{\rho}_1)\,\mathcal{K}_t(\boldsymbol{\rho}_1,\boldsymbol{\rho}_2)\,\phi_{t,\mu_2}(\boldsymbol{\rho}_2) &= \\ &= \int d\boldsymbol{\rho}_2\,\phi_{t,\mu_2}(\boldsymbol{\rho}_2)\int d\boldsymbol{\rho}_1\mathcal{K}_t(\boldsymbol{\rho}_2,\boldsymbol{\rho}_1)\,\phi_{t,\mu_1}(\boldsymbol{\rho}_1) \\ &= \Lambda_{t,\mu_1}\int d\boldsymbol{\rho}_2\,\phi_{t,\mu_1}(\boldsymbol{\rho}_2)\,\phi_{t,\mu_2}(\boldsymbol{\rho}_2).\end{aligned} \tag{35}$$

Comparing the r.h.s. of (34) and (35), we have

$$\langle\phi_{t,\mu_1}|\phi_{t,\mu_2}\rangle = \int d\boldsymbol{\rho}\,\phi_{t,\mu_1}(\boldsymbol{\rho})\,\phi_{t,\mu_2}(\boldsymbol{\rho}) = \delta_{\mu_1,\mu_2}. \tag{36}$$

Using the expansion (32), the functional integral (29) reduces to the integration over *all* complex coefficients $C_{t,\mu}$ in the expansion (32). However, due to the restriction that $U(\boldsymbol{\rho})$ is real, the pair $\{\mathrm{Re}C_{t,\mu}, \mathrm{Im}C_{t,\mu}\}$ can be "rotated" in such a way that instead of integrals over real and imaginary parts we get a single integral along the real axis,

$$p(\tau) = \frac{1}{2\pi}\int_{-\infty}^{\infty} dt\ e^{it/\tau}\,\frac{\int_{-\infty}^{\infty}\mathcal{D}\left\{\tilde{C}_{t,\mu}\right\}\exp\left(-\sum_\mu \Lambda_{t,\mu}\tilde{C}_{t,\mu}^2\right)}{\int_{-\infty}^{\infty}\mathcal{D}\left\{\tilde{C}_{0,\mu}\right\}\exp\left(-\sum_\mu \Lambda_{0,\mu}\tilde{C}_{0,\mu}^2\right)}. \tag{37}$$

To proceed further, we note that the real parts of all the eigenvalues $\Lambda_{t,\mu}$ are positive. Indeed, it follows from (31) and (33) that

$$\mathrm{Re}\Lambda_{t,\mu} = \left(\frac{1}{2k_0^4\Delta^2}\right)\frac{\iint d\boldsymbol{\rho}_1 d\boldsymbol{\rho}_2 \phi_{t,\mu}^*(\boldsymbol{\rho}_1)\kappa(\boldsymbol{\rho}_1,\boldsymbol{\rho}_2)\phi_{t,\mu}(\boldsymbol{\rho}_2)}{\iint d\boldsymbol{\rho}|\phi_{t,\mu}(\boldsymbol{\rho})|^2}. \tag{38}$$

Since κ is positively defined, $\mathrm{Re}\Lambda_{t,\mu}$ is positive. The last remark ensures the convergence of all the Gaussian integrals in (37). Thus, we obtain from (37)

$$p(\tau) = \frac{1}{2\pi}\int_{-\infty}^{\infty} dt\ e^{it/\tau}\left(\prod_\mu \frac{\Lambda_{0,\mu}}{\Lambda_{t,\mu}}\right)^{1/2}. \tag{39}$$

Since the product of the eigenvalues of an operator is equal to its determinant, (39) can be rewritten in the form

$$p(\tau) = \frac{1}{2\pi}\int_{-\infty}^{\infty} dt\ e^{it/\tau}\left(\frac{\det\hat{\mathcal{K}}_0}{\det\hat{\mathcal{K}}_t}\right)^{1/2}. \tag{40}$$

It is convenient to present the ratio of determinants in the integrand of (40) as a single determinant. This is achieved through the following sequence of steps

$$\begin{aligned}\frac{\det\hat{\mathcal{K}}_0}{\det\hat{\mathcal{K}}_t} &= \frac{1}{\det\hat{\mathcal{K}}_0^{-1}\det\hat{\mathcal{K}}_t} = \frac{1}{\det\left(\hat{\mathcal{K}}_0^{-1}\hat{\mathcal{K}}_t\right)} = \\ &= \frac{1}{\det\left[2k_0^4\Delta^2\hat{\kappa}^{-1}\left(\frac{1}{2}k_0^{-4}\Delta^{-2}\hat{\kappa} + it\hat{S}\right)\right]} = \\ &= \frac{1}{\det\left[1 + 2itk_0^4\Delta^2\,\hat{K}\hat{S}\right]},\end{aligned} \tag{41}$$

where we have used the explicit form (31) of the operator $\hat{\mathcal{K}}_t$. The operators $\hat{K}$ and $\hat{S}$ are the integral operators with the kernels $S(\boldsymbol{\rho}_1, \boldsymbol{\rho}_2)$ [(22)] and $K(\boldsymbol{\rho}_1, \boldsymbol{\rho}_2)$, respectively. We recall that the operator $\hat{K}$ is the inverse of the operator $\hat{\kappa}$ [(28)].

Upon transformation (41), the expression (40) for $p(\tau)$ takes the form

$$p(\tau) = \frac{1}{2\pi}\int_{-\infty}^{\infty} dt \frac{e^{it/\tau}}{\sqrt{\det\left[1+2itk_0^4\Delta^2\hat{K}\hat{S}\right]}}$$

$$= \frac{1}{\pi}\int_{0}^{\infty} dt \;\mathrm{Re}\left\{\frac{e^{it/\tau}}{\sqrt{\det\left[1+2itk_0^4\Delta^2\hat{K}\hat{S}\right]}}\right\}. \quad (42)$$

To proceed further, we need to analyze the properties of eigenfunctions and eigenvalues of the operators $\hat{K}$ and $\hat{S}$.

Properties of Operator $\hat{\boldsymbol{K}}$. It is easy to see that the eigenfunctions of $\hat{K}$ are plane waves. Indeed, since the kernel K depends only on the difference $(\boldsymbol{\rho}_1 - \boldsymbol{\rho}_2)$, we have

$$\hat{K}e^{i\boldsymbol{p}\boldsymbol{\rho}} = \int d\boldsymbol{\rho}_1 K(\boldsymbol{\rho}-\boldsymbol{\rho}_1)e^{i\boldsymbol{p}\boldsymbol{\rho}_1} = \tilde{K}(p)e^{i\boldsymbol{p}\boldsymbol{\rho}}, \quad (43)$$

so that eigenvalues of $\hat{K}$ are the Fourier components of the correlator $K(\rho)$. Thus, these eigenvalues are strongly suppressed if $p > R_c^{-1}$. For the particular case of Gaussian correlator we have

$$\tilde{K}(p) = \int d\rho K(\rho)\, e^{i\boldsymbol{p}\boldsymbol{\rho}} - \pi R_c^2 e^{-p^2 R_c^2/4}. \quad (44)$$

Properties of Operator $\hat{\boldsymbol{S}}$. Although the eigenfunctions, $\zeta_\mu(\boldsymbol{\rho})$, of the operator $\hat{S}$ are not plane waves, their width in the k-space is narrow (of the order of inverse spatial extent of the function ψ_0) as it can be seen from (22). To estimate the eigenvalues, λ_μ, of $\hat{S}$, defined as

$$\int d\boldsymbol{\rho}_1 S(\boldsymbol{\rho},\boldsymbol{\rho}_1)\xi_\mu(\boldsymbol{\rho}_1) = \lambda_\mu \xi_\mu(\boldsymbol{\rho}), \quad (45)$$

we first note that these eigenvalues satisfy the following sum rule

$$\sum_\mu \lambda_\mu = \frac{c}{2k}. \quad (46)$$

This rule follows from the identity

$$\sum_\mu \lambda_\mu = \iint d\boldsymbol{\rho}\, d\boldsymbol{\rho}_1 S(\boldsymbol{\rho},\boldsymbol{\rho}_1)\sum_\mu \xi_\mu^*(\boldsymbol{\rho})\xi_\mu(\boldsymbol{\rho}_1)$$

$$= \int d\boldsymbol{\rho}\; S(\boldsymbol{\rho},\boldsymbol{\rho}) = \frac{c}{2k}\int d\boldsymbol{\rho}\; |\psi_0(\boldsymbol{\rho})|^2, \quad (47)$$

in which the completeness of the set $\xi_\mu(\boldsymbol{\rho})$ is used, so that

$$\sum_\mu \xi^*_\mu(\boldsymbol{\rho})\xi_\mu(\boldsymbol{\rho}_1) = \delta(\boldsymbol{\rho} - \boldsymbol{\rho}_1). \tag{48}$$

The reason why (46) allows to estimate the eigenvalues is their specific distribution. Namely, the first $N \approx k^2 A$ eigenvalues are almost equal to each other, while the eigenvalues with numbers $\mu > N$ fall off rapidly, faster than $(N/\mu)^4$. This rapid fall off of λ_μ has a simple explanation. The eigenfunctions corresponding to large μ are close to plane waves, so that the value of λ_μ is determined by the integral of the rapidly oscillating plane wave over the area A and is small due to the cancellation. On the contrary, for $\mu < N$ the eigenfunction changes weakly within the area A. Thus, for $\mu < N$ we have $\lambda_\mu \approx \lambda_0 \sim c/(kN)$. For the particular case of the ring-shaped fluctuations $N \approx 2\pi\rho_0 w k^2$, so that $\lambda_0 \approx c/(\rho_0 w k^3)$.

Evaluation of the Integral (42) *Short-Range Disorder.* In this case $\tilde{K}(p) \approx \tilde{K}(0) \equiv \pi R_c^2$, so that the contribution to the integrand of (42) comes from N eigenvalues of the operator $\hat{S}$, i.e.,

$$\det\left[1 + 2itk_0^4\Delta^2\hat{K}\hat{S}\right] \approx \left[1 + 2i\pi t k_0^4\tilde{\Delta}^2\lambda_0\right]^N ,$$

where $R_c\Delta \to \tilde{\Delta}$ when $R_c \to 0$. Then the integration over t in (42) can be easily performed, yielding

$$\begin{aligned} p(\tau) &= \frac{1}{4\tau(N/2-1)!}\left(2\pi\,\tau\lambda_0\, k_0^4\tilde{\Delta}^2\right)^{2-N/2} \exp\left[-\left(2\pi\tau\lambda_0 k_0^4\tilde{\Delta}^2\right)^{-1}\right] \\ &\sim \exp\left[-\frac{N}{2}\ln\left(\frac{Q}{kl_e}\right) - \left(\frac{N}{8}\right)\frac{kl_e}{Q}\right], \end{aligned} \tag{49}$$

where we have used the large-N asymptotics of $N!$ and the fact that $\lambda_0 \approx c/(kN)$. Since the first term in the exponent is much larger than the second one, we recover with exponential accuracy the form of $P(\tau)$, obtained in the qualitative consideration.

Smooth Disorder. In the case of a smooth disorder, $kR_c > 1$, it is the fast decay of $\tilde{K}(p)$ [(44)] rather than λ_μ, that introduces a "cutoff" of the determinant in (42). In this case it is convenient to rewrite the determinant in (42) in the form

$$\begin{aligned} \det\left[1 + 2itk_0^4\Delta^2\hat{K}\hat{S}\right] &= \prod_\mu\left[1 + 2it\lambda_0\, k_0^4\Delta^2\tilde{K}(p_\mu)\right] \\ &= \exp\left[\sum_\mu \ln\left(1 + 2it\lambda_0\, k_0^4\Delta^2\tilde{K}(p_\mu)\right)\right]. \end{aligned} \tag{50}$$

The sum over μ in the exponent of (50) goes over both projections, p_x and p_y, of the momentum p. For ring-shaped fluctuations it is natural to consider the radial and azimutal components of p. Since $w \sim R_c$ and $\rho_0 \gg R_c$ the contribution to the sum comes only from a single radial component, while the sum over the angular component can be replaced by an integral. Thus we obtain

$$\det\left[1+2itk_0^4\Delta^2\hat{K}\hat{S}\right]=\exp\left[\rho_0\int_0^\infty dp\,\ln\Big(1+2it\lambda_0\,k_0^4\Delta^2\tilde{K}(p)\Big)\right]. \quad (51)$$

The main contribution to the integral (51) comes from the domain

$$\tau\lambda_0\,k_0^4\Delta^2\tilde{K}(p)\sim 1,$$

so that $p > R_c^{-1}$. It is instructive to perform further calculations for more general form of $\tilde{K}(p)$, namely $\tilde{K}(p) \sim \exp[-(pR_c)^n]$, which reduces to (44) when $n = 2$. Substituting this form into (51) and performing two subsequent integration by parts, we get

$$\begin{aligned}\det&\left[1+2itk_0^4\Delta^2\hat{K}\hat{S}\right]\\ &=\exp\left[\frac{i\rho_0}{R_c}\left(\frac{n}{n+1}\right)\int_0^\infty dw\;w^{(n+1)/n}\frac{e^{w-w_t}}{\left(e^{w-w_t}+i\right)^2}\right]\\ &=\exp\left[\frac{i\rho_0}{R_c}\left(\frac{n}{n+1}\right)w_t^{(n+1)/n}\int_{-\infty}^\infty dw\;\frac{e^{w}}{\left(e^{w}+i\right)^2}\right],\end{aligned} \quad (52)$$

where $w_t = \ln(2\pi t\lambda_0 R_c^2 k_0^4 \Lambda^2)$ In the second identity we made use of the fact that the function $\frac{e^{w-w_t}}{(e^{w-w_t}+i)^2}$ has a sharp maximum at $w = w_t$. The remaining integral in (52) can be evaluated exactly, $\int_{-\infty}^{\infty} dw\, \frac{e^w}{(e^w+i)^2} = -i$, so that (52) takes the form

$$\det\left[1+2itk_0^4\Delta^2\hat{K}\hat{S}\right]=\exp\left[\frac{\rho_0}{R_c}\left(\frac{n}{n+1}\right)w_t^{(n+1)/n}\right]. \quad (53)$$

Substituting this form into (42), we obtain

$$p(\tau)=\frac{1}{\pi}\int_0^\infty dt\,\cos(t/\tau)\exp\left[-\frac{n\rho_0}{2(n+1)R_c}\ln^{(n+1)/n}\left(2\pi t\lambda_0R_c^2k_0^4\Delta^2\right)\right]. \quad (54)$$

Evaluating the above integral with an exponential accuracy yields

$$p(\tau)\sim\exp\left[-\frac{n\rho_0}{2(n+1)R_c}\ln^{(n+1)/n}\left(2\pi\tau\lambda_0R_c^2k_0^4\Delta^2\right)\right]. \quad (55)$$

Since $p(\tau)$ and $P(\tau)$ have the same exponential dependence, the final expression for the prefactor, $P(\tau)$, takes the form

$$P(\tau)\sim\exp\left[-\frac{n\rho_0}{2(n+1)R_c}\ln^{(n+1)/n}\left(Q/kl_a\right)\right], \quad (56)$$

where τ in the argument of the logarithm was replaced by the trapping time $\omega^{-1}Q$. For the particular case of the Gaussian correlator (44) the probability that the lifetime with respect to additional leakage is longer than $\omega^{-1}Q$ is given by

$$P(\omega^{-1}Q) \sim \exp\left[-\frac{\rho_0}{3R_c}\ln^{3/2}(Q/kl_a)\right]. \tag{57}$$

Note, that for the short-range disorder, [(49)], the number N was the number of sections with the area $\sim k^{-2}$, which "cover" the ring-shaped trap. Correspondingly, for the smooth disorder the ratio ρ_0/R_c in the exponent is the number of squares with the side $\sim R_c$ that cover the trap. On the other hand, as it is seen from (56), the power of the logarithm is specific for the Gaussian correlator.

3.3 Optimal Ring

Combining the main exponent for the Gaussian correlator, (17), and the corresponding prefactor, (57), we obtain

$$\begin{aligned}\mathbf{N}_m(Q) &= P(\omega_c Q)e^{-S_m(Q)} \\ &\sim \exp\left[-\frac{m}{3kR_c}\ln^{3/2}Q - 3^{4/3}4^{-5/3}\pi\frac{kl}{(kR_c)^2}m^{-1/3}\ln^{4/3}Q\right].\end{aligned} \tag{58}$$

It is seen that the dependence of the main exponent and the prefactor on m are opposite. The prefactor, reflecting the "vulnarability" of the ring, falls off with m, while the main exponent favors large m value for which the Q-factor is higher. As a result of the competition between the two tendencies, $\mathbf{N}_m(Q)$ has a sharp maximum at the optimal value of m given by

$$m_{opt} = \frac{3\pi^{4/3}}{4^{5/4}}\left(\frac{l}{R_c}\right)^{3/4}\ln^{-1/8}Q. \tag{59}$$

Substituting $m = m_{opt}$ into (58), we arrive at the final result

$$\mathbf{N}_{opt}(Q) = \exp\left[-2^{-1/2}\pi^{3/4}\frac{(kl)^{3/4}}{(kR_c)^{7/4}}\ln^{11/8}Q\right]. \tag{60}$$

If we use a general form of the prefactor (56), then the changes in (60) amount to an additional factor $2^{-13/4}(3n)^{1/4}(n+1)^{-5/4}(11n+2)$. Also the power of $\ln Q$ in (60) modifies to $(5n+1)/4n$. Overall, these changes are inessential, so that the result, (60), is rather robust. It shows how the trapping is enhanced due to a smooth disorder, when the additional leakage is taken into account. Without the prefactor this enhancement manifested itself through the combination $(kR_c)^2$ in the denominator of the main exponent, S_m. With the prefactor, $(kR_c)^2$ is replaced by $(kR_c)^{7/4}$, so that the enhancement is weaker, but insignificantly.

As a final remark, we note that additional leakage, caused by the scattering out of the plane, can be incorporated into the theory in a similar fashion as the in-plane additional leakage. Corresponding changes are outlined in the end of Sec.2. Recall also, that for the smooth disorder the suppression of $\mathbf{N}_m(Q)$ due to additional leakage is dominated by the in-plane scattering processes.

4 Conclusion

In the present paper we studied a new type of solutions [66,76] of the wave equation, (1), in a weakly disordered medium. The solutions, dubbed as "almost localized" states, describe a wave which is confined primarily to a small ring. In an *open* sample, of size L much smaller than the two-dimensional localization length ξ, the almost localized states correspond to sharp resonances, residing in the high-Q ring-shaped cavities, as discussed throughout the paper. However, in a *closed* sample – which would require perfectly reflecting walls – the resonances turn into true eigenstates, whose almost entire weight is located at the rings. In this respect the "almost localized" states differ from the "prelocalized" states, extensively studied in the context of electronic transport [72–75].

We have provided a quantitative theory of the almost localized states and the associated random resonators, and pointed out their relevance for the phenomenon of random lasing. We stress, however, that these random resonators exist already in the *passive* medium, and gain is only needed "to make them visible". Moreover, the resonators are "self-formed", in the sense that no sharp features (like Mie scatterers or other "resonant entities") are introduced: the model is defined by (1), which describes a featureless dielectric medium with fluctuating dielectric constant.

Our study substantiates the intuitive image [43,44,58] of a resonant cavity as a closed-loop trajectory of a light wave bouncing between the point-like scatterers. The intuitive picture in [43,44,58] assumed that light can propagate along a loop of scatterers by simply being scattered from one scatterer to another. Such a picture, however, is unrealistic due to the scattering out of the loop. We have demonstrated that the scenario of light traveling along closed loops can be remedied. In our picture the "loops", i.e., the random resonators, can be envisaged as rings with dielectric constant larger than the average value. The reason why such rings are able to trap the light is that the constituting scatterers act as a *single entity*: only the coherent multiple scattering of light by *all* the scatterers in the resonator can provide trapping. We have also established that correlations in the fluctuating part of the dielectric constant highly facilitate trapping.

We acknowledge the support of the National Science Foundation under Grant No. DMR-0202790 and of the Petroleum Research Fund under Grant No. 37890-AC6.

References

1. E. Abrahams, P.W. Anderson, D.C. Licciardello, and V. Ramakrishnan, Phys. Rev. Lett. **42**, 673 (1979)
2. N. Garcia and A.Z. Genack, Rev. Lett. **66**, 1850 (1991)
3. A.Z. Genack and N. Garcia, Phys. Rev. Lett. **66**, 2064 (1991)
4. D.S. Wiersma , P. Bartolini, A. Lagendijk, and R. Righini, Nature (London) **390**, 671 (1997)
5. F. Scheffold, R. Lenke, R. Tweer, and G. Maret, Nature (London) **398**, 206 (1999)

6. D. S. Wiersma, J.G. Rivas, P. Bartolini, A. Lagendijk, and R. Righini, Nature (London) **398**, 207 (1999)
7. M.P. van Albada and A. Lagendijk, Phys. Rev. Lett. **55**, 2692 (1985)
8. P.E. Wolf and G. Maret, Phys. Rev. Lett. **55**, 2696 (1985)
9. E. Akkermans, P.E. Wolf, R. Maynard, and G. Maret, J. Phys. France **49**, 77 (1988)
10. F.J.P. Schuurmans, M. Megens, D. Vanmaekelbergh, and A. Lagendijk, Phys. Rev. Lett. **83**, 2183 (1999)
11. A.A. Chabanov, M. Stoytchev, and A.Z. Genack, Nature (London) **404**, 850 (2000)
12. A.A. Chabanov and A.Z. Genack, Phys. Rev. Lett. **87**, 153901 (2001)
13. A.A. Chabanov and A.Z. Genack, Phys. Rev. Lett. **87**, 233903 (2001)
14. V.S. Letokhov, Z. Éksp. Teor. Fiz. **53**, 1442 (1967) [Sov. Phys. JETP **26**, 835 (1968)]
15. N.M. Lawandy, R.M. Balachandran, A.S.L. Gomes, and E. Sauvain, Nature (London) **368**, 436 (1994)
16. A.Z. Genack and J.M. Drake, Nature (London) **368**, 400 (1994)
17. S. John and G. Pang, Phys. Rev. A **54**, 3642 (1996)
18. D.S. Wiersma and A. Lagendijk, Phys. Rev. E **54**, 4256 (1996)
19. M. Siddique, R.R. Alfano, G.A. Berger, M. Kempe, and A.Z. Genack, Opt. Lett. **21**, 450 (1996)
20. G.A. Berger, M. Kempe, and A.Z. Genack, Phys. Rev. E **56**, 6118 (1997)
21. R.M. Balachandran, N.M. Lawandy, and J.A. Moon, Opt. Lett. **22**, 319 (1997)
22. D.S. Wiersma, M.P. van Albada, and A. Lagendijk, Nature (London) **373**, 103 (1995)
23. W.L. Sha, C.-H. Liu, and R.R. Alfano, J. Opt. Soc. Am. B **19**, 1922 (1994)
24. W. Zhang, N. Cue, and K.M. Yoo, Opt. Lett. **20**, 961 (1995)
25. R.M. Balachandran and N.M. Lawandy, Opt. Lett. **20**, 1271 (1995)
26. M.A. Noginov, H.J. Caufield, N.E. Noginova, and P. Venkateswarlu, Opt. Commun. **118**, 430 (1995)
27. J. Martorell, R.M. Balachandran, and N.M. Lawandy, Opt. Lett. **21**, 239 (1996)
28. R.M. Balachandran, D.P. Pacheo, and N.M. Lawandy, Appl. Opt. **35**, 640 (1996)
29. W.L. Sha, C.-H. Liu, F. Liu, and R.R. Alfano, Opt. Lett. **21**, 1227 (1996)
30. P.C. de Olivera, J.A. McGreevy, and N.M. Lawandy, Opt. Lett. **22**, 700 (1997)
31. B.R. Prasad, H. Ramachandran, A.K. Sood, C.K. Subramanian, and N. Kumar, Appl. Opt.**36**, 7718 (1997)
32. G. Beckering, S.J. Zilker, and D. Haarer, Opt. Lett. **22**, 1427 (1997)
33. H.Z. Wang, F.L. Zhao, Y.J. He, X.G. Zheng, and X.G. Huang, Opt. Lett. **23**, 777 (1998)
34. Gijs van Soest, M. Tomita, and A. Lagendijk, Opt. Lett. **24**, 306 (1999)
35. G. Zacharakis, G. Heliotis, G. Filippidis, D. Anglos, and T.G. Papazoglou, Appl. Opt. **38**, 6087 (1999)
36. M. Shukri and R.L. Armstrong, Appl. Opt. **39**, 4300 (2000)
37. V.M. Markushev, V.F. Zolin, and C.M. Briskina, Sov. J. Quantum Electron. **16**, 281 (1986)
38. N.È. Ter-Gabrièlyan, V.M. Markushev, V.R. Belan, C.M. Briskina, and V.F. Zolin, Sov. J. Quantum Electron. **21**, 32 (1991)
39. C. Gouedard, D. Husson, C. Sauteret, F. Auzel, A. Migus, J. Opt. Soc. Am. B **10**, 2362 (1993)
40. G.R. Williams, S.B. Bayram, S.C. Rand, T. Hinklin, and R.M. Laine, Phys. Rev. A **65**, 013807 (2001)

41. F. Hide, B.J. Schwartz, M.A. Diaz-Garcia, and A.J. Heeger, Chem. Phys. Lett. **256**, 424 (1996)
42. D.J. Denton, N. Tessler, M.A. Stevens, and R.H. Friend, Adv. Mater. **9**, 547 (1997)
43. H. Cao, Y.G. Zhao, H.C. Ong, S.T. Ho, J.Y. Dai, J.Y. Wu, and R.P.H. Chang, Appl. Phys. Lett. **73**, 3656 (1998)
44. H. Cao, Y.G. Zhao, S.T. Ho, E.W. Seelig, Q.H. Wang, and R.P.H. Chang, Phys. Rev. Lett. **82**, 2278 (1999).
45. S.V. Frolov, Z.V. Vardeny, K. Yoshino, A. Zakhidov, and R.H. Baughman, Phys. Rev. B **59**, R5284 (1999)
46. S.V. Frolov, Z.V. Vardeny, A.A. Zakhidov, and R.H. Baughman, Opt. Commun. **162**, 241 (1999)
47. D.S. Wiersma, Nature **406**, 132 (2000)
48. D.S. Wiersma and S. Cavalieri, Nature **414**, 708 (2001)
49. H. Cao, Y. Ling, J.Y. Xu, C.Q. Cao, and P. Kumar, Phys. Rev. Lett. **86**, 4524 (2001)
50. R.C. Polson, A. Chipouline, and Z.V. Vardeny, Adv. Mater. **13**, 760 (2001)
51. A.Y. Zyuzin, Europhys. Lett. **26**, 517 (1994)
52. D.S. Wiersma M.P. van Albada, and A. Lagendijk, Phys. Rev. Lett. **75**, 1739 (1995)
53. H. Cao, J.Y. Xu, D.Z. Zhang, S.-H. Chang, S.T. Ho, E.W. Seelig, X. Liu, and R.P.H. Chang, Phys. Rev. Lett. **84**, 5584 (2000)
54. C. Vanneste and P. Sebbah, Phys. Rev. Lett. **87**, 183903 (2001)
55. C. Vanneste and P. Sebbah, Phys. Rev. B **66**, 164202 (2002)
56. X. Jiang and C. Soukoulis, Phys. Rev. E **65**, 025601(R) (2002)
57. H. Cao, Y. Ling, J.Y. Xu, and A.L. Burin, Phys. Rev. E **66**, 025601(R) (2002)
58. H. Cao, Y.G. Zhao, H.C. Ong, and R.P.H. Chang, Phys. Rev. B **59**, 15107 (1999)
59. H. Cao, J.Y. Xu, S.-H. Chang, and S.T. Ho, Phys. Rev. E **61**, 1985 (2000)
60. H. Cao, Y.G. Zhao, X. Liu, E.W. Seelig, and R.P.H. Chang, Appl. Phys. Lett. **75**, 1213 (2000)
61. Y. Ling, H. Cao, A.L. Burin, M.A. Ratner, X. Liu, and R.P.H. Chang, Phys. Rev. A **64**, 063808 (2001)
62. H. Cao, J.Y. Xu, Y. Ling, S.-H. Chang, S.T. Ho, E.W. Seelig, X Liu, and R.P.H. Chang, in *Photonic Crystals and Light Localization in the 21st Century*, edited by C.M. Soukoulis (Kluwer, Dordrecht, 2001)
63. R.C. Polson, M.E. Raikh, and Z.V. Vardeny, Comptes Rendus Physique, **3**, 509 (2002)
64. C.M. Soukoulis, X. Jiang, J.Y. Xu, and H. Cao, Phys. Rev. B **65**, 041103 (2002)
65. H. Cao, in *Optical Properties of Nanostructured Random Media*, ed. V.M. Shalaev (Springer-Verlag, 2002)
66. V.M. Apalkov, M.E. Raikh, and B. Shapiro, Phys. Rev. Lett. **89**, 016802 (2002)
67. T.S. Misirpashaev and C.W.J. Beenakker, Phys. Rev. A **57**, 2041 (1998)
68. X. Jiang and C.M. Soukoulis, Phys. Rev. Lett. **85**, 70 (2000)
69. G. Hackenbroich, C. Viviescas, B. Elattari, and F. Haake, Phys. Rev. Lett. **86**, 5262 (2001)
70. G. Hackenbroich, C. Viviescas, and F. Haake, Phys. Rev. Lett. **89**, 083902 (2002)
71. M. Patra, Phys. Rev. A **65**, 043809 (2002)
72. B.L. Altshuler, V.E. Kravtsov, I.V. Lerner, in *Mesoscopic Phenomena in Solids*, eds. B.L. Altshuler, P.A. Lee, and R.A. Webb (North Holland, Amsterdam, 1991)
73. B.A. Muzykantskii and D.E. Khmelnitskii, Phys. Rev. B **51**, 5480 (1995)
74. V.I. Fal'ko and K.B. Efetov, Phys. Rev. B **52**, 17413 (1995)

75. A.D. Mirlin, Phys. Rep. **326**, 259 (2000)
76. V.G. Karpov, Phys. Rev. B **48**, 4325 (1993)
77. V.G. Karpov, Phys. Rev. B **48**, 12539 (1993)
78. D.S. Wiersma, M.P. Albada, B.A. van Tiggelen, and A. Lagendijk, Phys. Rev. Lett. **74**, 4193 (1995)
79. G. van Soest, Ph.D. thesis, Universiteit van Amsterdam, The Netherlands, 2001 (unpublished)
80. S. Hikami, Phys. Rev. B **24**, 2671 (1981)
81. B.L. Altshuler, V.E. Kravtsov, and I.V. Lerner, Sov. Phys. JETP **64**, 1352 (1986)
82. B.I. Halperin and M. Lax, Phys. Rev. **148**, 722 (1966)
83. J. Zittartz and J.S. Langer, Phys. Rev. **148**, 741 (1966)
84. V.M. Apalkov, M.E. Raikh, and B. Shapiro Phys. Rev. Lett. **89**, 126601 (2002)
85. E. Brézin and G. Parisi, J. Phys. C **13** L307 (1980)
86. A.L. Efros and M.E. Raikh, in *Optical Properties of Mixed Crystals* eds. R.J. Elliott and I.P. Ipatova, (North Holland, Amsterdam, 1988)

Manifestations of Anderson Localization in Semiconductor Optics

Erich Runge[1] and Roland Zimmermann[2]

[1] MPI Physik komplexer Systeme, Noethnitzer Strasse 38, 01187 Dresden, Germany
[2] Institut für Physik der Humboldt-Universitt zu Berlin, Newtonstr. 15, D-12489 Berlin, Germany

1 Optics of Semiconductor Nanostructures

Most physicists world-wide are solid-state physicists; most solid-state physicists work more or less directly about semiconductors, and most studies on semiconductors consider nanostructures, where at least one characteristic system dimension is on the nanometer scale. Your judgment whether this provocative statement actually is true or not will depend on how broadly you define the terms *semiconductor* and *nanostructures*. But doubtlessly, the enormous economic impact of semiconductor nanostructures with, e.g., the laser diode as the work horse of the telecommunication industry and the CD player as well as the intellectual challenge provided, e.g., by the quantum Hall effect attract this huge interest and justify the manpower spend in this field of research.

In the present contribution, we argue that optical experiments with semiconductor nanostructures allow to study some of the less explored aspects of the Anderson model and its ramifications. We highlight two aspects of relevance for the following: (*i*) The conceptually simplest and, arguably, richest model system for quantum effects in nanometer-scale semiconductor structures is the quantum well (QW), which simply is a thin layer of one semiconductor sandwiched between another semiconductor material, see Fig. 1a. The interfaces are never perfect, which leaves us with a two-dimensional (2D) model system for the Anderson disorder Hamiltonian, as will be shown below in the context of (1). One-dimensional quantum wires (QWR) can be derived from quantum wells, e.g., by stripe etching. (*ii*) The very advanced fabrication technology allows to realize almost any 0D, 1D, or 2D structure which could be of interest. Optical spectroscopy provides the tools to study phenomena with temporal resolution Δt down to the picosecond and femtosecond range and/or μeV energy resolution $\hbar\Delta\omega$, subject only to the fundamental limitation $\Delta t\,\Delta\omega \gtrsim 1$. This can be supplemented by spatial resolution down to about 150 nm if, e.g., in so-called *near-field scanning optical microscopy* the light is brought to the sample and/or collected through a thin glas fiber tip which is scanned across the sample surface [1], see Fig. 1.

The fundamental optical process in a semiconductor is the generation or annihilation of an electron-hole pair. At low temperatures, these form a bound state, the so-called *exciton*. The wavefunction of the exciton in a QW with rough interfaces, see Fig. 1a, can approximately be written as product

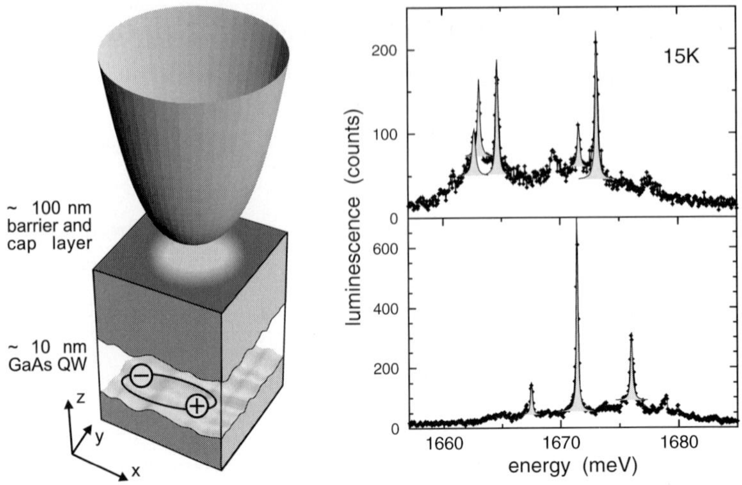

Fig. 1. *Left:* Schematic presentation of a glas fiber collecting near-field spectra from excitons in a semiconductor quantum well (QW) with rough interfaces. *Right:* Examples of two near-field spectra from a semiconductor nanostructure and their interpretation as a sum of many individual lines of Lorentzian shape

$\psi_\alpha(\boldsymbol{R})\,\varphi_{1s}(\boldsymbol{r}_e - \boldsymbol{r}_h)\,u_e(z_e)\,u_h(z_h)$, with factors describing, in this order, the center-of-mass (COM) motion in the QW plane, the (hydrogen-atom-like) relative function, and factors describing the vertical confinement of electron and hole. The Bloch parts involving the underlying atomic orbitals are left out, emphasizing the effective-mass approach used. Next, the expectation value of the Hamiltonian involving kinetic and potential energy as well as Coulomb attraction is formed with respect to all degrees of freedom except $\boldsymbol{R}$ [2]. This yields an effective one-particle Hamiltonian for the COM motion of the exciton as an entity:

$$\left(-\frac{\hbar^2\,\Delta}{2m_X} + v(\boldsymbol{R})\right)\psi_\alpha(\boldsymbol{R}) = \epsilon_\alpha\;\psi_\alpha(\boldsymbol{R})\,. \tag{1}$$

The explicit form of the effective potential $v(\boldsymbol{R})$ in terms of the underlying interface roughness on the atomic scale has been derived [2,3], but is of no relevance for the present argument. Here, it is sufficient to note that the energetic fluctuations $\sigma^2 = \overline{v(\boldsymbol{R})^2}$ are of the order $\sigma \sim 3 - 10\,\text{meV}$ and that the corresponding correlation function $g(\boldsymbol{R} - \boldsymbol{R}') = \overline{v(\boldsymbol{R})\,v(\boldsymbol{R}')}$ for most QW samples decays rapidly on a 10 - 100 nm scale. A typical value for the exciton mass m_X is $0.3\,m_0$. In summary, we describe the QW exciton by a 2D Anderson model [4,6–8] with correlated disorder.

A semiconductor-specific aspect of Anderson-localization physics is the importance of the so-called *optical matrix element*, which describes the exciton's coupling to the electromagnetic field. Defined as the exciton-photon transition matrix element, it is the overlap of the product wavefunction mentioned above taken at $\boldsymbol{r}_e = \boldsymbol{r}_h$ with the photon's plane-wave state [5]

$$M_\alpha(\boldsymbol{k}) = \mathcal{C} \int d\boldsymbol{R}\, e^{i\boldsymbol{k}\cdot\boldsymbol{R}}\, \psi(\boldsymbol{R})\,. \tag{2}$$

The prefactor $\mathcal{C}$ comprises the bare interaction vertex as well as the integrals of the other wavefunction factors over the remaining coordinates. For convenience, we shall often neglect the small photon momentum $\boldsymbol{k}$ and call the squared matrix element $r_\alpha = |\, M_\alpha(\boldsymbol{k}{=}0)\,|^2$ the excitonic *oscillator strength* or its *radiative rate* (inverse radiative life time).

A near-field absorption experiment measures the spectrum

$$I^{(\mathrm{abs})}(E) = \sum_\alpha r_\alpha\, \delta(E - \epsilon_\alpha)\,, \tag{3}$$

where the sum includes the exciton states α in the illuminated region. For studies of photoluminescence, the state occupation enters as an additional factor. Equation (3) has to be further modified for comparison with actual experiments by including the collection efficiency $f(h) < 1$, which can be very small if the exciton is located at a large distance h away from the center of the illumination/detection spot.

As it stands, the expression (3) demonstrates one of the key insights from optical studies of Anderson-model physics in semiconductors: Properties of the energy-level distribution and wavefunction characteristics – often parameterized via the optical matrix element – can, in general, not be disentangled.

Stimulated by experimental near-field spectra, we started [9,10] to study the averaged autoconvolution of spectra

$$R(\Delta E) = \frac{1}{\Omega} \int dE \;\; \overline{I^{(\mathrm{abs})}\Big(E + \frac{\Delta E}{2}\Big)\; I^{(\mathrm{abs})}\Big(E - \frac{\Delta E}{2}\Big)} \tag{4}$$

$$= \overline{\frac{1}{\Omega} \sum\nolimits_{\alpha\beta} r_\alpha\, r_\beta\, \delta\Big(\Delta E - (\epsilon_\alpha - \epsilon_\beta)\Big)} \tag{5}$$

hoping to see level repulsion when comparing to the autoconvolution of the averaged spectra

$$R_0(\Delta E) = \Omega \int dE' \;\overline{D(E')} \cdot \overline{D(E' - \Delta E)}\,. \tag{6}$$

The averaged absorption spectrum

$$D(E) = \overline{\frac{1}{\Omega}\, I^{(\mathrm{abs})}} = \overline{\frac{1}{\Omega} \sum\nolimits_\alpha r_\alpha\, \delta(E - \epsilon_\alpha)}\,. \tag{7}$$

is also-called *optical density*.

2 A Productive Challenge

When we presented our early results [9–11] based on (3–7) at the conference *Percolation, Interaction, and Localization: Simulations of transport in disordered systems* (PILS'98, October 6-9, Berlin, Germany), we were aware of the long-standing history of level repulsion and the classical results of random matrix theory (RMT) [12,13] – and we used the corresponding terminology. We were not aware that at that time the presence or absence of level repulsion in large-scale numerical simulations of the Anderson model had been established by Bernhard Kramer [14] and others as a very sensitive criterion for the position of the mobility edge and for the importance of finite-size effects [15]. In the conducting phase at energies E above the mobility edge, they observed – as predicted by RMT – a strong reduction of the probability to find energy levels separated by a small amount ΔE [8,14]. For large system size Ω, one finds in the absence of magnetic fields and spin-orbit scattering:

$$\overline{\frac{1}{\Omega^2} \sum_{\alpha,\beta} \delta\Big(E + \tfrac{\Delta E}{2} - \epsilon_\alpha\Big)\, \delta\Big(E - \tfrac{\Delta E}{2} - \epsilon_\beta\Big)} \propto \frac{\Delta E}{\overline{\Delta}} \ll 1\,, \tag{8}$$

where $\overline{\Delta}$ is the mean level spacing, $\propto 1/\Omega$.

In contrast, all states are localized at energies E below the mobility edge. A simple argument explains why level repulsion is not seen: The D-dimensional sample volume Ω is split in a *Gedankenexperiment* into M compartments, each of diameter larger than typical localization lengths: $\ell_{\mathrm{loc},\alpha}^D < \Omega/M$. One can expect to find a behavior resembling (8) within each compartment, but energy levels α, β in different compartments are certainly uncorrelated. This suggests for the joint level probability for $\Delta E \lesssim \overline{\Delta}$ the approximate form

$$\frac{M-1}{M}\,1 + \frac{1}{M}\Big(\frac{\Delta E}{\overline{\Delta}}\Big) \tag{9}$$

with

$$\frac{1}{M} \sim \frac{\ell_{\mathrm{loc}}^D}{\Omega}\,. \tag{10}$$

For large systems, nothing is left from the level repulsion!

We will refer to this argument as: *The* $1/M$ *argument.* It is certainly only of qualitative nature because issues such as wavefunctions localized near the compartment boundaries or the possible existence of rare states with exceptionally large $\ell_{\mathrm{loc},\alpha}$ are not accounted for. However, it conveys the relevant physics.

As told above, the community dominating at PILS'98 was quite aware of all this, when one of us (E.R.) presented our message: *There should be level repulsion in the near-field spectra.* Bernhard Kramer pointed out that within the Anderson model the QW excitons (2D!) should be localized [6] and thus, according to the $1/M$ argument, should *not* show level repulsion.

We took home from the conference what could be phrased as Bernhard Kramer's challenge to us: *How can one see the quantum physics of localization in the disordered QW exciton system?* (Note that 'seeing' can be taken almost literally as we primarily deal with optical experiments.) One can rephrase the challenge as: *How to circumvent the* $1/M$ *argument?*

3 Answers and Experimental Verifications

The challenge turned out to be a productive one: The following years brought several answers and some experimental verifications, which we hope Bernhard Kramer will enjoy to see summarized on the occasion of his 60th birthday!

3.1 Look Locally: Near-Field Level Spectroscopy

The simplest way to avoid the consequences of the $1/M$ argument is to keep M close to unity. This was implicit in our original proposal to study near-field spectra [9,10]. In particular, the area $\mathcal{A}_{\text{det}}$ of the excitation and/or detection spot plays the role of the volume Ω. The reduction of the probability to find states close in energy – called *level repulsion dip* henceforth – was thus predicted according to (9) to be $\propto 1/\mathcal{A}_{\text{det}}$. This has qualitatively be confirmed by Guest, Steel, Gammon et al. in [17] and lead to reasonable estimations for the localization length ($\sim 200\,\text{nm}$).

Later developments in *level correlation spectroscopy* [18–21] showed a rich phenomenology depending on, e.g., the particular QW sample and its interface quality. First of all, it turned out to be useful to replace the visual comparison of the averaged autoconvolution (4,5) and the autoconvolution of the average (6) by the consideration of the difference [31]:

$$R_c(\Delta E) = R(\Delta E) - R_0(\Delta E)\,. \tag{11}$$

Note that for systems with localized state, this implies the cancellation of the dominant contribution $\propto \Omega$, leaving, with our choice of prefactors, $R_c \propto \Omega^0$. Note further, that the level repulsion dip in $R(\Delta E)$ – defined via a reduced, but still positive probability – shows up in R_c as a negative values of R_c at small ΔE.

At least two additional features besides level repulsion have been found in the analysis of R_c. The first one could be called *classical level attraction.* It is best visualized in a scenario often used in the theory of the integer quantum Hall effect: small wave functions distributed evenly on a smooth potential landscape of large correlation length. The level correlation is then dominated by the correlation of the potential energy values

$$\frac{1}{\Omega}\int d\boldsymbol{R}\int d\boldsymbol{R}'\;\overline{\delta\Big(E+\frac{\Delta E}{2}-v(\boldsymbol{R})\Big)\,\delta\Big(E-\frac{\Delta E}{2}-v(\boldsymbol{R}')\Big)}\,, \tag{12}$$

which in general has a broad peak at $\Delta E \to 0$.

Figure 2a shows $R_c(E)$ for an actual experiment together with its interpretation (inset). With increasing level separation ΔE, one sees *(i)* a positive broadened δ contribution at the origin, resulting from $\alpha = \beta$ in (5), *(ii)* the negative quantum mechanical level repulsion analogous to the RMT behavior, *(iii)* the positive contribution from classical level attraction, and *(iv)* the negative contribution from the subtraction of the uncorrelated part R_0. Note that by definition the integral over $R_c(\Delta E)$ vanishes.

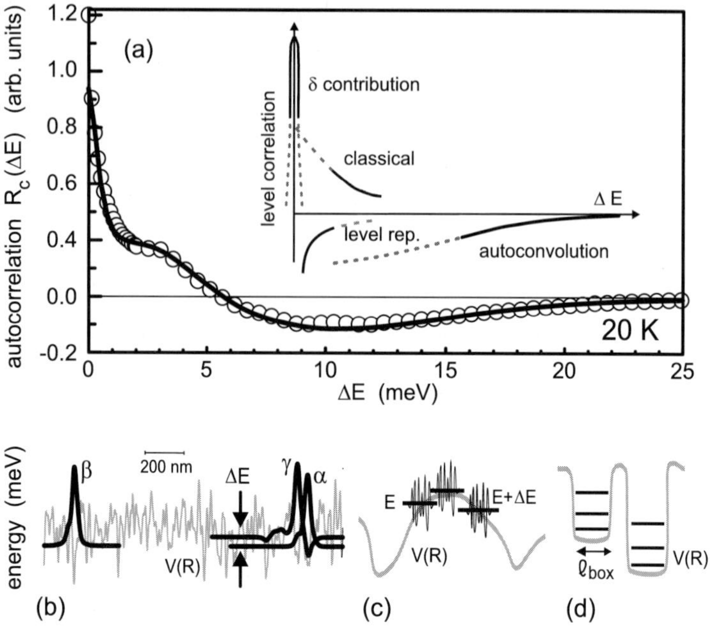

Fig. 2. **a** Experimental level correlation function for a 3-nm-wide GaAs/AlGaAs quantum well grown by molecular beam epitaxy on a GaAs (311)A substrate (after Intonti et al. [19]). *Open symbols:* data. *Solid line:* calculated level repulsion with Lorentzian broadening using the experimentally determined linewidth. *Inset:* Interpretation of the full level correlation as sum of various contributions, see text. **b** Visualization of quantum mechanical level repulsion between states α and γ which overlap in real space. In contrast, states α and β which are far separated in real space can have almost identical energy. **c** In a smooth potential landscape, the presence of a state at energy E increases the chance to find other states close-by in real space and therefore close-by in energy at $E + \Delta E$. **d** If many deep potential minima have approximately the same size ℓ_{box}, positive peaks in R_c at ΔE values corresponding to characteristic level spacings can occur

Panel (b) of Fig. 2 emphasizes the local character of level repulsion between localized wavefunctions, while panel (c) sketches the classical level attraction. Panel (d) illustrates the second additional feature, emphasized by von Freymann et al. in [21] but not seen in the particular autocorrelation of panel (a): If the considered potential profiles involve box-like minima with widths distributed narrowly around a characteristic width ℓ_{box}, one would expect peaks in the level correlation at energy differences determined by the particle-in-a-box energy levels $\Delta E \propto \hbar^2/(2m_X\ell_{\mathrm{box}}^2)$.

3.2 Look at Wavefunctions: Oscillator-Strength Distribution

The wavefunctions of random matrix theory for the N-dimensional orthogonal ensemble are evenly distributed points on the N-dimensional sphere. In the limit

$N \to \infty$, each component is Gaussian distributed (Berry's conjecture), and consequently its squares are distributed according to the so-called *Porter-Thomas distribution* (PTD) [12,13]

$$PTD: \quad P(r) = \frac{e^{-r/(2\overline{r})}}{\sqrt{2\pi\,\overline{r}\,r}} \,. \tag{13}$$

Within RMT, this holds in almost every basis.

Coming back to the QW exciton wavefunctions, we can *not* expect a PTD for the wavefunction values in the real-space representation because a localized wavefunction is almost zero almost everywhere. However, one can hope to find it for the wavefunction values in the $\boldsymbol{k}$-space presentation, which coincide with the $M_\alpha(\boldsymbol{k})$ introduced in (2). One appealing aspect of this quantity is that increasing the system size beyond the localization length does not change the $M_\alpha(\boldsymbol{k})$ and their distribution. Even more import is that in the exciton context the $r_\alpha = |M_\alpha(\boldsymbol{k}=0)|^2$ are measurable quantities, namely the weights of the peaks in the spectrum (3). Indeed, numerical simulations [8,22] showed that in the high-energy tail of the inhomogeneously broadened QW exciton line a PTD is found for the distribution of oscillator strengths.

For the one-dimensional QWR case with short-range disorder, the full energy-dependent distribution $P_E(r)$ of radiative rates r can be obtained with arbitrary accuracy from an integral recursion for the compound distribution of wavefunction value, normalization integral, and optical matrix element [24]. Its derivation [25] follows early work of Halperin [26] and is formulated for the tight-binding Anderson model, written in the site representation $n = 0, 1, \ldots, N$ recursively as.

$$\varphi_{n+1} = -\varphi_{n-1} + \frac{V_n + 2T - E}{T}\,\varphi_n \,. \tag{14}$$

The transfer matrix element T (band width equals $4T$) can be interpreted as resulting from the discretization with grid spacing Δ_x of the continuous model, (1), thus $T = \hbar^2/(2m_X\Delta_x)$, or as hopping from, e.g., one monomer to the next in a polymer semiconductor.

The left panel of Fig. 3 illustrates the strongly differing shapes of $P_E(r)$ depending on whether the considered energy is above, at, or below the band edge. Also included in the figure (right panel) is a first comparison with experimental oscillator-strength distributions obtained from near-field spectra of the sample studied in [19]. Two such spectra together with Lorentzian fits to some of the peaks were reproduced as Fig. 1b. The weight of the peaks ($\propto r_\alpha$) were separated into those above and below the maximum of the averaged intensity, yielding histograms $P^>(r)$ and $P^<(r)$. The corresponding average is performed for the theoretical results, taking additionally into account a Gaussian distribution of collection efficiencies [24]. The agreement of experiment and theory is quite satisfactory.

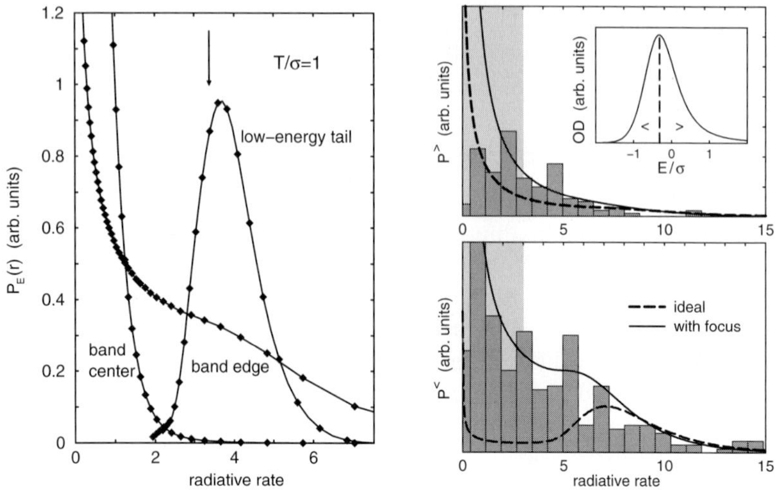

Fig. 3. *Left:* Distribution $P_E(r)$ of radiative rates of the tight-binding Anderson model for $T/\sigma = 1$ in the band center, at the lower band edge ($E = 0$), and in the low-energy tail where the optical density has dropped down to 10% of its maximum value, $E/\sigma = -1.4$ (r in tight-binding units); after [24,25] calculated with a moderate number of grid points (diamonds). The arrow marks the oscillator strength predicted by optimum fluctuation theory [27,28]. Courtesy of Helge Ludwig, Berlin. *Right:* Average distribution of oscillator strength separately for the optically active band region (*top*) and in the low-energy tail (bottom) with the position of the maximum in the optical density (OD) taken for the separation of these regimes, see inset. *Dashed curves*: $P^{\lessgtr}(r)$ as calculated; *full curves:* taking into account the finite collection efficiency (Gaussian focus); *histograms:* experimental data [24]. Light gray shading indicates that small peaks are not well resolved in the experiment and agreement with theory can not be expected here

3.3 Involve Transition Matrix Elements

The core of the $1/M$ argument is that for localized states the number of pairs $\{\alpha, \beta\}$ of spatially separated states ($\propto \Omega^2$) overwhelms the number of pairs of states which are close in real space ($\propto \Omega^1$). This is not the case, if the presence of transition matrix elements of an operator $\hat{A}$ force α and β to be close in real space, i.e., for quantities of the form:

$$\overline{\sum_{\alpha,\beta} \left|\langle\alpha|\hat{A}|\beta\rangle\right|^2 \delta\Big(E + \tfrac{\Delta E}{2} - \epsilon_\alpha\Big)\, \delta\Big(E - \tfrac{\Delta E}{2} - \epsilon_\beta\Big)} \,. \tag{15}$$

In the exciton context, we suggested that level repulsion of states connected by non-vanishing matrix elements may lead to measurable consequences in the low-temperature relaxation kinetics ($\hat{A}$: exciton-phonon coupling) [11]. This has not yet been confirmed by experiments. An example more along mainstream localization physics is the ac conductivity of a disordered system with $\hat{A}$ being the current operator [29].

3.4 Look at Fluctuations: Level Correlation in Rayleigh Scattering

Considered carefully, the $1/M$ argument suggests not the absence of level repulsion effects, but only their smallness $\propto 1/\Omega$ compared to the uncorrelated contribution. Fortunately, there are cases where the leading contribution in the double sum (5) over levels vanishes and the measurable signal are the fluctuations, which show up in the next order in $1/\Omega$. Such a case is Rayleigh scattering in a QW with microscopic disorder, i.e., the light intensity $I_{\boldsymbol{q}}(t)$ detected at time t after a short coherent excitation, with the three-dimensional emission direction characterized by the in-plane wavevector $\boldsymbol{q}$.

The initial excitation induces in the macroscopically homogeneous sample oscillations of the local dipoles localized at $\boldsymbol{R}_\alpha$. They emit at their respective frequency $\epsilon_\alpha/\hbar$ with oscillator strength r_α. The expectation value of the resulting electromagnetic field

$$\mathcal{E} \sim \sum_\alpha r_\alpha \, e^{i\boldsymbol{q}\cdot\boldsymbol{R}_\alpha} \, e^{-i\epsilon_\alpha t/\hbar} \tag{16}$$

is formally proportional to the excitation area, but it vanishes everywhere except in the reflected and transmitted direction. The diffusive light scattering in all other directions (Rayleigh signal) is proportional to the fluctuations of the electromagnetic field around its (vanishing) average, thus it is formally smaller by one order $1/\Omega$. It can be written as the Fourier transform of the optically weighted level-level correlation function (11) [5,30,31]

$$I^{\mathrm{Ray}}(t) = \int d\Delta E \; R_c(\Delta E) \, \cos(\Delta E t) \,. \tag{17}$$

We note explicitly that the level correlation is weighted by the oscillator strength, but that $r_\alpha \, r_\beta$ enter in a factorized form. This is conceptually completely different from the selection of pairs by the presence of transition matrix elements in Sect. 3.3. Again, we note in passing the connection to the ac conductivity, which according to the Kubo formula is a fluctuation correlation as well.

A level correlation function such as the one in Fig. 2a with a pronounced structure, e.g., a peak or a shoulder at a finite energy difference translates upon Fourier transformation into the time domain into an oscillation. Indeed, a faint oscillation has been reported by Savona et al. [32] as a first experimental evidence for level correlation features in the time-dependent Rayleigh signal.

3.5 Watch the Early Stages of Localization: Enhanced Backscattering of Quantum Well Excitons

We finish this section with the description of one aspect of Rayleigh scattering which is of such fundamental interest that it deserves its own subsection, namely the enhanced backscattering (EBS) and the early localization dynamics.

The enhanced backscattering of a disordered medium into a very narrow cone around the incoming direction is, arguably, the most spectacular manifestation of weak localization [33]. It has been observed directly in a wide variety of systems – including light scattering in suspensions of microspheres in water, in two-dimensional organic systems, and in ultra-cold atom gases, as well as scattering of ultrasonic acoustic waves and of elastic waves in crystals. Weak localization and EBS explains also the positive magneto-conductance of dirty metal wires at low temperatures [33].

In almost all these cases, delocalized waves were involved. Only very few claims of EBS from disorder localized states exist [34]. Therefore looking for EBS at resonant excitation of QW excitons is complementary to the usual light-scattering experiments. Our calculations showed that indeed EBS should be observed, however with an enhancement factor much smaller than the famous factor of two [35].

Two aspects are worthwhile to be mentioned explicitly: *(i)* In analogy to light-scattering in the localized regime [34], the profile of the backscattering peak does not show the characteristic non-analytic kink, but is rounded. *(ii)* The angular width first shrinks before the scattering profile approaches its final shape. This narrowing in $\boldsymbol{k}$ space reflects the spread of the exciton in real space, which stops when the localization length is approached. In this sense, studying the time-dependence of the EBS is tantamount to watching the early stages of localization. *(iii)* A strong forward peak develops with some time delay, finally obtaining equal weight. This is easily understood in the language of disorder-localized wavefunction [5], but hard to obtain in the usual diagrammatic perturbation theory (ladders plus maximally crossed diagrams).

Results for the time-dependent Rayleigh signal from a 1D system are shown in Fig. 4. Panel (a) shows clearly the mentioned features. Unfortunately, the large dielectric constant of the semiconductor implies in Snellius' law that even incident angles close to 90° correspond to rather small internal angles and small in-plane momenta $\boldsymbol{q}_{\text{in}}$. For the more realistic value of $\boldsymbol{q}_{\text{in}}$ in panel (b), the finite slope tilted towards the backscattering direction provides an experimentally accessible measure of EBS. This prediction could recently be confirmed even quantitatively [36].

4 Summary and Perspectives for Future Experiments

Bernhard Kramer's productive challenge at PILS'98 how one could see Anderson localization and its ramifications in quantum well spectroscopy has found various answers, each intriguing in its own ways. Several predictions lead to experiments, which profitted in many aspects from the high technological level described in the introduction.

Further experimental work is needed here, e.g., a comparison of the oscillator strength distribution with experiment more detailed than that of Fig. 3b is highly desirable. So is the confirmation of the predicted decrease of the EBS slope at late times, see Fig. 4b.

One can think of non-linear experiments or the interaction of the exciton with an additional charge. Here, obvious analogies to the so-called *level velocities* [12] exist. A third hallmark of random matrix theory besides level repulsion

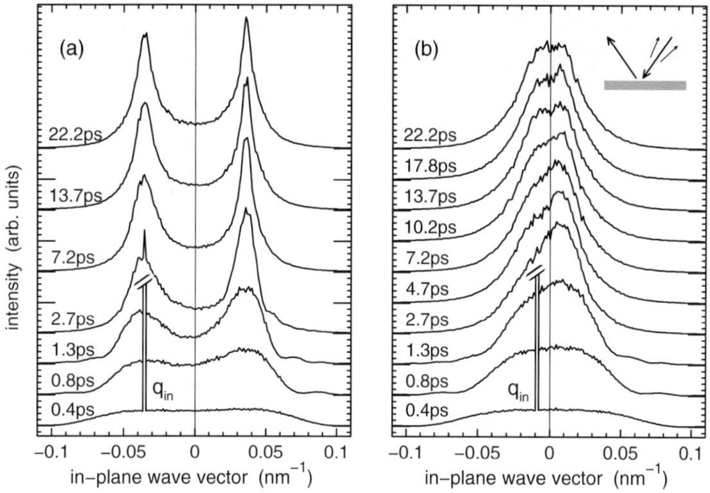

Fig. 4. Model calculations for the time-dependent Rayleigh scattering signal and enhanced backscattering (EBS) for a quantum wire with $\sigma = 3\,\mathrm{meV}$ and $m_X = 0.3\,m_0$. The imaginary part of the self-energy, which would lead to approximately exponential decay, is neglected. **a** Calculated with an unrealistically large photon momentum q_{in} for a demonstration of the underlying physics: the EBS peak develops with some delay after the excitation. Its angular width first shrinks, but then becomes constant. Simultaneously, a symmetric feature in forward direction develops. **b** Calculated with an incident momentum within the experimentally accessible external light cone [35]. Forward and backward scattering features overlap, but the symmetry-breaking finite slope at $k = 0$ is left as characteristic EBS signal. The huge signal of the specular reflected beam at q_{in} is suppressed in the curves for $t < 2\,\mathrm{ps}$

and wavefunction statistics is the *spectral rigidity*. We would not be surprised if manifestations of this feature will be found soon in excitonic quantum well physics as well.

References

1. H.F. Hess, E. Betzig, T.D. Harris, L.N. Pfeiffer, K.W. West: Science **264**, 1740 (1994)
2. E. Runge: 'Excitons in Semiconductor Nanostructures'. In: *Solid State Physics, Vol. 58*, ed. by H. Ehrenreich, F. Spaepen (Academic, San Diego 2002) pp. 149-305
3. R. Zimmermann, F. Grosse, E. Runge: Pure and Applied Chemistry **69**, 1179 (1997)
4. P.W. Anderson, Phys. Rev. **109**, 1492 (1958)
5. R. Zimmermann, E. Runge, V. Savona: 'Theory of resonant secondary emission: Rayleigh scattering versus luminescence'. In: *Quantum Coherence, Correlation and Decoherence in Semiconductor Nanostructures*, ed. by T. Takagahara (Elsevier Science, Amsterdam, to appear 2003)

6. E. Abrahams, P.W. Anderson, D.C. Licciardello, T.V. Ramakrishnan: Phys. Rev. Lett. **42**, 673 (1979)
7. B. Kramer, A. Mac Kinnon: Rep. Prog. Phys. **56**, 1469 (1993)
8. See contributions and references in: *Proceedings Localisation 1999*, July 30–Aug. 3, Hamburg, special issue Annalen der Physik (Leipzig) **8-SI7-9**
9. E. Runge, R. Zimmermann: phys. stat. sol. (b) **206**, 167 (1998)
10. E. Runge, R. Zimmermann: 'Optical properties of localized excitons in nanostructures: Theoretical aspects'. In: *Festkörperprobleme/Advances in Solid State Physics Vol. 38*, ed. by B. Kramer (Verlag Vieweg, Braunschweig 1998) pp. 251–263
11. E. Runge, R. Zimmermann: Annalen der Physik (Leipzig) **7**, 417 (1998)
12. T. Guhr, A. Müller-Groeling, H. Weidenmüller: Phys. Rep. **299**, 190 (1998)
13. M.L. Mehta: *Random Matrices*, 2nd edn. (Academic Press, San Diego 1990)
14. I.K. Zharekeshev, B. Kramer: Phys. Rev. Lett. **79**, 717 (1997)
15. K. Müller, B. Mehlig, F. Milde, M. Schreiber: Phys. Rev. Lett. **78**, 215 (1997)
16. E. Runge, R. Zimmermann: in [8], pp. SI-229-232
17. J.R. Guest, T.H. Stievater, D.G. Steel, D. Gammon, D.S. Katzer, D. Park: 'Nonlinear near-field spectroscopy and microscopy of single excitons in a disordered quantum well'. In: *Quantum Electronics and Laser Science Conference, OSA Technical Digest* (Optical Soc. of America, Washington DC 2000) p. 6; see also the reproduction of otherwise unpublished results in [2], p. 281
18. G. von Freymann, E. Kurtz, C. Klingshirn, M. Wegener: Appl. Phys. Lett. **77**, 394 (2000)
19. F. Intonti, V. Emiliani, C. Lienau, Th. Elsaesser, V. Savona, E. Runge, R. Zimmermann, R.Nötzel, K.H. Ploog: Phys. Rev. Lett. **87**, 076801 (2001); C. Lienau, F. Intonti, V. Emiliani, V. Savona, E. Runge, R. Zimmermann: Mat. Sci. Eng. B **91** 105 (2002)
20. A. Crottini, R. Idrissi Kaitouni, J.L. Staehli, B. Deveaud, X.L. Wang, M. Ogura: phys. stat. sol. (a) **190**, 631 (2002)
21. G. von Freymann, U. Neuberth, M. Deubel, M. Wegener, G. Khitrova, H.M. Gibbs: Phys. Rev. B **65**, 205327 (2002)
22. E. Runge, R. Zimmermann: phys. stat. sol. (b) **221**, 269 (2000)
23. C.E. Porter, ed.: *Statistical Theory of Spectra: Fluctuations* (Academic Press, New York 1965)
24. H. Ludwig, E. Runge, R. Zimmermann: submitted
25. H. Ludwig: Das diskrete Anderson-Modell und Exzitonen in ungeordneten Quantendrähten: Simulationen und exakte Resultate, Diploma Thesis, Humboldt Universität, Berlin (2002)
26. B.I. Halperin, Phys. Rev. **139**, 1A, A104 (1965)
27. B.I. Halperin, M. Lax: Phys. Rev. **148**, 722 (1966); J. Zittartz, J.S. Langer: *ibid.* 741
28. I.M. Lifshits, S.A. Gredeskul, L.A. Pastur: *Introduction to the Theory of Disordered Systems* (Wiley, New York 1988)
29. N.F. Mott, *Metal-Insulator Transitions* (Taylor&Francis, London 1990)
30. R. Zimmermann: Nuovo Cimento D **17**, 1801 (1995)
31. V. Savona, R. Zimmermann: Phys. Rev. B **60**, 4928 (1999)
32. V. Savona, S. Haacke, B. Deveaud: Phys. Rev. Lett. **84**, 183 (2000)
33. P. Sheng: *Scattering and Localization of Classical Waves in Random Media* (World Scientific, Singapore 1990)
34. D.S. Wiersma, P. Bartolini, A. Lagendijk, R. Righini: Nature **390**, 671 (1997)
35. V. Savona, E. Runge, R. Zimmermann: Phys. Rev. B **62**, R 4805 (2000)
36. W. Langbein, E. Runge, V. Savona, R. Zimmermann, Phys. Rev. Lett. **89**, 157401 (2002)

Electron Dynamics in AC-Driven Quantum Dots

C.E. Creffield[1,2] and G. Platero[1]

[1] Instituto de Ciencia de Materiales de Madrid (CSIC), Cantoblanco, E-28049, Madrid, Spain
[2] Dipartimento di Fisica, Università di Roma "La Sapienza", Piazzale Aldo Moro 2, I-00185 Roma, Italy

1 Introduction

A quantum dot (QD) is a structure in which electrons can be confined to small length scales, comparable to their Fermi wavelength. A set of electrons held in such a structure is conceptually similar to a set of atomic electrons bound to a nucleus, and for this reason quantum dots are sometimes termed "artificial atoms" [1]. Unlike real atoms, the physical properties of quantum dots can be easily varied, which gives theorists and experimentalists the opportunity to study novel quantum effects in a well-controlled system.

To extend the atomic analogy further, we can consider linking QDs together to form "artificial molecules". By allowing electrons to tunnel between the QDs, the electronic states on the QDs can hybridize, and form new states that extend over the whole system [2]. The degree of the tunneling determines the strength of this hybridization. If the tunneling is weak the electrons remain essentially localised on the QDs in analogy with ionic bonding states. Conversely, if the tunneling is strong then the electrons form delocalised states with a covalent character. Recent transport experiments using AC potentials [3] have been performed on double QD systems, and have indeed revealed the ionic or covalent character of the electronic states by measurement of the induced photo-current.

Ever since the pioneering work of Anderson [4], it has been known that random spatial disorder can cause electronic states to become localised in quantum systems. More recently it has been found that an AC driving field can produce a similar intriguing effect termed *dynamical localisation*, in which the tunneling dynamics of a particle can be destroyed. One of the first systems in which this effect was predicted is that of a particle moving in a double-well potential [5]. A physical realization of this could consist of two coupled QDs containing a single electron – the simplest type of artificial molecule possible. If this system is prepared with the electron occupying one of the QDs, we can expect it to tunnel across to the other QD on a time scale set by the Rabi frequency. However, if an AC field of the correct strength and frequency is applied to the system, the tunneling is destroyed, and the particle will remain trapped in the initial well.

Weak time-dependent fields are generally treated as small perturbations, which produce transitions between the eigenstates of the unperturbed quantum system. This approach, however, is not applicable to treat the strong driving fields required to produce dynamical localisation, and instead the technique of

Floquet analysis [6], which is valid in all regimes of driving, has proven to be extremely effective. In this approach, which we briefly outline in the next section, the important quantities to calculate are the *quasi-energies*, which play a similar role in driven systems to the eigenenergies in the undriven case. In particular, dynamical localisation occurs when two quasi-energies of states participating in the dynamics approach each other, and become either degenerate (a crossing) or close to degenerate (an avoided crossing). Using this formalism, analytic and numerical studies of the double-well system have shown [7–9] that in the limit of high frequencies, quasi-energy crossings occur when the ratio of the field strength to the frequency is a root of the Bessel function J_0.

Adding a second electron to the coupled QD system, however, introduces considerable complications. At the low electron densities typically present in QDs, strong correlations produced by the Coulomb interaction can significantly influence the electronic structure. One of the most dramatic consequences of this is the formation of *Wigner molecule* states [10]. Understanding the interplay between electron correlations and the driving field is, however, extremely desirable, as the ability to rapidly control electrons using AC fields [11] has immediate applications to quantum metrology [12] and quantum information processing. In particular, manipulating entangled electrons on short timescales is of great importance to the field of quantum computation [13].

We study this problem here by applying the Floquet formalism to systems of *interacting* particles. The first system we consider is that of two interacting electrons confined to a pair of coupled QDs. A consequence of the interaction is that the system only responds strongly to the field when the field frequency is in resonance with the Coulomb interaction energy. When this condition is satisfied we find that, as for the single-particle case, coherent destruction of tunneling (CDT) can again occur, but that is governed by the roots of the higher-order Bessel functions. We then go on to consider a *two-dimensional* QD in the Wigner regime, which may also be described by a lattice model of Hubbard-type [14]. Using the same approach we show that CDT can again occur when similar conditions are satisfied, and we clarify how an applied AC field can drive charge redistributions within a strongly correlated QD. Finally we summarize our results and give some brief conclusions.

2 Methods and Approaches

2.1 Introduction to Floquet Theory

We consider a general quantum system driven by a periodic electric field, described by a time-dependent Hamiltonian which we can divide in the following way:

$$H(t) = H_t + H_I + H_{AC}(t), \quad H_{AC}(t) = H_{AC}(t + nT). \tag{1}$$

Here H_t holds the tunneling terms, H_I holds the electron-electron interaction terms and $H_{AC}(t)$ describes the interaction of the system with the T-periodic driving field. The periodicity of the driving field allows us to use the Floquet theorem to write solutions of the Schrödinger equation as $\psi(t) = \exp[-i\epsilon_j t]\phi_j(t)$ where ϵ_j is called the quasi-energy, and $\phi_j(t)$ is a function with the same period as the driving field, called the Floquet state. This type of expression is familiar in the context of solid-state physics, where *spatial* periodicity permits an analogous rewriting of the spatial wavefunction in terms of quasi-momenta and Bloch states (Bloch's theorem).

The Floquet states provide a complete basis, and thus the time-evolution of a general state may be written as:

$$|\Psi(t)\rangle = \sum_j \left(c_j \mathrm{e}^{-i\epsilon_j t}\right) |\phi_j(t)\rangle, \tag{2}$$

which is formally analogous to the standard expansion in the eigenvectors of a time-independent Hamiltonian. Indeed, in the adiabatic limit, $T = 2\pi/\omega \to \infty$, the quasi-energies evolve to the eigenenergies, and the Floquet states to the eigenstates. It is important to note that in this expansion both the basis vectors (the Floquet states) and the expansion coefficients explicitly depend on time. The nature of this time-dependence is very different however, and the superposition of the T-periodicity of the Floquet states with the phase factors arising from the quasi-energies produces a highly complicated, quasi-periodic time-dependence in general. As the Floquet states have the same period as the driving field, they are only able to produce structure in the time-dependence on short time-scales. Consequently, the dynamics of the system on time-scales much larger than T is essentially determined by *just* the quasi-energies, and hence evaluating the quasi-energies provides a simple and direct way of investigating this behavior. In particular, when two quasi-energies approach degeneracy the time-scale for tunneling between the states diverges, producing the phenomenon of CDT.

As we shall see for the specific quantum systems we consider, it is frequently the case that the total Hamiltonian is invariant under the generalized parity operation: $x \to -x; t \to t + T/2$. As a result the Floquet states can also be classified into parity classes, depending whether they are odd or even under this parity operation. Quasi-energies belonging to different parity classes may cross as an external parameter (such as the field strength) is varied, but if they belong to the same class the von Neumann-Wigner [15] theorem forbids this, and the closest approaches possible are avoided crossings. Identifying the presence of crossings and avoided crossings in the quasi-energy spectrum thus provides a necessary (though not sufficient) condition for CDT to occur.

2.2 Perturbation Theory for Floquet States

Although the quasi-energies are extremely useful for interpretation of the time-dependence of a quantum system, they are usually difficult to calculate and numerical methods must be employed. When the driving field dominates the dynamics, however, it is possible to use a form of perturbation theory introduced by Holthaus [8], in which the time-dependent part of the problem is solved exactly, and the tunneling part of the Hamiltonian, H_t, acts as the perturbation. This was generalized to treat interacting systems in [16,17] and was found to be very successful in the high-frequency regime, where $\hbar\omega$ is the dominant energy-scale. We now give a brief outline of this method.

The Floquet states and their quasi-energies may be conveniently obtained from the eigenvalue equation:

$$\left(H(t) - i\hbar \frac{\partial}{\partial t} \right) |\phi_j(t)\rangle = \epsilon_j |\phi_j(t)\rangle, \tag{3}$$

where we consider the operator $[H(t) - i\hbar\partial/\partial t]$ to operate in an *extended* Hilbert space of T-periodic functions [18]. The procedure consists of dividing the Hamiltonian as in (1), and finding the eigensystem of the operator $[H_I + H_{AC}(t) - i\hbar\partial/\partial t]$, while regarding the tunneling Hamiltonian H_t as acting as a perturbation. Standard Rayleigh-Schrödinger perturbation theory can now be used to evaluate the order-by-order corrections to this result, requiring only that we define an appropriate inner product for the extended Hilbert space:

$$\langle\langle \phi_m | \phi_n \rangle\rangle_T = \frac{1}{T} \int_0^T \langle \phi_m(t') | \phi_n(t') \rangle dt'. \tag{4}$$

Here $\langle \cdot | \cdot \rangle$ denotes the usual scalar product for the spatial component of the wave-vectors, and $\langle \cdot | \cdot \rangle_T$ is the integration over the compact time coordinate. We shall show in later sections how this method can be used to obtain analytical forms which accurately describe the behavior of the quasi-energies for the systems we study.

2.3 Numerical Methods

To study the time-evolution of each system, we used a fourth-order Runge-Kutta method to evolve a given initial state in time – typically of the order of fifty periods of the driving field. Throughout the time-evolution, physical quantities such as the number occupation of a given site were measured, and it was ensured that the unitarity of the wavefunction was accurately preserved.

A number of different methods can be used to numerically calculate the quasi-energies of a quantum system, and a detailed description of them is given in [6]. One technique well-suited to our approach is to evaluate the unitary time-evolution operator for one period of the driving field $U(t+T,t)$, and then to diagonalize it. It may be easily shown that the eigenvectors of this operator are equal to the Floquet states, and its eigenvalues are related to the quasi-energies via $\lambda_j = \exp[-i\epsilon_j T]$. This method is particularly convenient for our purposes, as $U(T,0)$ can be obtained by propagating the unit matrix in time over one period of the field, using the same Runge-Kutta method described above.

3 The Driven Double Quantum-Dot

We consider a simplified model of this system, in which each QD is replaced by a single site. Electrons are able to tunnel between the sites, and we include the

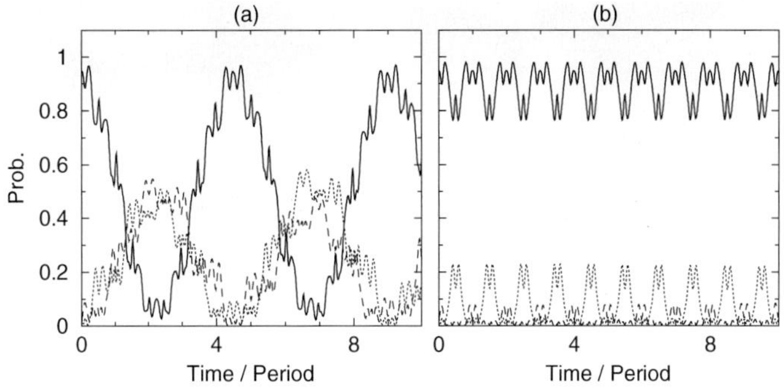

Fig. 1. Time evolution of the driven double QD system for $\tilde{U} = 8$ and $\omega = 4$. **a** electric potential, $E = 30.0$; **b** $E = 33.5$. *Thick solid line* = $p_{RL}(t)$, *dotted line* = $p_{LL}(t)$, *dashed line* = $p_{RR}(t)$

effect of interactions by means of a Hubbard-U term:

$$H = \tilde{t} \sum_{\sigma} \left(c_{1\sigma}^{\dagger} c_{2\sigma} + \text{H.c.} \right) + \sum_{i=1}^{2} \left(\tilde{U} n_{i\uparrow} n_{i\downarrow} + E_i(t) n_i \right) . \tag{5}$$

Here $\tilde{t}$ is the hopping parameter, and for the remainder of this work we shall take $\hbar$ equal to one, and measure all energies in units of $\tilde{t}$. $E_i(t)$ is the external electric potential applied to site i. Clearly only the potential difference, $E_1 - E_2$, is of physical importance, so we may choose to take the symmetric parametrization:

$$E_1(t) = \frac{E}{2} \cos \omega t, \qquad E_2(t) = -\frac{E}{2} \cos \omega t. \tag{6}$$

The Hilbert space of Hamiltonian (5) is six-dimensional, comprising three singlet states and a three dimensional triplet space. Measurements on semiconductor QDs have shown that the spin-flip relaxation time is typically extremely long [19], and so we have chosen not to include any spin-flip terms in the Hamiltonian. Consequently the singlet and triplet sectors are completely decoupled, and so if the initial state possesses a definite parity this will be retained throughout its time-evolution, and only states of the same parity need to be included in the basis.

To study the time evolution of the system, we used the ground state of the static Hamiltonian (a singlet) as the initial state, and evolved it in time as described in Sect. 2.3. Three probability functions were measured throughout the time evolution: $p_{LL}(t)$, $p_{RR}(t)$ and $p_{RL}(t)$, which are respectively the probability that both electrons are in the left QD, both are in the right QD, and that one electron is in each of the QDs. The Coulomb interaction favors separating the electrons, and thus for strong interactions the ground-state has a large value of p_{RL}, and relatively small values of p_{LL} and p_{RR}. We show in Fig. 1 the time evolution of these quantities for $\tilde{U} = 8$ and $\omega = 4$, at two different values of the electric potential. In both cases the detailed form of the time-evolution

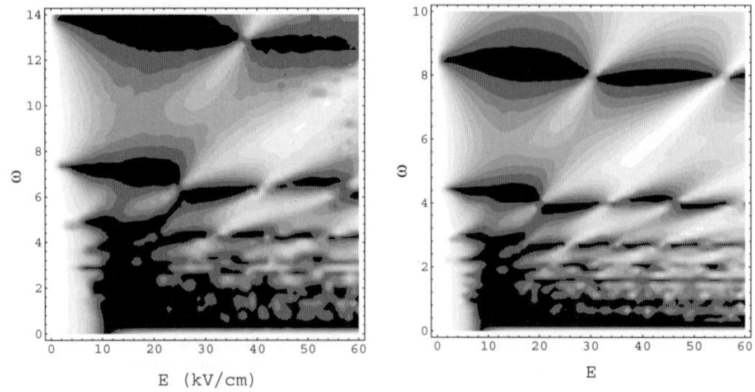

Fig. 2. p_{min}as a function of the strength E and energy $\hbar\omega$ of the AC field: **a** for a full simulation of a QD system ($\hbar\omega$ in units of meV) **b** for the two-site model with $\tilde{U} = 8$ (both axes in units of $\tilde{t}$)

is highly complicated, but it is clear that the system behaves in two distinct ways. In Fig. 1a the value of p_{RL} periodically cycles between its initial high value (indicating that each dot holds approximately one electron) to nearly zero, while the values of p_{LL} and p_{RR} correspondingly rise and fall at its expense. This behavior is very different from that shown in Fig. 1b, where p_{RL} never drops below a value of 0.78, and the other two probabilities oscillate with a very small amplitude. It thus appears that CDT is occuring in the second case, and that the system's time evolution is essentially frozen. We shall term the minimum value of p_{RL} attained during the time-evolution p_{min}, and use this to quantify whether CDT occurs, as a high value of p_{min}signifies that tunneling has been destroyed, while a low value indicates that the electrons are free to move between the QDs.

In Fig. 2b we present a contour plot of p_{min}as a function of both the frequency and strength of the AC field. Dark areas correspond to low values of p_{min} , and it can be seen that they form horizontal bands, indicating that the system is excited strongly by the AC field only at "resonant" values of ω. Close examination of this plot reveals that these bands occur at frequencies $\omega = \tilde{U},\ \tilde{U}/2,\ \tilde{U}/3\ldots$, at which the system can absorb an integer number of photons to overcome the Coulomb repulsion between electrons, thereby enabling tunneling processes such as $|\uparrow,\downarrow\rangle \rightarrow |0,\uparrow\downarrow\rangle$ to occur. We can additionally observe that these bands are punctuated by narrow zones in which CDT occurs. Their form can be seen more clearly in the cross-section of p_{min}given in Fig. 3a, which reveals them to be narrow peaks. These peaks are approximately equally spaced along each resonance, the spacing increasing with ω. In Fig. 2a we show another contour plot of p_{min}, this time obtained from a full simulation of the detailed physical model of two interacting electrons confined to a pair of coupled GaAs QDs (for more details on this simulation see [16]). The striking similarity between these results clearly indicates that our simple, effective model (5) indeed captures the essential processes occurring in the full system.

We emphasize that these results are radically different to those obtained for non-interacting particles. In this case an analogous plot of delocalisation shows a

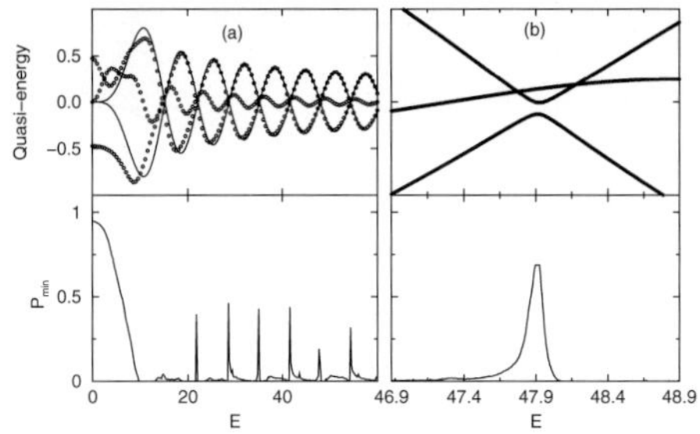

Fig. 3. **a** Quasi-energy spectrum for the two-site model for $\tilde{U} = 8$ and $\omega = 2$, *circles* = exact results, *lines* = perturbation theory, **b** magnified view of exact results for a single avoided crossing. Beneath are the corresponding plots of p_{min}

fan-like structure [9], in which localisation occurs along lines given by $\omega = E/x_j$, where x_j is the j-th root of the Bessel function $J_0(x)$. As a test of our method, we repeated our investigation with the inter-electron Coulomb repulsion $\tilde{U}$ set to zero, and found that the fan structure was indeed reproduced.

In Fig. 3a we show the Floquet quasi-energies as a function of the field strength for $\omega = 2$, one of the resonant frequencies visible in Fig. 2b. We see that the system possesses *two* distinct regimes of behavior, depending on whether the driving potential is weaker of stronger than $\tilde{U}$. For weak fields $E < \tilde{U}$, as studied previously in [20], the Floquet spectrum consists of one isolated state (which evolves from the ground state) and two states which make a set of exact crossings. Although in this regime p_{min} shows little structure, these crossings do in fact influence the system's dynamics. To demonstrate this, we show in Fig. 4 the Floquet quasi-energies in the weak-field regime for the case of $\tilde{U} = 16$, and plot beneath it the minimum value of p_{LL} attained during the time-evolution, where this time the state $\mid \uparrow\downarrow, 0\rangle$ has been used as the initial state. It can be seen that for this choice of initial condition, the crossings of the quasi-energies again produce CDT and freeze the initial state – despite the Coulomb repulsion between the electrons.

This surprising result may be understood as follows. For large values of $\tilde{U}$, the singlet eigenstates of the undriven system consist of the ground state, separated by the Hubbard gap $\tilde{U}$ from two almost degenerate excited states. For small values of the driving potential, the two excited states remain isolated from the ground state, and constitute an effective two-level system with a level-splitting of $\Delta \simeq 4\tilde{t}^2/\tilde{U}$. Thus if the system is prepared in an initial state which projects mainly onto the excited states, its dynamics will be governed by the two-level approximation [7–9], and CDT will occur at the roots of J_0. We show in Fig. 4a the quasi-energies obtained from the two-level approximation, which give excellent agreement with the actual results with *no* adjustable parameters. As E becomes comparable to the Hubbard gap, however, the two excited states are no longer isolated from the ground state, and all three levels must be taken into account. This can be seen in the progressive deviation of the quasi-energies from the two-level approximation as the electric potential approaches $\tilde{U}$.

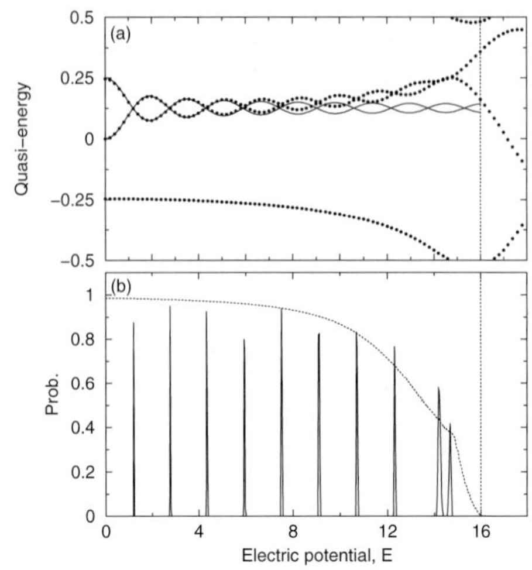

Fig. 4. a Quasi-energy spectrum for the two-site model for $\tilde{U} = 16$ and $\omega = 2$, *circles* = exact results, *solid line* = two-level approximation, $\epsilon_\pm = \pm(\Delta/2)J_0(2E/\omega)$. **b** *Solid line* = minimum value of $p_{LL}(t)$, starting from the initial configuration $|\uparrow\downarrow, 0\rangle$. The *dotted line* denotes p_{min}, which shows little structure. In both plots the *vertical dotted line* marks the transition to the strong-driving regime

When the electric potential exceeds $\tilde{U}$, the system displays a very different behavior, in which p_{min} remains close to zero except at a series of narrow peaks, corresponding to the close approaches of two of the quasi-energies. A detailed examination of these approaches (see Fig. 3b) reveals them to be *avoided crossings* between the Floquet states which evolve from the ground state and the higher excited state, and have the same generalized parity. The remaining state, of opposite parity, makes small oscillations around zero, but its exact crossings with the other two states do not correlate with any structure in p_{min} .

To interpret this behavior in the strong-field regime, we now obtain analytic expressions for the quasi-energies via the perturbation theory described in Sect. 2.2. The first step is to solve the eigenvalue equation (3) in the absence of the tunneling component H_t. In a real-space representation the interaction terms are diagonal, and so it can be readily shown that an orthonormal set of eigenvectors is given by:

$$
\begin{aligned}
|\epsilon_0(t)\rangle &= (\exp[i\epsilon_0 t],\ 0,\ 0) \\
|\epsilon_+(t)\rangle &= \left(0,\ \exp\left[-i(\tilde{U} - \epsilon_+)t + i\frac{E}{\omega}\sin\omega t\right],\ 0\right) \\
|\epsilon_-(t)\rangle &= \left(0,\ 0,\ \exp\left[-i(\tilde{U} - \epsilon_-)t - i\frac{E}{\omega}\sin\omega t\right]\right)
\end{aligned} \tag{7}
$$

Imposing T-periodic boundary conditions reveals the corresponding eigenvalues (modulo ω) to be $\epsilon_0 = 0$ and $\epsilon_\pm = \tilde{U}$. These eigenvalues represent the zeroth-order approximation to the Floquet quasi-energies, and for frequencies such that

$\tilde{U} = n\,\omega$ all three eigenvalues are degenerate. This degeneracy is lifted by the perturbation H_t, and to first-order, the quasi-energies are obtained by diagonalizing the perturbing operator $P_{ij} = \langle\langle \epsilon_i | H_t | \epsilon_j \rangle\rangle_T$. By using the well-known identity:

$$\exp\left[-i\beta \sin \omega t\right] = \sum_{m=-\infty}^{\infty} J_m(\beta) \exp\left[-im\omega t\right], \tag{8}$$

to rewrite the form of $|\epsilon_{\pm}(t)\rangle$, the matrix elements of P can be obtained straightforwardly, and its eigenvalues subsequently found to be $\epsilon_0 = 0$ and $\epsilon_{\pm} = \pm 2 J_n(E/\omega)$. Fig. 3a demonstrates the excellent agreement between this result (with $n = 4$) and the exact quasi-energies for strong and moderate fields, which allows the position of the peaks in p_{min} to be found by locating the roots of J_n. Similar excellent agreement occurs at the other resonances. For weak fields, however, the interaction terms do not dominate the tunneling terms and the perturbation theory breaks down, although we are still able to treat the system phenomenologically by using the effective two-level approximation.

4 The Square Quantum Dot

4.1 What Is a Wigner Molecule?

As we remarked in the Introduction, the Coulomb interaction between the electrons can significantly affect the electronic structure of a QD. Such strongly correlated problems are notoriously difficult to treat, and the addition of a time-dependent field complicates the problem even further. When the mean inter-electron separation exceeds a certain critical value, however, a surprising simplification occurs, as the Coulomb interaction dominates the kinetic energy and drives a transition to a quasi-crystalline arrangement which minimizes the total electrostatic energy. In analogy to the phenomenon of Wigner crystallization in bulk two-dimensional systems [1,22] such a state is termed a *Wigner molecule* [10]. As the electrons in the Wigner state are sharply localised in space, the system can be naturally and efficiently discretized by placing lattice points just at these spatial locations. A many-particle basis can then be constructed by taking Slater determinants of single-particle states defined on these lattice sites, from which an effective Hamiltonian of Hubbard-type can be generated to describe the low-energy dynamics of the system [14]. A major advantage of this technique over standard discretization [23,24] schemes, in which a very large number of lattice points is taken to approximate the continuum limit, is that the dimension of the effective Hamiltonian is much smaller (typically by many orders of magnitude), which permits the investigation of systems which would otherwise be prohibitively complex. This approach has proven to be extremely successful in treating a variety of static problems, including one-dimensional QDs [14], two-dimensional QDs with polygonal boundaries [25,26], and electrons confined to quantum rings [27,28]. We further develop this method here by including a time-dependent electric field, and study the temporal dynamics of the system as it is driven out of equilibrium.

4.2 Model and Methods

We consider a system of two electrons confined to a square QD with a hard-wall confining potential – a simple representation of a two-dimensional semiconductor QD. Such a system can be produced by gating a two-dimensional electron gas confined at a heterojunction interface, and by placing a gate split into four quadrants over the heterostructure [29], the potentials at the corners of the QD can be individually regulated. In Fig. 5a we show the ground-state charge-density obtained from the exact diagonalization of a square QD [25], for device parameters placing it deep in the Wigner molecule regime. It can be seen that the charge-density is sharply peaked at four points, located close to the vertices of the QD. This structure arises from the Coulomb interaction between the electrons, which tends to force them apart into diagonally opposite corners of the dot. As there are two such diagonal states, degenerate in energy, we can understand the form of the ground-state by considering it to be essentially a superposition of these two states (with a small admixture of higher energy states). The four points at which the peaks occur define the sites on which the effective lattice-Hamiltonian operates, as shown in Fig. 5b.

We take an effective lattice-Hamiltonian of the form:

$$H = \sum_{\langle i,j \rangle, \sigma} \left[\tilde{t} \left(c^{\dagger}_{i\sigma} c_{j\sigma} + \text{H.c.} \right) + \tilde{V} \; n_i n_j \right] + \sum_i \left[\tilde{U} \; n_{i\uparrow} n_{i\downarrow} + E_i(t) n_i \right] . \tag{9}$$

Here $\tilde{V}$ represents the Coulomb repulsion between electrons occupying neighboring sites, and $\tilde{U}$ is the standard Hubbard-U term, giving the energy cost for double-occupation of a site. As before, $E_i(t)$ denotes the electric potential at site i, which in general can have a static and a time-dependent component. In experiment, static offsets can arise either from deviations of the confining potential of the QD from the ideal geometry, or by the application of gating voltages to the corners of the QD. Applying corner potentials in this way would substantially enhance the stability of the Wigner molecule state, and could also be used to ensure that the multiplet of states included in the effective lattice-model is well-separated from the other excited states of the QD system. In this work,

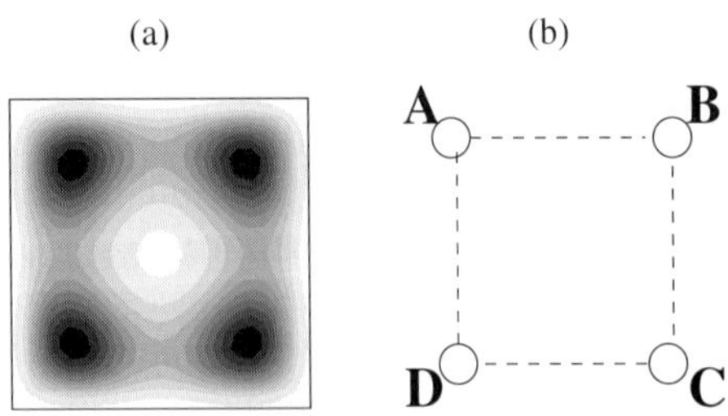

Fig. 5. **a** Ground-state charge-density for a two-electron square QD. GaAs material parameters are used, and the side-length of the QD is 800 nm, placing it in the Wigner regime. The *dark areas* indicate peaks in the charge-density. **b** Lattice points used for the effective lattice-Hamiltonian

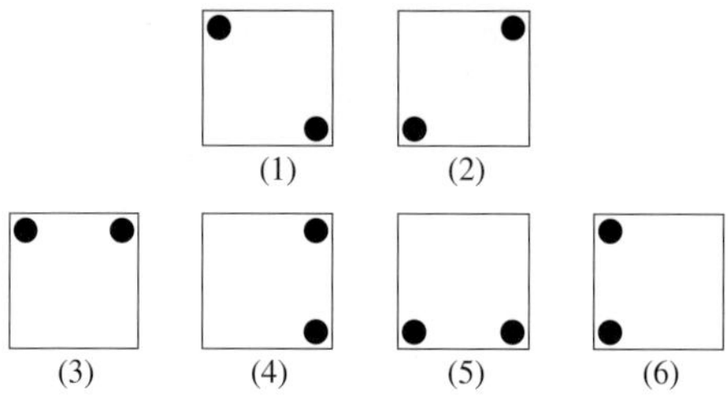

Fig. 6. Schematic representation of the two-particle basis states for the singlet sector of the Hamiltonian. The ground state of the QD is approximately a superposition of states (1) and (2)

however, we do not explicitly consider the effects of static gates, and we neglect the influence of small, accidental offsets encountered in experiment as we expect them to have only minor effects, and indeed may even stabilize CDT [30]. For convenience, we consider applying an AC field aligned with the x-axis of the QD, which can be parameterized as:

$$E_A = E_D = \frac{E}{2} \cos \omega t, \qquad E_B = E_C = -\frac{E}{2} \cos \omega t, \tag{10}$$

where A,B,C,D label the sites as shown in Fig. 5b. We emphasize that although we have the specific system of a semiconductor QD in mind, the effective-Hamiltonian we are using can describe a wide range of physical systems, including 2×2 arrays of connected QDs [31], and our results are thus of general applicability.

As for the case of the double QD, we include no spin-flip terms in (9) and so the singlet and triplet sectors are again decoupled. We choose to use initial states with singlet symmetry, which corresponds to the symmetry of the system's ground state. Simple state counting reveals that the singlet sector has a dimension of ten, and can be spanned by the six states shown schematically in Fig. 6, together with the four states in which each site is doubly-occupied.

5 Results

5.1 Interacting Electrons, Double Occupancy Excluded

We begin our investigation by taking the Hubbard-U term to be infinitely large – that is, to work in the sub-space of states with no double occupation. Our Hilbert space is thus six-dimensional, and we use the states shown in Fig. 6 as a basis. We show in Fig. 7 the time-dependent number occupation of the four sites at two different values of E, in both cases using the state (6) as the initial state, and setting the AC frequency to $\omega = 8$. In Fig. 7a E has a value of 100.0, and it can be clearly seen that the electrons perform driven Rabi oscillations between the left side of the QD and the right. Accordingly, the occupation number of the sites varies continuously between zero and one. In Fig. 7b, however, we see that

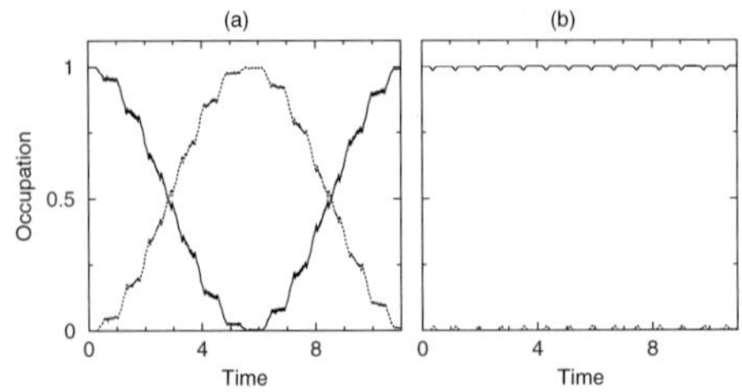

Fig. 7. Time development of the system for $\tilde{U}$ infinite, $\tilde{V} = 80$ and $\omega = 8$: **a** electric potential, E = 100.0 **b** E = 115.7. *Solid line* indicates the occupation of sites A and D, the *dotted line* the occupation of sites B and C

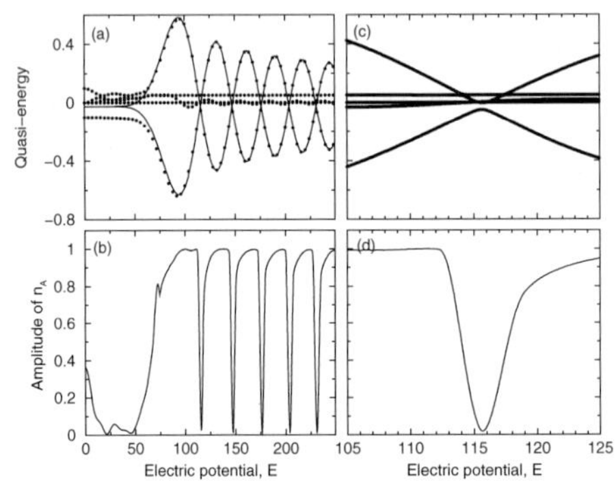

Fig. 8. a Quasi-energies of the system for $\tilde{U}$ infinite, $\tilde{V} = 80$ and $\omega = 8$: *circles* = exact results, *lines* = perturbative solution $[\pm 2J_{10}(E/\omega)]$. **b** Amplitude of oscillation of the occupation of site A. **c** Detail of quasi-energy spectrum, showing an avoided crossing. **d** Detail of amplitude of oscillations

changing the electric potential to a value of $E = 115.7$ produces dramatically different behavior. The occupations of sites A and D only vary slightly from unity, while sites B and C remain essentially empty throughout the time-evolution. Only a small amount of charge can transfer per period of the driving field between the left and right sides of the system, producing the small spikes visible in this figure. The amplitude of these features is extremely small, however, indicating that the tunneling between left and right sides has been almost totally destroyed.

To confirm that CDT is occurring, we present in Fig. 8 a comparison of the amplitude of the oscillations of n_A with the quasi-energy spectrum, as a function of the electric potential E. Similarly to the double QD system, we can see in Fig. 8a that the quasi-energies have two different regimes of behavior. The first of these is the weak-field regime, $E < \tilde{V}$, at which the driving field does not dominate the dynamics. In this regime the quasi-energy spectrum, and correspondingly, the amplitude of oscillations shows little structure. The second regime occurs at strong values of potential, $E > \tilde{V}$, for which the quasi-energy spectrum clearly shows a sequence of close approaches. In Fig. 8c we show an enlargement of one of these approaches which reveals it to be an *avoided cros-*

sing. Employing the perturbative method described in Sect. 2.2 demonstrates that the two quasi-energies involved in these avoided crossings are described by $\pm 2J_n(E/\omega)$, where n is equal to $\tilde{V}/\omega$. We may thus again think of n as signifying the number of photons the system needs to absorb to overcome the Coulomb repulsion between the electrons occupying neighboring sites. The results in Fig. 8b and Fig. 8d clearly show that the locations of the avoided crossings correspond exactly to quenching of the oscillations in n_A, and so confirm that CDT indeed occurs at these points.

5.2 Interacting Electrons, Double-Occupancy Permitted

We now take the most general case, and consider the competition between the $\tilde{U}$ and $\tilde{V}$ terms. Setting $\tilde{U}$ to a finite value means that the four doubly-occupied basis states are no longer energetically excluded from the dynamics, and accordingly we must take the full ten-dimensional basis set.

Although it is difficult to obtain precise estimates for the values of parameters of the effective Hamiltonian, it is clear that in general $\tilde{U} > \tilde{V}$. Accordingly we choose the parameters $\tilde{U} = 160$, $\tilde{V} = 16$ to separate the two energy-scales widely for our investigation. We again set the frequency of the AC field to $\omega = 8$, and in Fig. 9a we show the quasi-energy spectrum obtained by sweeping over the field strength. It is immediately clear from this figure that for electric potentials $E < \tilde{U}$ the form of the spectrum is extremely similar to the infinite-$\tilde{U}$ case. Performing perturbation theory confirms that, as in the previous case, the behavior of the quasi-energies is given by $\pm 2J_n(E/\omega)$ where $n = \tilde{V}/\omega$. We show in Fig. 9b the amplitude of the oscillations of n_A when the system is initialized in state (6),

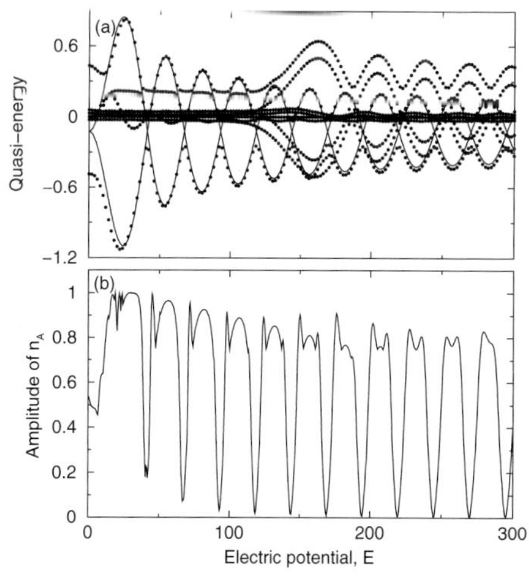

Fig. 9. **a** Quasi-energies of the system for $\tilde{U} = 160$ and $\tilde{V} = 16$, $\omega = 8$: *circles* = exact results, *lines* = perturbative solution $[\pm 2J_2(E/\omega)]$. **b** Amplitude of oscillation of the occupation of site A, with (6) as the initial state

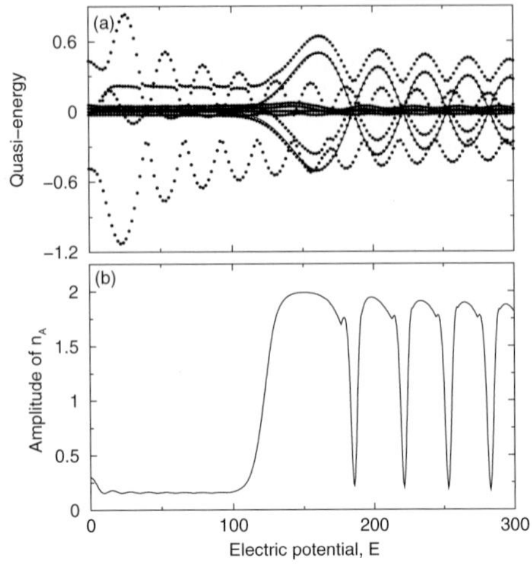

Fig. 10. a Quasi-energies of the system for $\tilde{U} = 160$ and $\tilde{V} = 16$, $\omega = 8$: *circles* = exact results, *lines* = perturbative solution $[\pm 2J_{18}(E/\omega)]$. **b** Amplitude of oscillation of the occupation of site A, with site A doubly-occupied as the initial state

which demonstrates that at the locations of the avoided crossings the tunneling parallel to the field is again quenched.

When the electric potential exceeds the value of $\tilde{U}$, however, new structure appears in the quasi-energy spectrum. A group of four quasi-energies, that for weaker fields cluster around zero, become "excited" and make a sequence of avoided crossings as the field strength is increased. Perturbation theory predicts that these high-field quasi-energies are given by $\pm 2J_m(E/\omega)$, where $m = (\tilde{U} - \tilde{V})/\omega$, and thus these avoided crossings arise when the absorption of m photons equates to the electrostatic energy difference between the two electrons being on neighboring sites, and doubly-occupying one site. This then indicates that this structure arises from the coupling of the AC field to the doubly-occupied states.

To probe this phenomenon, we time-evolve the system from an initial state consisting of *two* electrons occupying site A. In Fig. 10b it can be seen that for electric potentials weaker than $\tilde{U}$ the amplitude of the oscillations in n_A remains small, and shows little dependence on the field. As the potential exceeds $\tilde{U}$, this picture changes, and the AC field drives large oscillations in n_A, and in fact mainly forces charge to oscillate between sites A and B. At the high-field avoided crossings, however, the tunneling between A and B is suppressed, which shuts down this process. Instead, the only time-evolution that the system can perform consists of *undriven* Rabi oscillations between sites A and D, perpendicular to the field. As these oscillations are undriven they have a much longer time-scale than the forced dynamics, and thus during the interval over which we evolve the system the occupation of A only changes by a small amount, producing the very sharp minima visible in Fig. 10b, centered on the roots of $J_m(E/\omega)$.

As the tunneling perpendicular to the field is undriven, it is straightforward to evaluate the time evolution of the initial state, if we assume that the left side

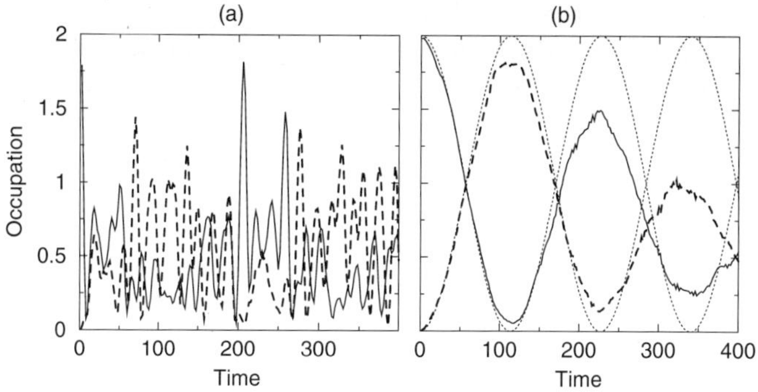

Fig. 11. Time development of the $\tilde{U} = 160$ and $\tilde{V} = 16$ system for $\omega = 8$: **a** electric potential, E = 200.0 (no CDT) **b** E = 185.8 (CDT). *Thick solid line* indicates the occupation of site A, the *thick dotted line* the occupation of site D. *Dotted lines* in **b** show the Rabi oscillations of the isolated two-site system, (11)

of the QD is completely decoupled from the right side. The occupation of sites A and D is then given by:

$$n_A(t) = 1 + \cos \Omega_R t, \quad n_D = 1 - \cos \Omega_R t, \tag{11}$$

where $\Omega_R = 4t^2/(\tilde{U} - \tilde{V})$. In Fig. 11 we display the occupations of sites A and D as a function of time, for two values of electric potential. At the first value, $E = 200$, tunneling between the left and right sides of the QD is not quenched, and accordingly the occupation of the two sites varies rapidly between zero and two as the electrons are driven by the AC field around the system. The second value, $E = 185.8$, corresponds to the first high-field avoided crossing. It can be clearly seen that the charge oscillates between sites A and B, with a frequency of Ω_R. These Rabi oscillations are damped, however, indicating that the isolation between the left and right sides of the QD is not perfect. In this sense we can regard the two sites B and C as providing an environment, causing the quantum system composed of sites A and D to slowly decohere in time. When the tunneling between the left and right sides of the QD is strong, for example at $E = 200$, this decoherence occurs very rapidly. By moving to an avoided crossing, however, and suppressing the tunneling, the rate of mixing between the two sides of the QD can be considerably reduced, and is just limited by the separation in energy between the two quasi-energies. Tuning the parameters of the driving field therefore gives us a simple and controllable way to investigate how a two-electron wavefunction can decohere in a QD.

6 Conclusions

In this work we have studied the interplay between Coulomb interactions and an AC driving field in two different configurations of QDs. For the case of the double QD system we have found that the system presents two different types of behavior, depending on whether the driving field dominates the interaction energy. For weak driving, we have found the surprising result that a suitable AC field can nonetheless freeze the time-evolution of a doubly-occupied QD *despite* the Coulomb interaction, and that this can be understood by means of an effective two-level model. At high field strengths the system is only driven strongly by the field at frequencies for which an integer number of quanta, n, is equal to the interaction energy. When this condition is satisfied CDT is again able to occur at certain well-defined parameters of the field, and by using Floquet theory we have shown that these points correspond to the roots of $J_n(E/\omega)$.

Strong electronic correlations allowed us to use an effective lattice model of just four sites to treat the square QD, by taking advantage of the natural discretization of the system in a Wigner molecule state. In the effective model, the inter-electron Coulomb interaction is described by two parameters, $\tilde{U}$ and $\tilde{V}$, and the dynamics of the system consists essentially of tunneling from corner to corner, along the perimeter of the QD. We find again that when the frequency of the driving field is in resonance with the Coulomb gap (that is, $m\omega = \tilde{V}$ or $m\omega = (\tilde{U} - \tilde{V})$) charge is able to circulate freely around the system, except at sharply defined field strengths at which tunneling parallel to the field is destroyed. Floquet theory again proved an excellent tool to understand this behavior, and revealed that these points correspond to the roots of $J_m(E/\omega)$.

We have thus shown that AC-fields may not only be used as a spectroscopic tools to probe the electronic structure of QD systems, but can also be used to dynamically control the time-evolution of the system. Possible applications of CDT range from stabilising the leakage of trapped electrons in physical realisations of quantum bits, to acting as "electron tweezers" by destroying or restoring tunneling between regions of a mesoscopic device. The tunability of the CDT effect, and its ability to discriminate between doubly-occupied and singly-occupied states, make it an excellent means for rapid manipulation of the dynamics of strongly correlated electrons in mesoscopic systems.

Acknowledgments

This work was supported by the Spanish DGES grant MAT2002-02465, by the European Union TMR contract FMRX-CT98-0180 and by the European Community's Human Potential Programme under contract HPRN-CT-2000-00144, Nanoscale Dynamics.

References

1. M.A. Kastner, Phys. Today **46**, 24 (1993); R.C. Ashoori, Nature (London) **379**, 413 (1996)
2. R.H. Blick, D. Pfannkuche, R.J. Haug, K. von Klitzing and K. Eberl, Phys. Rev. Lett. **80**, 4032 (1998)
3. T.H. Oosterkamp, T. Fujisawa, W.G. van der Wiel, K. Ishibashi, R.V. Hijman, S. Tarucha and L.P. Kouwenhoven, Nature (London) **395**, 873 (1998)

4. P.W. Anderson, Phys. Rev. **109**, 1492 (1958)
5. F. Grossmann, T. Dittrich, P. Jung and P. Hänggi, Phys. Rev. Lett. **67**, 516 (1991)
6. M. Grifoni and P. Hänggi, Phys. Rep. **304**, 229 (1998
7. J.H. Shirley, Phys. Rev. **138**, B979 (1965)
8. M. Holthaus, Z. Phys. B **89**, 251 (1992)
9. F. Grossmann and P. Hänggi, Europhys. Lett. **18**, 571 (1992)
10. K. Jauregui, W. Häusler and B. Kramer, Erophys. Lett. **24**, 581 (1993)
11. B.E. Cole, J.B. Williams, B.T. King, M.S. Sherwin and C.R. Stanley, Nature (London) **410**, 60 (2001); D. Vion, A. Aasime, A. Cottet, P. Joyez, H. Pothier, C. Urbina, D. Esteve and M.H. Devoret, Science **296**, 886 (2002)
12. P.I. Tamborenea and H. Metiu, Phys. Rev. Lett. **83**, 3912 (1999)
13. C.H. Bennett and D.P. DiVincenzo, Nature (London) **404**, 247 (2000)
14. J.H. Jefferson and W. Häusler, Phys. Rev. B **54**, 4936 (1996)
15. J. von Neumann and E.P. Wigner, Phys. Z **30**, 467 (1929)
16. C.E. Creffield and G. Platero, Phys. Rev. B **65**, 113304 (2002)
17. C.E. Creffield and G. Platero, Phys. Rev. B, **66**, 235303 (2002)
18. H. Sambe, Phys. Rev. A **7**, 2203 (1973)
19. T. Fujisawa, D.G. Austing, Y. Tokura, Y. Hirayama and S. Tarucha, Phys. Rev. Lett. **88**, 236802 (2002)
20. P. Zhang and X.-G. Zhao, Phys. Lett. A **271**, 419 (2000)
21. E.P. Wigner, Phys. Rev. **46**, 1002 (1934)
22. B. Tanatar and D.M. Ceperley, Phys. Rev. B **39**, 5005 (1989)
23. S. Akbar and I.-H. Lee, Phys. Rev. B **63**, 165301 (2001)
24. P.A. Schulz, P.H. Rivera, and N. Studart, Phys. Rev. B **66**, 195310 (2002)
25. C.E. Creffield, W. Häusler, J.H. Jefferson and S. Sarkar, Phys. Rev. B **59**, 10719 (1999)
26. C.E. Creffield, J.H. Jefferson, S. Sarkar and D.L.J. Tipton, Phys. Rev. B **62**, 7249 (2000)
27. W. Häusler, Physica B **222**, 43 (1996)
28. M. Koskinen, M. Manninen, B. Mottelson and S.M. Reimann, Phys. Rev. B **63**, 205323 (2001)
29. D.G. Austing, T. Honda and S. Tarucha, Semicond. Sci. Technol. **12**, 631 (1997)
30. J.T. Stockburger, Phys. Rev. E **59**, R4709 (1999)
31. C.A. Stafford and S. Das Sarma, Phys. Rev. Lett. **72**, 3590 (1994); R. Kotlyar and S. Das Sarma, Phys. Rev. B **55**, R10205, (1997)

Phonons and Phononic Gaps in Continuous Inhomogeneous Media

E.N. Economou and R.S. Penciu

Research Centre of Crete, FORTH, Heraclion, Crete, Greece

1 Introduction

Sound wave propagation in continuous inhomogeneous media plays an important role in various physical phenomena (e.g. earthquakes [1]) as well as in technological applications (such as oil exploration [2], non-destructive evaluation [3], ultrasound in medicine [4], underwater acoustics [5], etc). The relatively large number of material parameters (mass density, ρ, and velocities of the longitudinal, c_{l}, and transverse, c_{t}, components of the sound wave) and the full vector character of the wave give a rich physics.

Over the last ten years there is a growing interest in this field among physicists and physical chemists because of two recent developments, namely the emphasis in the study of sub-micron and nano-structures [6,7] and the appearance of the so called phononic crystals [8]-[13] in analogy with the photonic crystals [14].

Various submicron and nanostructures [15] have attracted recently widespread attention due to their connection to nanotechnology and their affinity to molecular biology and biotechnology. The dynamic behavior of these interesting systems depends among others on the nature of elementary sound excitations, i.e. phonons [16], which arise in such structures. The wavelength of these phonons is of the same order of magnitude as the structure size. For structures of nanometers, the phonons have the frequency in the GHz range, and they can be experimentally revealed by Brillouin spectroscopy. The latter involves inelastic light scattering by the phonons and determines the complicated dispersion relation of these phonons, $\omega_s = f_s(\boldsymbol{q})$, for each mode s, where $\boldsymbol{q} = \boldsymbol{k}_{\mathrm{f}} - \boldsymbol{k}_{\mathrm{i}}$ ($\boldsymbol{k}_{\mathrm{f}}$ and $\boldsymbol{k}_{\mathrm{i}}$ are the wave vectors of the incident and the scattered light, respectively) (see, e.g., Fig. 1b and d). These dispersion relations can be achieved also indirectly, by transmission vs frequency measurements.

The phononic crystals are composite structures with periodically modulated material parameters. The phonons in such structures behave similarly to electrons moving through the periodic potential in a crystal and their propagation may be described by a phononic band structure. The interest in studying the phononic crystals is related to the question of the existence or not of spectral gaps over frequency ranges for which the phonon propagation is forbidden in all directions. The conditions for the existence of gaps depends mainly on the density and velocity contrast of the components of the composite, the volume fraction of one of the two components, the lattice structure, the topology (cermet vs network) and the scatterers geometrical shape. By breaking the periodicity of the phononic crystal, it is possible to create highly localized defect modes within the phononic gap, which are analogous to the localized impurity states in a semiconductor. As John and Rangajan [17] and Economou and Zdetsis [18] suggest, the existence of both spectral gaps and localized eigenstates are due to *destructive interference* of *multiple scattered* waves.

A large variety of phonons appears in composite structures as those described above. The most usual is the acoustic phonon with linear dispersion relation, $\omega = c|\boldsymbol{k}|$, where ω is its frequency, c is the velocity of sound and $\boldsymbol{k}$ is the wavevector. In many cases, this acoustic phonon propagates mainly through the host component, avoiding as much as possible the inclusions, and its velocity is almost equal with that in the host, c_o. In other cases, the acoustic wave *"sees"* the actual inhomogeneous medium as a frequency dependent, homogeneous effective medium. Then, the sound velocity, c_e, can be determined by using the coherent potential approximation (CPA).

Another kind of modes which is simple to describe, is the mode related to the oscillations of each scatterer. For simplicity, we consider here the case of spherical inclusions. A single spherical particle in vacuum exhibits many vibrational eigenmodes. When the sphere is embedded in the solvent each eigenmode of sphere oscillation is coupled with solvent acoustic waves and, consequently, energy is leaking out of the sphere and is radiated away. As a result, each sphere eigenmode acquires a finite lifetime, τ. If the coupling is weak, the leakage of energy is slow, the lifetime is long ($\omega\tau \gg 1$) and we can still talk about approximate sphere eigenmodes (also called resonances). On the other hand, if $\omega\tau \leq 1$, the lifetime is so short that the concept of sphere eigenmode is no longer relevant. The simplest way to obtain theoretically the approximate eigenmodes of a single solid sphere embedded in the solvent is to send a plane acoustic wave which upon impinging on the sphere sets in a forced oscillation. When the frequency of the plane wave coincides with an approximate sphere eigenmode, there is a resonance response which appear as peaks in the scattering cross section of the incoming plane acoustic wave by the solid sphere. Their position gives the frequencies of the approximate sphere eigenmodes and their half-widths give the inverse of their lifetimes. To assure that a peak associated with approximate eigenmodes is not partly hidden in the background of purely geometrical nature, one can calculate also the scattering amplitude by a hard sphere (i.e. one with $\rho = \infty$, c_l, $c_\mathrm{s} = finite$) of the same radius and subtract it from the calculated scattering amplitude by the spherical particle.

In the composite system, these eigenmodes of each solid sphere couple with the acoustic waves in the solvent and through these waves are connected with the eigenmodes of the neighboring spheres. Thus, the system sustains in general complicated propagating modes, which are hybrids of all the individual eigenmodes and the acoustic waves as well. In some cases, the hybridization involves mainly the equal frequency approximate eigenmodes of each individual particle giving thus rise to a global mode where the wave propagates by coherently hopping from particle to particle, similarly to electronic propagation in solids by a linear combination of atomic orbitals (LCAO), the approximate sphere eigenmodes being the analogues of the atomic orbitals. These modes are localized inside each particle and because their frequency, ω, is almost independent of $\boldsymbol{k}$, we call them optic-like phonons.

If the hybridization involves individual eigenmodes of different frequency and some times the acoustic phonon also, the global eigensolutions of the wave equation cannot be analyzed simply in terms of the host acoustic phonons, or CPA, or isolated particle modes, and more sophisticated computational techniques involving multiple scattering effects are required.

In the case of inelastic light scattering by phonons in periodic systems, when the momentum lost by the light, $\hbar\boldsymbol{q}$, is large, it is transferred partly to the phonon, $\hbar\boldsymbol{k}$, and partly to the periodic lattice, $\hbar\boldsymbol{G}$ (where $\boldsymbol{G}$ is a reciprocal lattice vector). The corresponding modes are called *"Bragg"* induced modes.

In conclusion, there are two basic channels of wave propagation in a composite: one is through the utilization mainly of the acoustic phonon and the other is by hopping coherently among the scatterers by a linear combination of the resonances. It follows from these qualitative considerations, that a phononic gap would appear when both channels of propagation are absent. This happens in those frequency regions (if any) which (a) are not so close to single particle resonances (in order to eliminate the first channel of propagation) and (b) exhibit large scattering cross-section, σ_{h}, by a single hard sphere ($\sigma_{\mathrm{h}}/\pi r_{\mathrm{s}}^2 \gg 1$), in order to block the second channel of propagation. Indeed detailed calculations strongly support this conclusion.

Several material combinations of particles/host have been studied (e.g. glass spheres in water (H. P. Schriemer et al [11]), steel spheres in epoxy [9], coated lead spheres in epoxy [12], steel cylinders in air (R. Martinez-Sala et al [10]), mercury cylinders in aluminum alloy (M. Torres et al 2001 [10]), tungsten carbide beads in water [20], air bubbles in water [21], as well as colloidal systems consisting of suspended particles (SiO_2 or PMMA) in various solvents (R. S. Penciu et al [19]) and melts of copolymers [22]).

The fabrication of most of these systems was guided by the goal of creating the widest possible gaps around the desired midgap frequency(ies). On the other hand the colloidal systems and the diblock copolymers were studied because they are very interesting physical systems, whose dynamics are controlled to some extent by phonons and details in their structure may be revealed by photon spectroscopy.

In the case of strongly disordered systems, where $\boldsymbol{k}$ is not well defined (except at very long wavelengths), experiments are concentrated in measuring transmission, backscattering, diffusion coefficient, mean free path, or velocity (phase, group, energy) vs frequency. Recently the question of interest is whether one can obtain flat acoustic lenses by employing phononic crystals.

2 Theoretical Methods

The elastic wave equation in an homogeneous isotropic medium is:

$$(\lambda + 2\mu)\nabla(\nabla \cdot \boldsymbol{u}) - \mu\nabla \times \nabla \times \boldsymbol{u} + \rho\omega^2 \boldsymbol{u} = 0 \; , \tag{1}$$

where $\boldsymbol{u}$ is the displacement vector at the point $\boldsymbol{r}$, ρ is the mass density, and λ and μ are the so called Lamé coefficients which are related with the bulk modulus B by the relation $B = \lambda + \frac{2}{3}\mu$. The quantity μ is the shear modulus, which in fluids is zero. The displacement $\boldsymbol{u}$ can always be decomposed into a longitudinal, $\boldsymbol{u}_\mathrm{l}$, component ($\nabla \times \boldsymbol{u}_\mathrm{l} = 0$), and a transverse, $\boldsymbol{u}_\mathrm{t}$, component ($\nabla \cdot \boldsymbol{u}_\mathrm{t} = 0$) (Note that in fluids only the longitudinal component exists):

$$\boldsymbol{u}(\boldsymbol{r}) = \boldsymbol{u}_\mathrm{l}(\boldsymbol{r}) + \boldsymbol{u}_\mathrm{t}(\boldsymbol{r}) \ , \tag{2}$$

satisfying the usual wave equation:

$$\nabla^2 \boldsymbol{u}_\alpha + \frac{\omega^2}{c_\alpha^2} \boldsymbol{u}_\alpha = 0 \ , \ \alpha = \mathrm{l}, \mathrm{t} \ , \tag{3}$$

where c_l, c_t are the longitudinal and transverse sound velocities:

$$c_\mathrm{l} = \sqrt{\frac{\lambda + 2\mu}{\rho}} \quad \text{and} \quad c_\mathrm{t} = \sqrt{\frac{\mu}{\rho}} \ . \tag{4}$$

In a homogeneous isotropic medium the elastic wave can, in general, be purely longitudinal or purely transverse. In a composite system, at the interfaces between the two components, mixing of the longitudinal and transverse components takes place and the eigenmodes become very complicated.

Various theoretical techniques which have been employed for calculating quantities of interest, are shortly presented below.

Single Particle Scattering. This is the simplest method which identifies the position of the approximate sphere eigenmodes (in the host environment) as peaks in the single sphere scattering cross-section and their life times as the inverse of the widths of the peaks. Furthermore, it provides strong evidence for the appearance or not of gaps and their locations; the gaps are expected to appear in regions between two consecutive widely separated resonances, if these regions exhibit high scattering cross section by a hard sphere.

Plane Wave Expansion. This is a fast and easy to apply method for calculation of the band structure of infinite periodic systems. It is based on the expansion of the periodic coefficients in the wave equation (i.e. $\boldsymbol{u}(\boldsymbol{r})$, $\rho(\boldsymbol{r})$, $\lambda(\boldsymbol{r})$ and $\mu(\boldsymbol{r})$) in Fourier series. This method is proper when both the scatterers and the host are either fluids or solids.

Multiple Scattering Formalism. This method of calculation is applicable to both infinite periodic media and finite periodic and aperiodic systems and for all phase combinations. It is based on the observation that the total scattered wave, Ψ, out of an aggregate of individual scatterers is the sum of the individual scattered waves, ψ_{sn}, from each scatterer n. However, ψ_{sn} is a known linear functional of the total incident wave on n, ψ_{in}: $\psi_{sn} = F(\psi_{in})$. But ψ_{in} is the sum of the scattered waves from all the other scatterers p $(p \neq n)$ plus the external wave, ψ_{e}, if any. Hence, we obtain a system of equation of the form:

$$\psi_{sn} = F\left(\sum_{p\neq n} \psi_{sp} + \psi_{\mathrm{e}}\right) \quad , \quad n = 1, 2, ... \tag{5}$$

The implementation of the multiple scattering method requires the expansion of each ψ_{sn} and ψ_{in} in spherical harmonics (assuming spherical shape of the scatterers) and expansion of the spherical harmonic centered at the particle p in terms of the spherical harmonics centered at the particle n. This expansion has to be truncated (creating thus a source of numerical errors); the higher the frequency the larger the number of terms to be kept in the expansion. The only disadvantage of this method is that the computational time increases very quickly with the total number of terms kept in the expansion.

Transfer Matrix Formalism. The transfer matrix formalism allows us to determine the wave at the lattice plane $n+1$, if the wave at the lattice plane n is known. The basic idea is elegantly presented in the book by Landau & Lifshitz, Quantum Mechanics, §25; the generalization and implementation of the method for electromagnetic and acoustic waves can be found in papers by Stefanou et al [23] and by Psarobas et al [13,16].

Coherent Potential Approximation (CPA). This method is a mean field approximation which replaces the actual inhomogeneous medium by a properly determined, frequency dependent, homogeneous effective medium. The condition for determining the effective medium is the following: *The local replacement of the effective medium by various configurations of the actual medium should produce on the average no scattered wave.*

Putting this average scattered wave equal to zero gives an equation from which one calculates the parameters of the effective medium. Thus, the effective bulk modulus, B_{e}, and the effective mass density, ρ_{e}, at low ω are given by the formulas:

$$B_{\mathrm{e}}^{-1} = \phi B_{\mathrm{i}}^{-1} + (1+\phi) B_{\mathrm{o}}^{-1} \tag{6}$$

and

$$\rho_{\mathrm{e}} = \rho_{\mathrm{o}}[\phi(\rho_{\mathrm{i}} - \rho_{\mathrm{o}}) + 2\rho_{\mathrm{i}} + \rho o]/[2\phi(\rho_{\mathrm{o}} - \rho_{\mathrm{i}}) + 2\rho_{\mathrm{i}} + \rho_{\mathrm{o}}] \ , \tag{7}$$

where the indexes "i" and "o" denote the inclusions and the host, respectively, while ϕ is the volume fraction occupied by the inclusions.

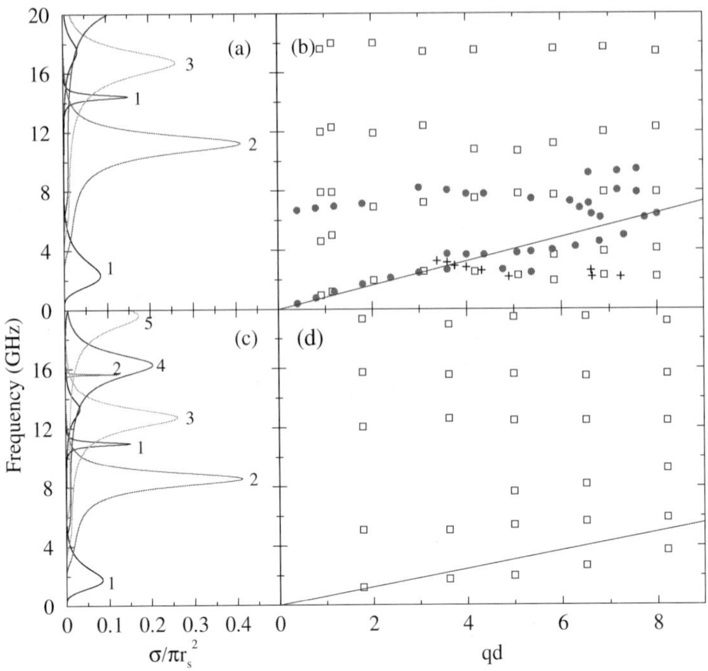

Fig. 1. b Experimental (*open squares*) and theoretical (*solid circles*) phonon dispersion relations in a silica polycrystalline colloidal suspension ($d = 250$ nm, $\phi = 0.62$). The *crosses* denote the theoretical results of low light scattering intensity. **d** Experimental (*open squares*) phonon dispersion relations for the glassy silica ($d = 328$ nm, $\phi = 0.64$) colloidal suspension. The *solid lines* indicate the linear dispersion of the acoustic phonon in the solvent. **a** and **c** The partial scattering cross sections of an acoustic plane wave by an individual silica sphere of diameter $d = 250$ nm (**a**) and $d = 328$ nm (**c**). The numbers on the top of the peaks in the partial scattering cross section indicate the corresponding l component

3 Results

3.1 Silica Spheres in Cyclohexane/Decalin

In this section we give an example of nanostructures with rich phonon spectrum revealed by Brillouin spectroscopy, and we theoretically identify the nature of the phonons. The systems consists of silica spheres colloidal suspensions in cyclohexane/decalin mixture. In Fig. 1 we compare the crystalline state ($d = 250$ nm in panel (b)) with the glassy state ($d = 328$ nm in panel (d)). In the panels (a) and (c) we plot the sound scattering cross section by the single spheres with diameter $d = 250$ nm and $d = 328$ nm, respectively. The strong and very well separated peaks are revealed by subtracting from the scattering amplitude that of the hard sphere. The frequencies of the highest peaks ($l = 2$ and 3 in panel (a), and $l = 3, 4$ and 5 in panel (c)) match very well those of the flat experimental modes of the crystalline and glassy samples (open squares in Fig. 1b and d).

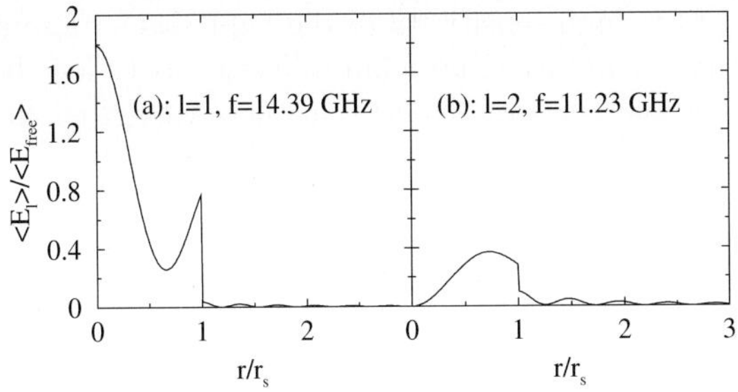

Fig. 2. The reduced partial energy density distribution vs distance from the center of the silica particle for $f = 14.39$ GHz (**a**) and $f = 11.23$ GHz (**b**), respectively, keeping only the contribution of the spherical harmonic responsible for the peak

One can observe the approximate $1/d$ dependence of the resonance frequencies. Since the resonances appear when half the wavelength, $l = c/2f$, is comparable to the diameter of the particle, we expect the resonance frequencies to be of the order of $5000/(2 \times 250 \times 10^{-9}) = 10$ GHz.

Note that this connection, $d \sim c/f$, creates in many cases practical difficulties in the fabrication of photonic or phononic crystals, since the lattice spacing must be of order of c/f. In the next subsection we shall consider a case where the relation $d \sim c/f$ does not hold.

The flatness of these modes in the q-space suggests that they are localized mostly inside each colloidal particle. In order to check the localization of these modes, we computed the partial energy density distributions in Fig. 2a and b for frequencies $f = 11.23$ GHz (corresponding to the $l = 2$ peak) and $f = 14.39$ GHz (corresponding to the second $l = 1$ peak). One can observe that the energy is well concentrated inside the sphere with a small fraction leaked into the solvent; for the sharper peak at $f = 14.39$ GHz the localization is better. However, this mode has not been detected experimentally probably because of its weak intensity. The propagation of these modes is done by hopping.

The lowest frequency mode, at low q, corresponds to an acoustic phonon propagating mainly in the host liquid. These modes are multiply scattered by the silica particles and hence become softer (more retarded) with increasing particle volume fraction.

For the polycrystalline colloidal suspensions (Fig. 1b), the low frequency modes match well the theoretical ω vs k results for an infinite fcc lattice obtained by using the multiple scattering (MS) method.

As one can observe from the dispersion data, the present polycrystalline system exhibits clear *"Bragg"* induced modes (high q, low ω open squares in Fig. 1b). In order to find these modes theoretically, one should extend the MS band structure calculations to higher Brillouin zones (up to the seventh). Modes (crosses in Fig. 1b) were obtained which are in good agreement with the experimental data although their theoretical light scattering intensity was weaker than what was expected to be experimentally detectable. This discrepancy was attributed to the very high sensitivity of the light scattering intensity to $\boldsymbol{q}$ and to the details of the structure. The main difference between the glassy and the crystalline states is that, as one expects, in the former the *"Bragg"* modes are absent at low ω high q (Fig. 1d).

3.2 Air Bubbles in Liquids

In this subsection we review briefly recent results (Kafesaki et al. [21]) for the propagation of sound waves in bubbly liquids. In Fig. 3, the scattering cross section by a single spherical air bubble in water is plotted against the reduced frequency, $\omega r_\mathrm{s}/c_\mathrm{o}$, where r_s is the radius of the bubble and c_o is the sound velocity in water. The huge low frequency resonance, known as Minnaert resonance [5], at $\omega_0 r_\mathrm{s}/c_\mathrm{i} = \sqrt{3\rho_\mathrm{i}/\rho_\mathrm{o}}$, corresponds to the $l = 0$ spherical harmonic and is due to an isotropic compression and expansion of the bubble against the surrounding water. (Note that the ratio $\sqrt{3\rho_\mathrm{i}/\rho_\mathrm{o}}$ makes the resonance frequency much lower than what one would expect from the condition $\lambda_i \sim d$, which gives usually the order of magnitude of the resonance frequencies). The high off-resonance background is also due to the $l = 0$ spherical harmonic. In addition, there are extremely sharp resonances corresponding to $l \neq 0$ spherical harmonics at frequencies satisfying the following equation

$$\frac{j_l'(q_\mathrm{i})}{j_l(q_\mathrm{i})} = \frac{\rho_\mathrm{i}}{\rho_\mathrm{o}} \frac{c_\mathrm{i}}{c_\mathrm{o}} \frac{n_l'(q_\mathrm{o})}{n_l(q_\mathrm{o})}\ , \text{where} \quad q_\alpha = \omega r_\mathrm{s}/c_\alpha\ , \quad \alpha = \mathrm{o}, \mathrm{i}\ . \tag{8}$$

As has been pointed out before such a frequency dependence of the single scattering cross section provides ideal conditions for the appearance of wide spectral gaps in a multiple scattering environment, because in the frequency region between resonances neither the host material allows propagation, nor coherent hopping to neighboring scatterers utilizing the resonances can take place. Notice that the mean free path, l, at the Minnaert resonance, as obtained

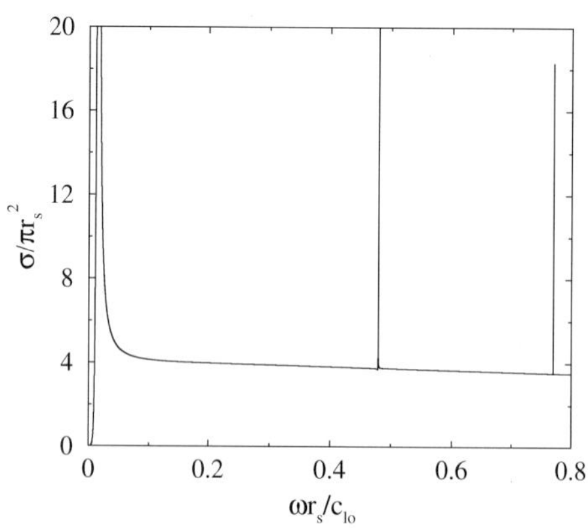

Fig. 3. Reduced scattering cross-section by a spherical air bubble in water vs the reduced frequency of the incoming plane sound wave

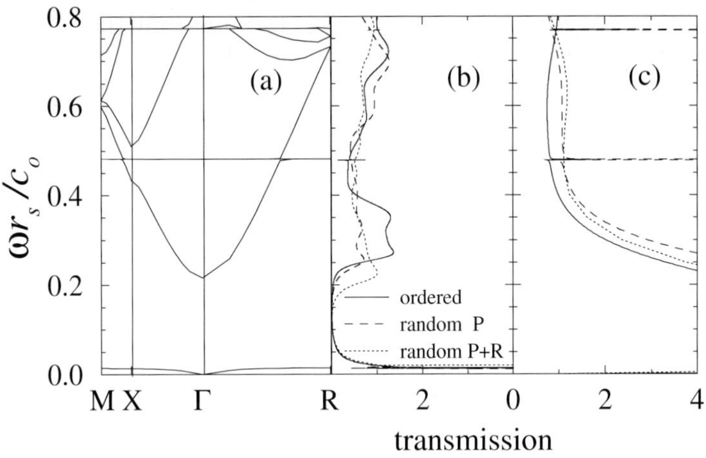

Fig. 4. a Dispersion relation along the MXΓR directions for a *sc* periodic composite consisting of air bubbles (of radius r_s) in water; ω is the frequency, c_o is the wave velocity in the water, and air volume fraction $\phi = 1\%$. **b** Generalized transmission coefficient through a finite cluster consisting of 147 air bubbles in water in ordered (*solid line*), positional (P) or bubble radius (R) disordered arrangement and volume fraction $\phi = 1\%$. **c** The same as in panel (b) but with the multiple scattering switched off

from the approximate formula $l = 1/n\sigma$ (where n is the average concentration of the bubbles), equals to the average nearest neighbor distance already at an air volume fraction, $\phi \approx 0.0002$. This means that multiple scattering effects are important even for such low ϕ. Thus, acoustic waves in bubbly liquids, provide an almost ideal case for examining in detail the localization question.

We applied first our multiple scattering method to sound waves in infinite periodic bubbly water in order to check our results against those of Fig. 3 as well as against those of Ruffa [20] and Kushwaha et al. [21]. In Fig. 4a we show the first nine bands for a sc arrangement of bubbles with volume fraction $\phi = 0.01$. The lowest band is essentially confined between $\omega = 0$ and the Minnaert frequency. Then a wide gap follows and above this gap, a complex wide band appears dominated by branches with an ω/k slope close to c_o; this shows that the propagation for $\omega r_s/c_o \geq 0.2$ takes place mostly through the water with the air bubbles playing a less important role except that of superimposing very flat bands at the frequencies of the higher resonances. From the lowest band (of $\omega \leq \omega_0$) one can easily obtain both the phase and group velocity. It should be pointed out that the long wavelength sound velocity can be obtained very easily and accurately by employing mean field theories, more specifically the coherent potential approximation (CPA) which gives $c = 468, 43$, and 35 m/s for a bubbly water of bubble volume fraction $\phi = 5.84 \times 10^{-4}, 0.1$, and 0.2 respectively (Ruffa [21]). These extremely low values are due to the fact that $c = \sqrt{B_e/\rho_e}$, where the effective bulk modulus, $B_{\rm e}$, is close to that of the air, (6), while the effective density is close to that of water, (7. In Fig. 4b we present the transmission of sound through a cluster of 147 nonoverlapping spherical bubbles vs frequency, for periodic or random (in position and/or size) arrangement. In all cases the air volume fraction is $\phi = 0.01$. We see a narrow acoustic band extending a little beyond the frequency of the strong resonance which gives rise to the maximum transmission. Then, in agreement with Fig. 4a, a wide gap follows which corresponds to $T \approx 0$ transmission for both the ordered and disordered case. It is impressive that this gap survives almost intact the positional disorder. When, in addition, we allowed the radius of each bubble to vary randomly between 0.75 to 1.25 of $r_{\rm s}$ the gap still survived although it was reduced by about 20% mostly from the upper side. Note that this gap is due entirely to multiple scattering since it disappears when multiple scattering is switched off (Fig. 4c). There are noticeable differences in the transmission through the ordered and the disordered cluster: There is higher transmission in the disordered case than in the ordered one (i) at the upper part of the gap due probably to a tail of localized eigenstates, and (ii) around Bragg points, where the disorder tends to diminish destructive interference. In the rest of the spectrum the disorder decreases the transmission as expected.

4 Current Trends

Veselago [24] has studied theoretically the question of electromagnetic (EM) wave propagation in dispersive materials for which both $\mu(\omega)$ and $\epsilon(\omega)$ are negative and real over a frequency range. He predicted peculiar and unexpected properties such as negative refractive index, refraction in the "wrong" half plane, etc.

Recently, Pendry [25] suggested the design of an artificial structure, which over a limited range of frequencies would exhibit negative effective $\mu(\omega)$ and $\epsilon(\omega)$. Shelby at al [26] constructed the structure suggested by Pendry and have demonstrated experimentally the negative refraction. Pendry suggested also that such artificial materials can produce flat lenses and "superlenses" in the sense that the sharpness of focusing can surpass the diffraction limit of the wavelength. Furthermore, Notomi [27] pointed out that photonic crystals can, under certain conditions, behave as negative refractive index materials. Foteinopoulou et al [28] demonstrated numerically how negative refraction takes place in such properly chosen photonic crystals; Chiyan Luo et al [29] have obtained numerically focusing by means of a flat slab of photonic crystal with sharpness equal to 0.67λ. However, as pointed out by Haldane [30], photonic crystals do not seem to allow enough freedom for reducing the sharpness dramatically. This is where phononic crystals may enter the picture: The fact that strong phononic resonances appear at frequencies much lower than the one resulting from the condition $\lambda \sim d$, opens up the possibility of producing sonic lenses with focusing sharpness much lower than that of the wavelength. This possibility is currently under intense investigation.

Acknowledgment

Support by EU grants FMRX-CT96-0042 and HPRN-CT-2000-00017 is acknowledged.

References

1. F.A. Dahlen and J. Tromp, *Theoretical Global Seismology*, Princeton Univ. Press, Princeton (1998)
2. Gregory A. Clark, Science and Technology Review , p. 12, (May 1999)
3. J.H. Page, M.L. Cowan and D.A. Weitz, in Proc. 17th International Congress on Acoustics, Vol II (CD version): Measurement techniques and transducers – Nondestructive testing (17th ICA, Rome, 2001), pp. 1–2
4. P. Atkinson and J.P. Woodcock, *Doppler Ultrasound and Its Use for Clinical Measurements*, Academic Press, London (1982)
5. T.G. Leighton, *The Acoustic Bubble*, Academic Press (1994)
6. A.P. Gast and W.B. Russel, Phys. Today **51**, No. 12, 24 (1998); G.A, McConnell, A.P. Gast, J.S. Huang, and S.D. Smith, Phys. Rev. Lett. **71**, 2102 (1993); R. Sigel, S. Pispas, D. Vlassopoulos, N. Hadjichristidis, and G. Fytas, Phys. Rev. Lett. **83**, 4666 (1999); B. Loppinet et al., Langmuir **16**, 6480 (2000)
7. P.N. Segre, S.P. Mecker, P.N. Pusey, and W.C.K. Roon, Phys. Rev. Lett. **75**, 958 (1995)
8. M.M. Sigalas and E.N. Economou, J. of Sound and Vibration, **158** 377 (1992); E.N. Economou and M. Sigalas, J. Acoust. Soc. Am. **95**, 4, (1994); M. Kafesaki and E.N. Economou, Phys. Rev. B **52**, 13317 (1995); Phys. Rev. B **60**, 11993 (1999); M. Kafesaki, R.S. Penciu, and E.N. Economou, Phys. Rev. Let. **84**, 6050 (2000)
9. M.S. Kushwaha et al., Phys. Rev. Lett. **71**, 2022 (1993); Phys. Rev. B **49**, 2313 (1994); M.S. Kushwaha, Int. Journ. of Modern Physics, B **10**, 977 (1996) and Recent Res. Devel. Applied Phys. **2**, 743 (1999)
10. R. Martinez-Sala et al., Nature **378**, 241 (1995); J.V. Sanchez-Perez et al, Phys. Rev. Lett.**80**, 5325 (1998); F.R. Montero de Espinosa et al., Phys. Rv. Lett.**80**, 1208 (1998); M. Torres et al., Phys. Rev. Lett.**82**, 3054 (1999); D. Caballero et al., Phys. Rev. **E60**, R6316 (1999); D. Garcia-Pablos et al., Phys. Rev. Lett.**84**, 4349 (2000); M. Torres et al., Phys. Rev. Lett.**86**, 4282 (2001)
11. J.H. Page et al., Phys. Rev.**E52**, 3106 (1995); Science **271**, 634 (1996); H.P. Schriemer et al., Phys. Rev. Lett.**79**, 3166 (1997); Z. Liu et al., Phys. Rev.**B62**, 2446 (2000)
12. Z. Liu et al., Science **289**, 1734 (2000); E.N. Economou, Physics World, 26, (Nov. 2000)
13. I.E. Psarobas et al., Phys. Rev. **B62**, 5536 (2000); R. Sainidou et al., Rhys. Rev. **B66**, 024303 (2002)
14. E. Yablonovitch, Phys. Rev. Lett. **58**, 2059 (1987); S. John, Phys. Rev. Lett. **58**, 2486 (1987); E. Yablonovitch and T.J. Gmitter, Phys. Rev. Lett. **63**, 1450 (1989); K.M. Ho et al., Phys. Rev. Lett. **65**, 3152 (1990); E. Yablonovitch et al., Phys. Rev. Lett. **67**, 2295 (1991); C.M. Soukoulis, ed., *Photonic Crystals and Light Localization in the 21*st *Century*, Kluwer Academic Puplishers, Dordrecth (2001); J.D. Joannopoulos et al., *Photonic Crystals*, Princeton Univ. Press, Princeton (1995); K. Sakoda, *Optical Properties of Photonic Crystals*, Springer-Verlay, Berlin (2001)

15. T.C.B. McLeish, ed., *Theoretical Challenges in the Dynamics of Complex Fluids*, Kluwer, London (1997); M. Muthakumar et al., Science **227**, 1095 (1996); G.A. McCornell et al., Phys. Rev. Lett. **71**, 2102 (1993); R. Seghrouchni et al., Europh. Lett. **42**, 271 (1998); W.E.K. Poon et al., eds, *Observation, Prediction and Simulation of Phase Transition in Complex Fluids*, Kluwer, Dordrecht (1995)
16. L. Schwartz and T.J. Plona, J. Appl. Phys. **55**, 3971 (1984); Zhengdong Cheng et al., Phys. Rev. Lett., **85**, 1560 (2000); I.E. Psarobas et al., Phys. Rev. **B65**, 064307-1 (2002); A.F. Bakker and C.P. Lowe, J. of Chem. Phys. **116**, 5867 (2002)
17. S. John, and R. Rangarajan, Phys. Rev. B **38**, 10 101 (1988)
18. E.N. Economou, and A. Zdetsis, Phys. Rev. B **40**, 1334 (1989)
19. Jos Derksen and Willem van de Water, Phys. Rev. **A45**, 5660 (1992); Ling Ye et al., Phys. Rev. **B48**, 2805 (1993); J. Liu et al., Phys. Rev. Lett. **65**, 2602 (1990); D.O. Reise and G.H. Wegdam, Phys. Rev. Lett. **82**, 1676 (1999); R.S. Penciu et al., Phys. Rev. Lett. **85**, 4622 (2000); R.S. Penciu et al., Europhys. Lett. **58**, 699 (2002)
20. Suxia Yang, J.H. Page, Zhengyou Liu, M.L. Cowan, C.T. Chan, and Ping Sheng, Phys. Rev. Lett. **88**, 104301 (2002)
21. Zhen Ye and A. Alvarez, Phys. Rev. Lett. **80**, 3503 (1998); M.S. Kushwaha et al, Phys. Lett. **A248**, 252 (1998); E. Hoskinson and Zhen Ye, Phys. Rev. Lett. **83**, 2734 (1999); A.A. Ruffa, J. Acoust. Soc. Am. **91**, 1 (1992); M. Kafesaki et al., Phys. Rev. Lett. **84**, 6050 (2000)
22. A.M. Urbas et al., to be published.
23. N. Stefanou et al., J. Phys.: Condensed Matter **4**, 7389 (1992)
24. V.G. Veselago, Soviet Physics USPECHI **10**, 509 (1968)
25. J.B. Pendry et al., IEEE Trans. MTT, **47**, 2075 (1999); J.B. Pendry, Phys. Rev. Lett. **85**, 3966 (2000)
26. R.A. Selby, D.R. Sith, S. Schultz, Science, **292**, 77 (2001)
27. M. Notomi, Phys. Rev. **B62**, 10696 (2000)
28. S. Foteinopoulou et al., ar Xiv: cond-mat/0210346v1, 16 Oct. 2002
29. Chiyan Luo et al., Phys. Rev. **B65**, 201104 (R) (2002)
30. F.D.M. Haldane, ar Xiv: cond-mat/0206420v1, 22 June 2002

Quantum Tunneling of an Acoustic Polaron in One-Dimensional Electron-Lattice System

Yoshiyuki Ono, Takashi Ebinuma, and Toshiyuki Ozawa

Department of Physics, Faculty of Science, Toho University, Miyama 2-2-1, Funabashi, Chiba 274-8510, Japan

1 Introduction

More than two decades ago, Donovan and Wilson [1] found peculiar charge carriers with an anomalously high mobility in photo-conduction experiments on polydiacetylene which belongs to conjugated polymers and could be regarded as a quasi-one-dimensional system; for example, in the single crystal of polydiacetylene toluene sulphonate, one of typical polydiacetylene, the minimum distance between neighboring polymer chains is 0.75 nm [2] which is much larger than the bond length between carbon atoms ($\sim$ 0.13 nm). The characteristics of the photo-excited charge carriers are summarized as follows; (1) they have a rather high mobility which is estimated to be no less than 20 $\mathrm{m}^2/(\mathrm{sec}\cdot\mathrm{V})$, and (2) their drift velocity shows saturation at a value of the order of the sound velocity of the system in the applied electric field with the strength 10^2 to 10^6 V/m. For a possible explanation of this charge carrier, Wilson [3] proposed an acoustic polaron which is a composite of an electron near the conduction band bottom and lattice distortions induced by the presence of the electron and giving rise to an effective attractive potential confining the electron to a locally distorted lattice region. In fact, starting from Su, Schrieffer and Heeger's (SSH) model [4] which is one of the standard theoretical models for one-dimensional coupled electron-lattice systems, Wilson [3] could derive theoretically the following properties of an acoustic polaron within the continuum approximation which is justified in the weak coupling limit; (1) the polaron has a saturation velocity equal to the sound velocity v_s of the system, (2) the extent of the polaron (referred to as "width" in the following) decreases with increasing velocity and tends to vanish as the velocity v approaches the saturation velocity (= the sound velocity, v_s), and (3) the energy of the moving polaron diverges as $(v_\mathrm{s}-v)^{-3}$ when the polaron velocity v approaches v_s. These properties of the acoustic polaron were confirmed later by numerical simulations [5] treating the SSH model as it is without assuming the weak coupling limit; details of these simulations will be explained in the following section.

In the above-mentioned studies, however, the dynamical properties of the acoustic polaron as a quasi-classical particle have been stressed. The electron is treated quantum mechanically, but the lattice system is regarded as classical. Therefore, it is not astonishing that the acoustic polaron behaves like a classical particle, even if it has an inner structure reflecting the fact that it is a composite object consisting of two different degrees of freedom, an electron and lattice distortions. Nevertheless, we cannot exclude the possibility to find a quantum nature of the polaron. As a typical quantum behavior we consider the tunneling effect, i.e., the issue dealt with in this article is whether an acoustic polaron shows quantum tunneling through a potential barrier or not. The knowledge of this kind of properties will be inevitable in understanding transport processes by the acoustic polarons.

2 "Classical" Properties of an Acoustic Polaron

In this section, the "classical" properties of an acoustic polaron obtained by numerical simulations are summarized [5]. Numerical calculations were carried out based on the SSH model,

$$\begin{aligned} H_{\mathrm{SSH}} = & -\sum_n (t_0 - \alpha y_n) \left(\mathrm{e}^{\mathrm{i}\gamma A} c_n^\dagger c_{n+1} + \mathrm{h.c.} \right) \\ & + \frac{K}{2} \sum_n y_n^2 + \frac{M}{2} \sum_n \dot{u}_n^2 , \end{aligned} \tag{1}$$

which is extended to included the effect of an external uniform electric field through the time-dependent vector potential A; the electric field E is given by $E = -\dot{A}$. Here $c_n^\dagger$ and c_n are field operators creating and annihilating an electron at the site n; since we are treating a single electron, the spin indices as well as electronic Coulomb interactions are disregarded. t_0 is the electronic transfer integral in a regular lattice. The bond variable y_n is related to the lattice displacements $\{ u_n \}$ as $y_n = u_{n+1} - u_n$. α describes the coupling between the electron and lattice due to the change of the transfer integral caused by the change of the bond length. The last two terms in (1) represent the lattice potential and kinetic energies, respectively, K being the spring constant and M the mass of an ion unit at a site. The parameter γ is defined as $\gamma = ea/\hbar$ with e the elementary charge and a the lattice constant. In the following, we use the inverse of the bare optical mode frequency $\omega_Q (\equiv \sqrt{4K/M})$ as a unit of time when we perform dynamical simulations. The standard values of parameters for the polydiacetylene are, $t_0 = 0.64$ eV, $K = 2.4 \times 10^{20}$ eV/m^2, $\alpha = 0.38 \times 10^{10}$ eV/m and $a = 4.9 \times 10^{-10}$ m, $\omega_Q (= \sqrt{4K/M}) = 1.47 \times 10^{13}$ sec^{-1} and the sound velocity $v_{\mathrm{s}} = 3.6 \times 10^3$ m/sec. It will also be clear that $E_0 \equiv \hbar\omega_Q/ea$ is a natural unit of the external electric field. The appearance of the vector potential in the phase factor of the transfer integral is due to the Peierls substitution [25]. In order to avoid the end point effect, we assume a periodic boundary condition. In this situation it is not convenient to introduce a scalar potential related to a uniform electric field. Therefore we use a time-dependent uniform vector potential to introduce an electric field which can make the polaron move in a natural way. By this method we can avoid introducing a not necessarily justifiable assumption on the shape of a moving polaron.

A brief explanation of the method of calculation is given in the following. Before starting a dynamical simulation, we have to prepare a static acoustic polaron in the absence of the external electric field. The electronic wave function and the lattice distortions corresponding to the lowest energy state are determined to minimize the total energy of the system; the minimization condition leads to a set of self-consistent equations for the electron wave function φ and static lattice distortions $\{ y_n\}$,

$$\varepsilon_0\varphi(n) = -(t_0 - \alpha y_n)\varphi(n+1) - (t_0 - \alpha y_{n-1})\varphi(n-1), \tag{2}$$

$$y_n = -\frac{2\alpha}{K}\varphi(n+1)\varphi(n) + \frac{2\alpha}{NK}\sum_{n'}\varphi(n'+1)\varphi(n'), \tag{3}$$

where ε_0 indicates the ground state energy eigenvalue for the electron, and the last term in (3) is due to the periodic boundary condition; note that the periodic boundary condition yields an additional condition $\sum_n y_n = 0$, which means that all of y_n's cannot be independent. It should be noted here that, if we scale the electronic energy by t_0 and treat $\alpha y_n/t_0$ as an effective bond variable instead of y_n itself, the only relevant coupling parameter is expressed as $\lambda = \alpha^2/Kt_0$, which is dimensionless. It is clear that the dimensionless coupling constant λ determines the spatial structure of the acoustic polaron. In the case of a polydiacetylene, λ is equal to 0.094 and the half width of the acoustic polaron is about 20 times the lattice constant. In general, the smaller the dimensionless coupling constant, the wider the polaron width.

The above self-consistent equations can be solved by iterations. The initial position of the polaron might be controlled by an appropriate choice of the initial set of y_n's in the iterative calculations. For the initial set of y_n's, the electronic ground state is calculated from (2), and the obtained wave function is substituted in (3). The latter procedure yields a new set of y_n's. If the difference between the original and new sets of y_n's is not sufficiently small, the same procedures are repeated until the quantity $\sum_n(y_n^{\rm new} - y_n^{\rm old})^2/\sum_n(y_n^{\rm old})^2$ becomes smaller than a certain small constant which is typically chosen to be 10^{-12}.

Once we have prepared the static polaron as an initial state, it can be accelerated by introducing an external electric field. In the dynamical simulations, the electric field is switched on and off slowly with a duration time τ as shown in the following,

$$E(t) = \begin{cases} E_{\rm max}\dfrac{1}{2}\left[1-\cos\left(\dfrac{\pi t}{\tau}\right)\right], & 0 \le t < \tau \\ E_{\rm max}, & \tau \le t < t_{\rm off} \\ E_{\rm max}\dfrac{1}{2}\left[1+\cos\left(\dfrac{\pi(t-t_{\rm off})}{\tau}\right)\right], & t_{\rm off} \le t < t_{\rm off}+\tau \\ 0, & t \ge t_{\rm off}+\tau \end{cases} \tag{4}$$

where $E_{\rm max}$ represents the field strength, $t_{\rm off}$ the time of switching off which is set infinite in some cases. The slow switching is employed in order to avoid disturbances due to sudden switching [7].

In order to follow the dynamics of the polaron in the presence of an external force, we have to find the time dependence of both the electronic wave function and the lattice displacements. The time development of the electronic wave function $\psi(n,t)$ and the lattice displacements $\{u_n(t)\}$, related to the bond variables

$\{y_n(t)\}$ as $y_n(t) = u_{n+1}(t) - u_n(t)$, is calculated by solving the time-dependent Schrödinger equation for the former and the classical equation of motion for the latter which involves the electronic energy as the Born-Oppenheimer potential. These equations are explicitly expressed in the following form,

$$\begin{aligned} \mathrm{i}\hbar\frac{\partial}{\partial t}\psi(n,t) &= -(t_0 - \alpha y_{n-1}(t))\mathrm{e}^{\mathrm{i}\gamma A}\psi(n-1,t) \\ &\quad -(t_0 - \alpha y_n(t))\mathrm{e}^{-\mathrm{i}\gamma A}\psi(n+1,t), \end{aligned} \tag{5}$$

$$\begin{aligned} M\ddot{u}_n(t) &= \alpha\{[\mathrm{e}^{\mathrm{i}\gamma A}\psi^*(n,t)\psi(t,n+1) + c.c.] \\ &\quad -[\mathrm{e}^{\mathrm{i}\gamma A}\psi^*(n-1,t)\psi(n,t) + c.c.]\} \\ &\quad +K\left[u_{n-1}(t) - 2u_n(t) + u_{n+1}(t)\right]. \end{aligned} \tag{6}$$

Both equations are numerically solved by introducing discrete time with a mesh Δt, which is set to be sufficiently smaller than the typical time constant of the system ω_Q^{-1} so that the change of the lattice displacements during Δt is small enough. The formal solution of the Schrödinger equation (5) is given by [8]

$$\begin{aligned} \psi(t+\Delta t) &= \mathrm{T}\exp\left[-\frac{\mathrm{i}}{\hbar}\int_t^{t+\Delta t} H(s)\mathrm{d}s\right]\psi(t), \\ &= \exp\left[\left(-\frac{\mathrm{i}}{\hbar}H(t) + \mathcal{T}\right)\Delta t\right]\psi(t), \end{aligned} \tag{7}$$

where $H(t)$ is the time dependent electronic Hamiltonian, the time dependence being introduced through A and $y_n(t)$, T the time ordering operator, and the super operator $\mathcal{T}$ is defined [8] by the following operation for arbitrary functions of time $F(t)$ and $G(t)$,

$$F(t)\mathrm{e}^{\mathcal{T}\Delta t}G(t) = F(t+\Delta t)G(t). \tag{8}$$

In the present model it is most convenient to decompose $H(t)$ into two parts,

$$H(t) = H_{\mathrm{odd}\to\mathrm{even}}(t) + H_{\mathrm{even}\to\mathrm{odd}}(t), \tag{9}$$

and to use the Suzuki-Trotter's fractal decomposition of exponential operators [8]. The first and second terms of (9) represent the electron transfers from odd sites to even sites and from even sites to odd sites, respectively. In the present calculation, we employ the simplest form of the fractal decomposition which is correct up to the second order of Δt and Δt is set to be $0.0005\omega_Q^{-1}$. The equation of motion for the lattice system is solved by replacing the second order differential equations by the lowest order difference equations. This rather crude approximation is justified because the motion of the lattice system is very slow due to the large mass of the ions. Needless to say, the initial conditions for these time dependent equations are those determined so as to minimize the total energy of the system in the absence of the external field, i.e. $\psi(n,0) = \varphi(n)$ and $u_n(0) = u_n^{(0)}$, the latter being obtained from the static stable bond variables $\{y_n\}$ under the postulation that the total center of mass of the ionic system is

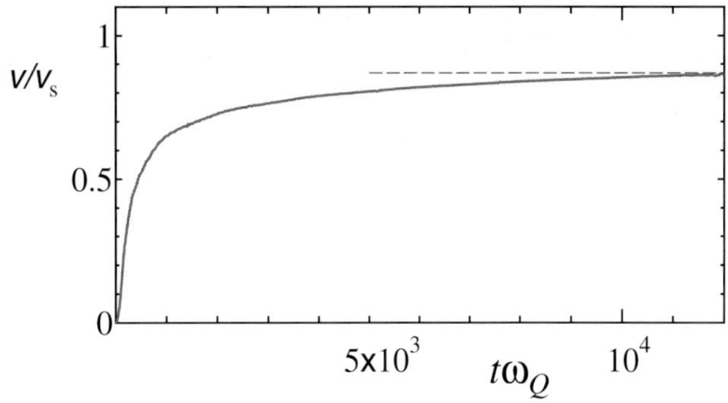

Fig. 1. Time dependence of the polaron velocity [5]. The values of parameters used in this calculation are as follows; the dimensionless coupling constant $\lambda = 0.094$, the system size $N = 200$, the field strength $E_{\max} = 0.005E_0$, the time mesh $\Delta t = 0.0025\omega_Q^{-1}$ and the duration time of switching $\tau = 200\omega_Q^{-1}$

fixed at the center of the system. Because of the periodic boundary condition the number of independent bond variables is $N - 1$ and therefore it is not possible to determine N values for $\{u_n\}$. This ambiguity can be lifted by postulating the center of mass of the lattice system to be fixed.

The position of a polaron is calculated as the center of mass of the electron density profile which is equal to $|\psi(n,t)|^2$. About how to incorporate the periodic boundary condition in the numerical determination of the center of mass, see e.g. [9]. Although we are adopting a rather small time mesh Δt in order to ensure the numerical accuracy, the electronic density profile is taken out every $100\Delta t$ since the change of the system is rather slow. Therefore the position of the polaron $x(t)$ is obtained every $100\Delta t$. The velocity of the polaron, $v(t)$, is estimated from the following formula,

$$v(t) = \frac{x(t + 50\Delta t) - x(t - 50\Delta t)}{100\Delta t} . \tag{10}$$

In Fig. 1, we show an example of the time dependence of the polaron velocity. The data were obtained by using the parameter values appropriate for the polydiacetylene, the time mesh $\Delta t = 0.0025\omega_Q^{-1}$, the system size $N = 200$, the field strength $E_{\max} = 0.005E_0$, the duration time for the switching $\tau = 200\omega_Q^{-1}$, and $t_{\text{off}} = \infty$. It will be seen that the velocity saturates at about $0.9v_s$ which is quite near the value v_s obtained by Wilson [3] within the continuum approximation appropriate in the weak coupling limit. In the same simulation we can find the time dependence of the polaron width (i.e. extent), which is most naturally defined as the second cumulant of the electronic density profile [5]. The width is found to decrease monotonically with time as far as the acceleration is continued. This is due to the monotonic increase of the polaron velocity as shown in Fig. 1.

The relation between the polaron velocity and width is obtained by eliminating the time variable from the time-dependences of the velocity and the width. The result is depicted in Fig. 2. It will be seen that the polaron width tends to

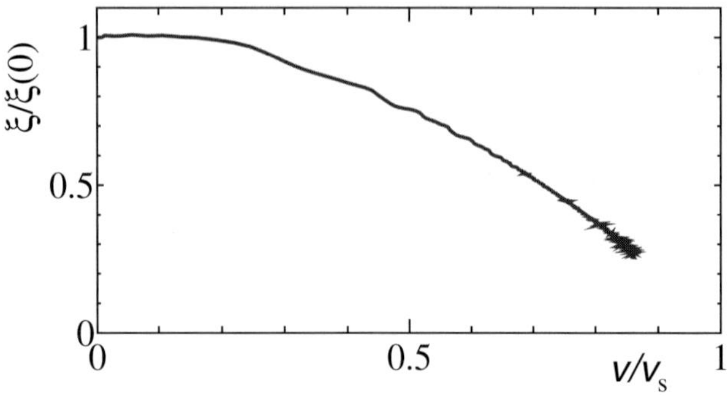

Fig. 2. The relation between the width and the velocity of an acoustic polaron obtained by eliminating the time variable from the data shown in Fig. 1 and the data for the width [5]. The width tends to vanish when the velocity approaches the saturation velocity $\sim v_s$

vanish when the velocity approaches the saturation velocity ($\sim v_s$). The decrease of the width with increasing the velocity could be understood by carefully studying the analytic argument in the continuum approximation by Wilson [3]. In this approximation, the coupled dynamical equations for the lattice displacement and the electronic wave function can be solved by assuming traveling solitary wave type solutions for both of them, with the same argument which is a linear combination of the space and time variables. In this situation the second order time derivative of the lattice displacement can be replaced by v^2 times the second order spatial derivative, resulting in the factor $(1 - v^2/v_s^2)$ in front of the second order spatial derivative when combined with the force term originating from the lattice potential energy. In the equation of motion for the displacement, the electronic density multiplied by the electron-lattice coupling constant appears as the Born-Oppenheimer potential term. These considerations lead to the effective electron-lattice coupling proportional to $(1 - v^2/v_s^2)^{-1}$ when the polaron is moving with a velocity v. Namely the coupling is effectively enhanced due to the motion of the polaron, which explains the decrease of the width with increasing velocity and also the tendency that the width goes to zero when the velocity approaches the sound velocity.

It will be worthwhile to see the behavior of the energy when the polaron is moving. The system energy can be decomposed into three parts as easily seen from the definition of the model Hamiltonian; the three parts are the electronic energy ε_e, the lattice potential energy ε_{lp} and the lattice kinetic energy ε_{lk}. The total energy ε_{tot} is defined as the sum of these three terms. The time dependence of the energy can be calculated within the above-mentioned simulation, and we can obtain the relation between the energy and the velocity by eliminating the time variable as done in getting Fig. 2. The results are summarized in Fig. 3. All the three parts of the energy show diverging behaviors as the velocity approaches the sound velocity, and the resulting total energy also shows a divergence. The decrease of the electronic energy and the increase of the lattice potential energy

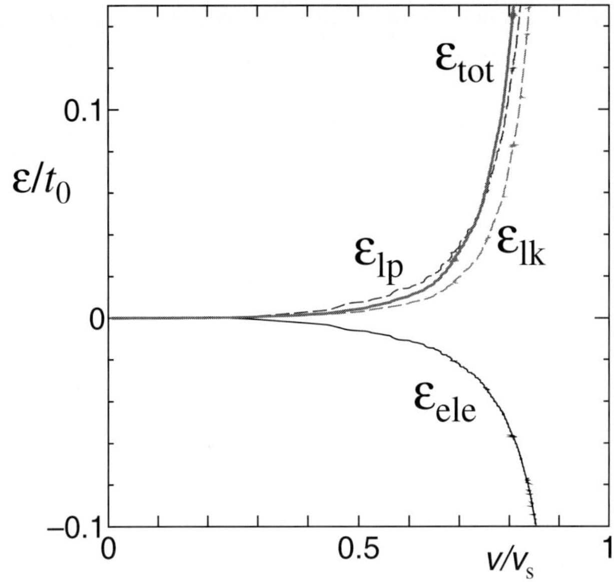

Fig. 3. The correlation between the different parts of the system energy and the velocity of a polaron [5]. The parameters are the same as in Fig. 1 and the data have been obtained similarly as in Fig. 2

will be understood if we note the enhancement of the effective coupling due to the motion of the polaron. According to the analytic calculation within the continuum approximation by Wilson [3], both of $\varepsilon_{\mathrm{lp}}$ and $\varepsilon_{\mathrm{lk}}$ diverge as $(1 - v^2/v_{\mathrm{s}}^2)^{-3}$ and the electronic energy ε_{e} as $(1 - v^2/v_{\mathrm{s}}^2)^{-2}$. The results shown in Fig. 3 are found to be consistent with these behaviors. The velocity dependence of the energy $\varepsilon_{\mathrm{tot}}$ in the small v region can be used to estimate the effective mass of the polaron. By fitting the $\varepsilon_{\mathrm{tot}}$-$v$ curve to a third order polynomial of v^2 in the region $v < 0.5v_{\mathrm{s}}$, we obtain the polaron effective mass from the coefficient of v^2, which is found to be about 100 times the bare electron mass in the case of polydiacetylene [5].

3 Quantum Tunneling of the Acoustic Polaron

The dynamical behavior discussed in the previous section involves only the motion of the polaron as a whole and the quantum nature is not considered explicitly. In this section, we address the issue whether the acoustic polaron can tunnel through a barrier which cannot be passed in the classical sense. In order to investigate this problem, a potential barrier located at a single site is introduced in the Hamiltonian, i.e. we consider the following Hamiltonian,

$$H = H_{\mathrm{SSH}} + H_{\mathrm{imp}}, \tag{11}$$

where the first term is defined in (1) and the second term is given by

$$H_{\text{imp}} = V c_i^{\dagger} c_i, \tag{12}$$

with i indicating the site where the potential barrier is located and V the barrier height. As will be understood from this model, we consider a thin (atomic-scale) barrier; this is because the tunneling is expected to be easily realized in the case of a thin barrier.

In order to see the possibility of quantum tunneling of an acoustic polaron through the barrier, we carry out a dynamical simulation where a freely moving polaron collides with the barrier. In the following the system size is fixed to be $N = 500$ and the barrier potential is located at the site $i = 350$. Furthermore the initial position of the polaron is set at the site 100 so that the initial distance from the barrier may be $N/2$ which is the largest distance when we apply the periodic boundary condition. The polaron is put in motion by a slowly switched external electric field as used in the previous section. Throughout the calculations discussed in this section, the switching duration time τ is fixed to be $250\omega_Q^{-1}$ and furthermore, in order to exclude the effect of the external field on the collision, the time of switching off, t_{off} in (4), is chosen to be $500\omega_Q^{-1}$. We have confirmed that, when the field is completely shut down (i.e. at $t = t_{\text{off}} + \tau$), the position of the polaron is sufficiently far away from that of the barrier and that the electronic density at the barrier site is less than 10^{-8} at the instant. Different ratios between the kinetic energy of the polaron and the potential height are realized by fixing the former and changing the latter. The kinetic energy of the polaron is defined as the difference in energy between the moving and the static state, and can be controlled by adjusting the strength E_{max} of the accelerating electric field. Although the total energy involves the lattice vibrational energy which is not localized around the polaron, this vibrational contribution can be suppressed by the slow switching to some extent.

Here it should be noted that the effective barrier height felt by the polaron is not equal to V involved in the Hamiltonian. This is because the potential energy is equal to the product of V and the electronic density at the barrier site. Since the polaron has a finite width, the local density at the barrier site becomes maximum when the peak of the electronic density profile arrives at the barrier site. Therefore what should be compared with the kinetic energy T of the polaron is not V itself but the product of V and the maximum value of the density profile just before the collision, ρ_{max}. If we regard the polaron as a classical particle, the conditions whether or not it can pass the barrier are expressed as $T > V\rho_{\text{max}}$ or $T < V\rho_{\text{max}}$, respectively.

The results of collision simulations are summarized in Fig. 4, where the maximum value of the potential strength V, below which more than ten percent of electronic charge related to the acoustic polaron can go through the potential barrier, is plotted (painted circles) for different values of the dimensionless coupling constant λ in the case with a fixed kinetic energy of the polaron ($T = 6.4 \times 10^{-3} t_0$). In the same figure, the classical criterion for the polaron transmission through the barrier potential, i.e. the value of the potential strength

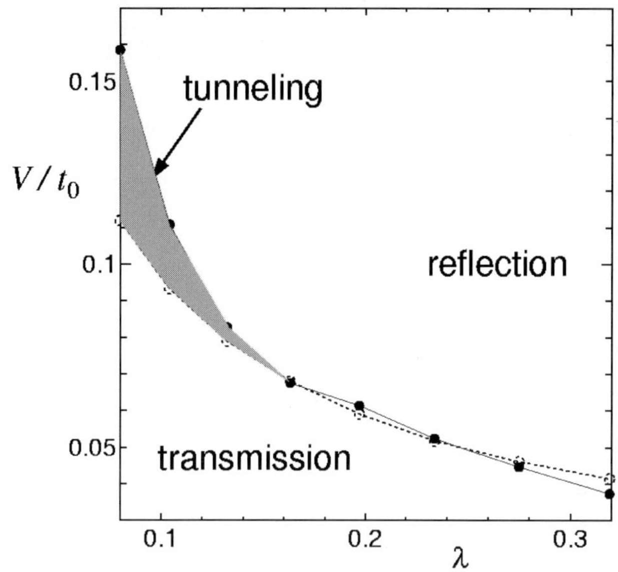

Fig. 4. The transmission and reflection of an acoustic polaron. This figure summarize the results of collision simulations for various values of the dimensionless coupling constant λ. The kinetic energy of the polaron just before the collision, T, is fixed at $T = 6.4 \times 10^{-3} t_0$. The data (*open circles*) connected by a dotted line are obtained by the condition $T = V\rho_{\max}$ with $\rho_{\max}$ the peak value of the electronic density profile just before the collision. The data (*painted circles*) connected by solid line are obtained from the criterion that, when the potential is smaller than that value, more than ten percent of electronic charge can go through the barrier

for which the kinetic energy is equal to the maximum of the effective potential felt by the polaron, is plotted (open circles). It is clearly seen that these two data are different particularly in the weak coupling region, the former being larger than the latter. This means that the quantum tunneling is realized within the shaded region of Fig. 4. As mentioned in the previous section, the dimensionless coupling constant determines the shape (or more explicitly the width) of the polaron; the weaker the coupling, the larger the width. It is reasonable to see quantum nature when the confinement of an electron is weak and its wave character can survive. In contrast to the weak coupling region, the confinement of the electron is rather strong in the strong coupling region, and therefore in this case the polaron behaves like a classical particle.

In order to see what is occurring in the shaded region of Fig. 4, we show in Fig. 5 a stereographic presentation of the space and time dependences of the electronic density profile in the case with $\lambda = 0.08$ and $V = 0.154t_0$ for which the maximum value of the effective potential is equal to $8.8 \times 10^{-3} t_0$ and larger than the polaron kinetic energy $T = 6.4 \times 10^{-3} t_0$. The partial transmission and reflection are clearly confirmed. The same data are drawn in the left-hand-side figure of Fig. 6 as snapshots at several different instants. The separate

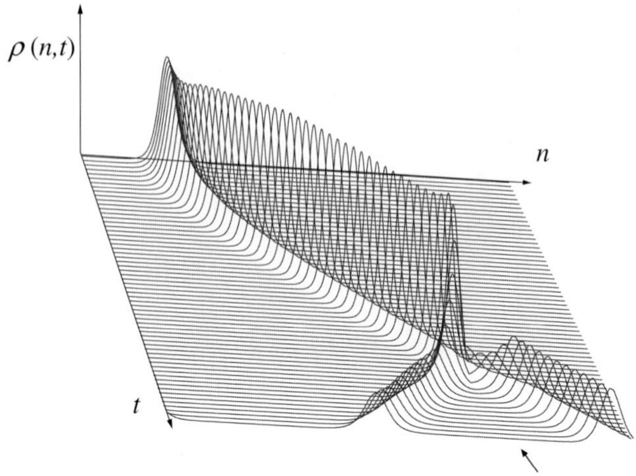

Fig. 5. The stereographic presentation of the space and time dependences of the electronic density profile in the case with $\lambda = 0.08$ and $V = 0.154t_0$. The kinetic energy of the polaron is fixed at $T = 6.4 \times 10^{-3}t_0$. The barrier is located at the site $n = 350$ (indicated by an *arrow*)

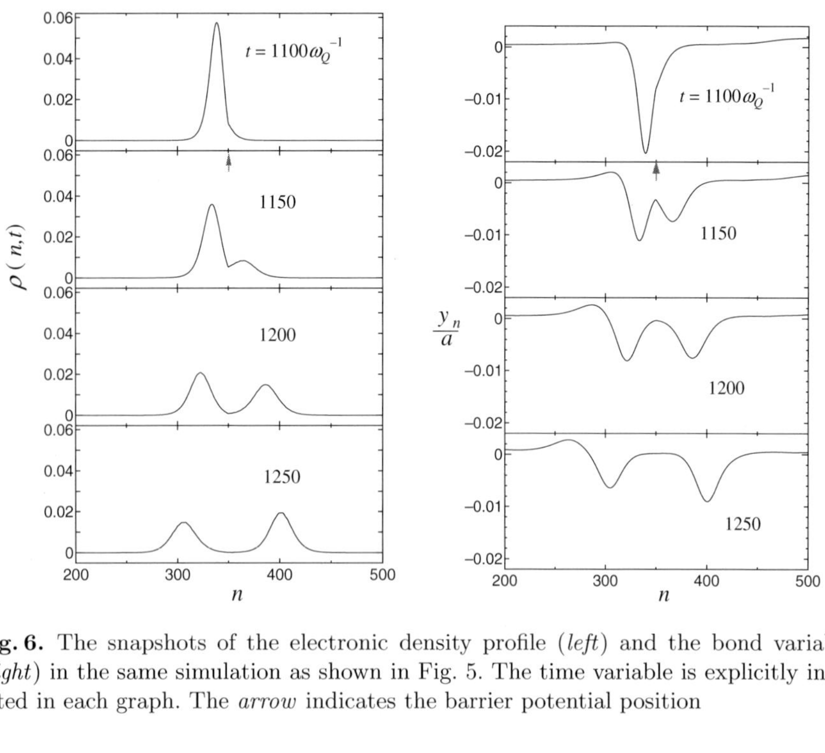

Fig. 6. The snapshots of the electronic density profile (*left*) and the bond variable (*right*) in the same simulation as shown in Fig. 5. The time variable is explicitly indicated in each graph. The *arrow* indicates the barrier potential position

integrations of the density profile in the left and right halves of the system at $t = 1250\omega_Q^{-1}$ (the bottom of Fig. 6) yield the reflection and transmission coefficients, respectively. In the case of Fig. 6 the reflection coefficient is about 0.45.

Since the lattice distortion is determined by the local density of the electron, the fact that the electron density profile is divided into two parts implies that the region of the lattice distortion is also divided into two parts. This can be seen by examining the snapshots of the bond variable $y_n(t)$ which are shown in the right-hand-side figure of Fig. 6. Implication of this behavior of the lattice distortion will be discussed later.

4 Concluding Remarks

In the present work, the dynamics of an acoustic polaron in a one-dimensional electron-lattice system has been reviewed and its quantum tunneling has been investigated by numerical simulations for the collision between the polaron and a barrier potential fixed at a site. In the simulations the electron has been treated quantum mechanically but the lattice system has been regarded as classical. When the repulsive impurity potential is strong enough, the polaron is almost completely reflected, and when the potential is very weak, it goes over the potential barrier. In the case with very strong electron-lattice coupling, the polaron is reflected by or penetrates the barrier potential depending on whether the kinetic energy of the polaron is smaller or larger than the maximum value of the effective potential felt by the polaron. In the weak coupling regime, however, as far as the maximum value of the effective potential is not very large compared to the kinetic energy of the polaron, we have found partial transmission and reflection of the polaron. This is nothing but the quantum tunneling of the polaron.

The boundary value of the barrier potential strength, above which the polaron cannot go through the impurity in a classical sense, has been estimated from the comparison of the kinetic energy of the polaron with the maximum value of the effective potential obtained as a product of the potential strength and the electron density profile before the collision. Although the details are not shown here, numerical simulations in the case, where the potential is sufficiently weak and the polaron can go over the barrier in a classical sense, indicate that the shape change of the polaron during passing through the barrier can reduce the effective potential particularly when the electron-lattice coupling is weak. Thus the boundary value of the impurity potential strength may be larger than that obtained by the above-mentioned naive estimate. Nevertheless the real boundary value of the impurity potential strength seems not much different from that obtained by the simple method, and in a certain region in the λ-V plane (the shaded region in Fig. 4) we find partial transmission and reflection of the polaron.

The partial transmission yields apparent fractional polarons which have fractional numbers of electrons. The stability of these fractional polarons could not be discussed within the present treatment since whether they are stable or not

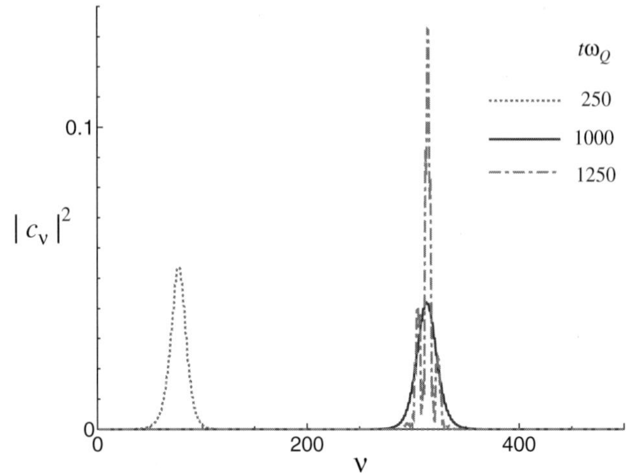

Fig. 7. Examples of the weights of different eigenfunctions at different instants when the electronic wave function of a moving polaron is expanded in terms of eigenfunctions of the electronic Hamiltonian with the lattice distortions at corresponding instants. The simulation was done for the case with $\lambda = 0.08$ and $V = 0.154t_0$, the kinetic energy of the polaron before the collision being set to be $6.4 \times 10^{-3}t_0$. The instants at which the snapshots are taken are $t\omega_Q = 250$ (dotted line), 1000 (*solid line*) and 1250 (*dash-dotted line*) as indicated in the graph

depends strongly on the method of observation of the polaron. Our speculation is that once a polaron is detected at some place, the polaron wave function will converge again into a single localized state with a unit charge and that fractional polarons would not live long.

Here it would be worthwhile to consider the expansion of the electronic wave function ψ for a moving polaron in terms of eigenfunctions for the electronic Hamiltonian with the corresponding lattice distortions. For a given configuration of lattice distortions, the eigenfunctions $\{\varphi_\nu\}$ can be obtained by numerically diagonalizing the electronic Hamiltonian. The expansion coefficients $\{c_\nu\}$ are determined by taking the scalar products of the wave function for the moving polaron with those eigenfucntions; $c_\nu = \langle\varphi_\nu|\psi\rangle$. The weight of each eigenfunction can be obtained as the square of the absolute value of the corresponding coefficient, $|c_\nu|^2$. In Fig. 7, examples of such weights at different times in one run of the simulation are depicted. The parameters used for this run are $\lambda = 0.08$ and $V = 0.154t_0$ (the polaron kinetic energy is fixed at $T = 6.4 \times 10^{-3}t_0$). For these parameters, partial transmission and reflection of the polaron are realized. At $t = 250\omega_Q^{-1}$ the polaron has not yet attained the final stationary velocity. The instants $t = 1000\omega_Q^{-1}$ and $1250\omega_Q^{-}1$ are before and after the collision, respectively. We have also carried out similar analyses for the cases with $V = 0.05t_0$ and $0.20t_0$, other parameters being unchanged. For the former case, the polaron is transmitted almost completely (99 %), and for the latter, it is almost completely reflected (99 %). The behaviors of $|c_\nu|^2$ at $t = 250\omega_Q^{-1}$ and $1000\omega_Q^{-1}$ are not changed as easily expected. The behavior at $t = 1250\omega_Q^{-1}$ is much different from that shown in Fig. 7. It is almost the same as that at $t = 1000\omega_Q^{-1}$. This behavior is not astonishing if we note that the motion of the polaron does not change much in these cases except for the change of direction of motion in the case with $V = 0.20t_0$. We should note that the eigenfucntions in the presence of the lattice distortion are not simple plane waves. Furthermore the degeneracy, which exists between right- and left-going waves or between sine and cosine waves in the absence of potentials due to lattice distortions and the barrier potential, is lifted in the presence of potentials since functions with different symmetry suffer different phase shifts.

The analysis depicted in Fig. 7 shows that the electronic wave function for the moving polaron is essentially a wave packet. However, in contrast to the normal wave packet, the polaron does not diffuse out because of the self-trapping mechanism due to the electron-lattice coupling. This suggest that the polaron might be used in studying the basic problems related to tunneling phenomena such as the tunneling time [11–13]. Although the definition of the tunneling time is not very clear, the visualized dynamics of a polaron might shed light on the tunneling time problem. Therefore the study of polaron dynamics is useful not only in discussing the transport phenomena in quasi-one dimensional systems but also in understanding basic problems of quantum mechanics.

This paper is dedicated to Bernhard Kramer for celebrating his sixtieth birthday. The work is partially supported by Grant-in-Aid for Scientific Research (C) (No. 14540365) from Japan Society for the Promotion of Science.

References

1. K.J. Donovan, E.G. Wilson: Philos. Mag. B **44**, 9 (1981)
2. E.G. Wilson: J. Phys. C **13**, 2885 (1980)
3. E.G. Wilson: J. Phys. C **16**, 6739 (1983)
4. W.P. Su, J.R. Schrieffer, A. J. Heeger: Phys. Rev. Lett. **42**, 1698 (1979); Phys. Rev. B **22**, 2099 (1980)
5. Y. Arikabe, M. Kuwabara, Y. Ono: J. Phys. Soc. Jpn. **65**, 1317 (1996)
6. R.E. Peierls: *Quantum Theory of Solids.* (Clarendon Press, Oxford, 1955)
7. Y. Ono, M. Kuwabara: 'Acceleration of Soliton in Polyacetylene by External Electric Field – Effect of Switching'. In: *AIP Conf. Proc. No. 256, Slow Dynamics in Condensed Matter*, ed. by K. Kawasaki, M. Tokuyama, T. Kawakatsu (AIP, New York, 1992), pp. 296–297
8. M. Suzuki: Proc. Jpn. Acad. **69**, 161 (1993)
9. Y. Ono, A. Terai: J. Phys. Soc. Jpn. **59**, 2893 (1990)
10. M. Kinoshita, Y. Hirano, M. Kuwabara, Y. Ono: J. Phys. Soc. Jpn. **66**, 703 (1997)
11. E.H. Hauge, J.A. Stovneng: Rev. Mod. Phys. **61**, 917 (1989)
12. R. Landauer, Th. Martin: Rev. Mod. Phys. **66**, 217 (1994)
13. N. Yamada: Phys. Rev. Lett. **83**, 3350 (1999)

Part III

Localization and Electron Correlations in Quantum Transport

Critical Chaotic Spectra of One and Two Interacting Electrons in Quasiperiodic Chains

Spiros Evangelou[1], Dimitris Katsanos[1], and Bernhard Kramer[2]

[1] Department of Physics, University of Ioannina, Greece
[2] Universität Hamburg, I. Institut für Theoretische Physik, Jungiusstr. 9, 20355 Hamburg, Germany

1 Introduction

In the last years we have seen a growing interest in the description of the quantum motion for few interacting electrons in a disordered potential [1–5]. Although this is by no means a true many-body problem, in the sense that it is a finite density and not just a few electrons which is required for a complete understanding of the combined interaction-disorder effects, it has shed some light on this very hard problem. On the other hand, the statistics of energy levels has become a valuable tool for extracting the essential information of the localization of eigenfunctions (without actually computing them) for non-interacting electrons in disordered media, making also possible to identify the localization-delocalization transition [6,7]. The level-statistics is a general tool since the spectra are always independent of the choice of basis while localization due to disorder or quasiperiodicity is usually expected to occur in the "position" basis. In this area a lot of emphasis was focused on understanding the critical level-statistics at the Anderson transition. This statistics turned out to be intermediate between Wigner (extended states) and Poisson (localized states) [6,7] being intimately connected to the fractal structure of the corresponding critical wave functions [8].

The critical nearest-level distribution function $P(S)$ for small energy spacings is metallic-like (Wigner) and for large energy spacings is insulating-like (Poisson) [6]. A simple analytical form which interpolates between Wigner and Poisson is the semi-Poisson curve $P(S) = 4S\exp(-2S)$ which appears in the following cases: pseudo-integrable billiard models of quantum chaos [9], $3D$ critical disordered systems when averaged over boundary conditions [10], $2D$ disordered systems with spin-orbit coupling (in this case the symplectic form of semi-Poisson) [11], critical points due to quasiperiodicity [12] and also two interacting electrons in a random potential [13]. The semi-Poisson distribution [9] can be easily obtained by considering the spacing statistics when extra levels are placed exactly in the middle between levels randomly distributed according to the integrable Poisson's law. It should be stressed here that for the level-statistics at criticality sensitivity to boundary conditions has been always found even in the thermodynamic limit [10]. This is a rather generic feature of criticality due to the presence of multifractal wavefunctions which are channel-like and thus very sensitive to boundary conditions.

The studies of few interacting electrons in disordered chains led to, at least, one important result. This concerns the unusual localization length enhancement of an electronic pair due to the interactions [1]. In the energy level-statistical description the presence of interactions between two electrons in addition to a random potential in one dimension quite never gives full quantum chaos (metallic state) but introduces instead a kind of weak, also named critical, chaos (quasi-metallic state) at the scale of the one-particle localization length [13]. The level-statistics for two electrons in a finite system of size equal to the one-particle localization length ξ_1 can be described by the intermediate semi-Poisson law [13], the same as the distribution obtained at the critical point of non-interacting disordered systems averaged over boundary conditions [10]. This quasi-metallic behavior, which has become known as "delocalization due to interactions", simply means a significant enhancement of the localization length in an otherwise localized one-dimensional disordered system.

The main idea of this article is to use two "toy" quasiperiodic models in order to explore the critical behavior at the delocalization-localization transition for one and two interacting electrons. For this purpose as alternatives to ordinary disordered lattices one could also use certain long-range mean-field type models [14]. We prefer the two simpler quasiperiodic systems [15,16] which can be regarded as unique models for studying accurately certain aspects of the critical behavior, even in the presence of many-body interactions. These are: (i) The Harper model [15] which is described by a simple one-dimensional tight-binding equation with quasiperiodic site potential $V_n = \lambda \cos(2\pi\sigma n)$ on each lattice site n, where σ is an irrational number which represents the "disorder" of the system. Despite the fact that it is one-dimensional the Harper model displays an Anderson delocalization-localization transition when the potential strength λ varies. The critical point is at $\lambda_c = 2$ and shares most of the features of a real $3D$ Anderson transition in disordered systems. The related critical behavior is characterized by fractal wave functions, fractal spectra, anomalous diffusive spreading in time of a δ-function wavepacket [17,18], power-law statistics of the gapwidths [19] and semi-Poisson statistics of the bandwidths [12]. The study of two-interacting electrons in this system has also revealed certain facts, such as absence [20] or very weak localization length enhancement due to interactions in the localized region ($\lambda > 2$)[21], a rather different behavior from what it is expected in the localized regime of ordinary disordered systems [1]. (ii) The second is the Fibonacci model where two values $+V$ or $-V$ associated with the site potential are arranged in a self-similar manner. This is different from Harper model since it has neither delocalized nor localized but only critical states [16]. It must be emphasised that the absence of translational invariance in these quasiperiodic models makes any analytical treatment almost impossible, so in the rest we rely on numerical computations in order to identify their key spectral features. We must stress that the quasiperiodicity which can be characterized by the golden mean irrational σ, with $\sigma^{-1} = (\sqrt{5}-1)/2$, in the Harper model determines the incommensurability between the potential and the lattice spacing and in the Fibonacci model the numbers of successive type A and B type units.

The study of one and two-body spectra in the Harper and Fibonacci models is interesting also for many physical reasons. For example, in order to identify the possible consequences of the electron-electron interaction in understanding the motion of Bloch electrons in two dimensional periodic potential at strong perpendicular magnetic fields, which is appropriate for the current-currying states in the quantum Hall effect [22]. Moreover, the problem of interacting fermions in self-similar potentials is also relevant for experiments in quasicrystals where the conductivity is known to exhibit unusual behavior, where it reduces with increasing temperature or disorder [23]. Our main aim, however, is to identify effects of the electronic interactions on the one-particle critical behavior.

The article addresses the following questions:

(i). What is the one-electron spectral density of states in the Harper model as a function of the quasiperiodic strength λ, in the Fibonacci model as a function of the potential strength V and how it changes for two electrons by switching on the interaction U?

(ii). Concerning spectral fluctuations is the critical semi-Poisson statistics, which is known to appear in the Harper model at criticality for averages over flux, shared by other one-electron quasiperiodic models, such as the diagonal and off-diagonal Fibonacci chains?

(iii). What is the nearest-level-statistics obtained for the fluctuations around averages over the "incommensurability" parameter σ, also in relation to fractal power-laws known to occur in the gap distribution before any "unfolding"? How does this change after "unfolding" procedures also for two interacting electrons?

(iv). What is the structure of the corresponding Hilbert space Q-matrix which describes many-body correlations in the case of critical one electron states?

We mostly concentrate on the spectral fluctuations for one and two interacting electrons in large finite quasiperiodic systems. Our purpose is to describe the fluctuations in the density of states by the intermediate scale-invariant critical semi-Poisson statistics. The outline of the article is as follows: In Sect. 3 we define the Harper and Fibonacci models, also in the presence of interactions. We compute their energy levels to obtain "meaningful" densities of states for one and two interacting electrons by introducing an "ensemble averaging" for all values of the incommensurability parameter σ. Subsequently we determine the level-statistics by studying the corresponding fluctuations in the introduced averages over σ. For one electron we find the nearest-level spacings $P(S)$ from the corresponding bandwidth distribution, which is an alternative procedure equivalent to "unfolding" over flux averages. We find that this distribution is in agreement with the semi-Poisson law also for the Fibonacci model. When the "unfolding" is done over various σ's the semi-Poisson is also relevant but only when we are restricted to some small energy windows. Our results by keeping all the energy levels for various σ's gave $P(S)$ having tails which correspond to power-laws known to occur for the gap-width distributions [19]. For two interacting electrons we demonstrate that the spectral fluctuations can be also reasonably well described by the semi-Poisson law. In Sect. 4 we obtain results for the many-body correlation function Q which consists of one-particle critical states. At the critical point $\lambda_c = 2$ we find Q which displays multifractal properties. Finally, in Sect. 5 we present some conclusions from our numerical study.

2 One and Two Interacting Electron Spectra

2.1 The Model

We consider the one-dimensional tight-binding Hubbard Hamiltonian with quasiperiodic site potential, unity nearest-neighbor hoppings and local Hubbard interaction of the form

$$\begin{aligned} H = & \sum_{n,s=\pm} V_n c^\dagger_{n,s} c_{n,s} + \sum_{n,s=\pm} (c^\dagger_{n+1,s} c_{n,s} + H.c.) \\ & + U \sum_n c^\dagger_{n,+} c_{n,+} c^\dagger_{n,-} c_{n,-} , \end{aligned} \tag{1}$$

where $c_{n,s}$ is the annihilation operator of an electron with spin $s = \pm$ on the nth site in the presence of the on-site quasiperiodic potential $V_n = \lambda \cos(2\pi\sigma n)$, σ is an irrational number specifying the incommensurability and U the Hubbard interaction strength. The one-electron Hamiltonian can be written in the site basis $|n\rangle$, $n = 1, 2, ..., N$ with the irrational σ chosen as the inverse golden mean approximated by the rational approximants M/N, where M, N are successive Fibonacci numbers. For the Fibonacci model the on-site binary alloy type potential is $V_n = +V$ for A type atoms or $-V$ for B type, and is created by making the replacements $A \to AB$ and $B \to A$, so that the successive Fibonacci chains are AB, ABA, $ABAAB$, $ABAABABA$, etc., having sizes equal to the Fibonacci numbers 2, 3, 5, 8, ... We have also studied a bond-type Fibonacci model obtained by arranging the hopping strengths t_A or t_B in a quasiperiodic fashion. This model respects chiral symmetry [24] so that the corresponding spectrum is symmetric around the zero energy $E = 0$ mode.

In order to describe two interacting electrons the singlet subspace with total spin $S = 0$ suffices. This implies two electrons of opposite spins so that the Hamiltonian can be written in the $N(N+1)/2$ two-electron symmetric basis states $(|n_1, n_2\rangle + |n_2, n_1\rangle)/\sqrt{2}$ for $n_1 < n_2$ and $|n_1, n_1\rangle$ for $n_1 = n_2$, where $n_1, n_2 = 1, 2, ..., N$ are the positions of each electron in the N-site chain. For two electrons in one dimension the diagonalization requires half the size of the Hilbert space for one electron in a squared lattice which excludes the diagonal plus the diagonal. The equivalent square lattice problem has unit hopping and site potential $V_{n_1} + V_{n_2}$ which is symmetric with respect to the main diagonal ($n_1 = n_2$) where the on-site interaction U is added. We have set up and diagonalized numerically the matrix corresponding to (1) for a chain of N sites and hard wall boundary conditions, for both Harper and Fibonacci type site potentials.

2.2 The Energy Levels

The computed energy levels for one and two interacting electrons in the considered quasiperiodic models are displayed in Figs. 1 and 2. For the Harper model

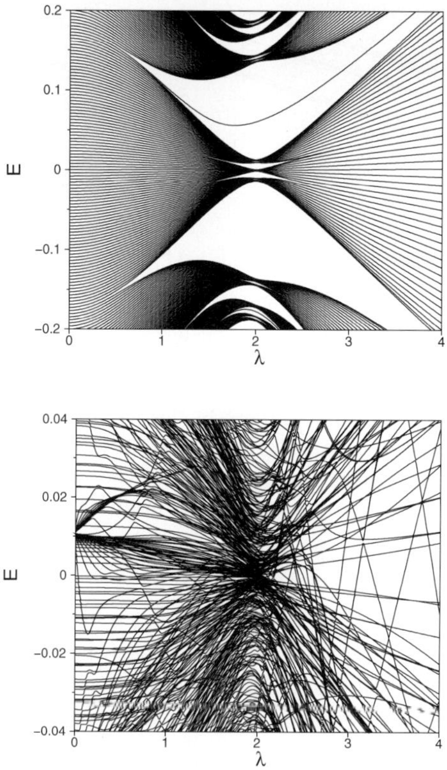

Fig. 1. a Energy levels close to $E = 0$ for the one-electron in the Harper model as a function of the quasiperiodic potential strength λ for $\sigma = M/N = 987/1597$ and $N = 1597$. In the delocalized regime ($\lambda < 2$) and its dual localized regime ($\lambda > 2$) trivial level-statistics can be seen. The unusual bandwidth statistics at the critical point $\lambda_c = 2$ is found to resemble closely the critical semi-Poisson distribution [12]. **b** The energy levels close to $E = 0$ for two electrons interacting with strength $U = 1$ for $M/N = 55/89$ with $N = 89$ as a function of the quasiperiodic potential strength λ. The two electron spectra for $\lambda \leq 2$ show some signs of chaotic level-repulsion due to the interaction

the one-electron spectra (Fig. 1a) clearly show the well-known duality between the delocalized regime ($\lambda < 2$) and the localized regime ($\lambda > 2$) where no spectral fluctuations exist. This is, of course, in sharp contrast with the delocalized and localized regimes in disordered systems where Wigner and Poisson fluctuations are observed, respectively. The quasiperiodic models are much simpler. The Harper model exhibits ballistic motion (instead of diffusive) in the delocalized regime and correlated localized behavior (instead of random) in the localized regime with interesting statistics obtained only at the critical point $\lambda_c = 2$. In [12] it was established that the statistics of the energy levels after "unfolding" over various flux values is described by the intermediate semi-Poisson critical law. The obtained spectra shown in Fig. 1b for two interacting electrons in the

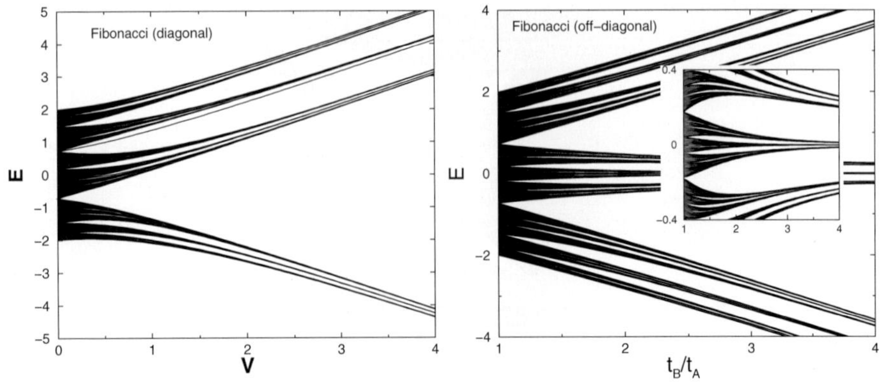

Fig. 2. The energy spectra for one-electron in the Fibonacci model for size $N = 1597$: **a** for the diagonal model as a function of V, and **b** for the off-diagonal model as a function of the hopping ratio t_B/t_A. In the inset the chiral spectra of case (b) are magnified around the $E = 0$ zero mode which is present only for odd size N

Harper model allow two additional conclusions to be drawn: First, the critical point $\lambda_c = 2$ seems to be relevant also in the presence of interactions and, second, chaotic signs of level repulsion are visible when switching on the interactions U.

The one-electron spectra for the diagonal and off-diagonal Fibonacci models are shown in Figs. 2a,b, respectively. In this case we observe critical spectra for all non-zero values of V (diagonal model) or t_B/t_A (off-diagonal model), with features similar to the critical point $\lambda_c = 2$ spectra of the Harper model (Fig. 1a). For the off-diagonal Fibonacci model the spectrum is symmetric around the origin due to the sublattice chiral symmetry where the lattice consists of two interconnected sublattices [24]. Moreover, for the large V or large t_B/t_A Fibonacci model the spectra diminish and we observe a slow approach to the localized limit. In this limit the spectra are no longer singularly continuous but become point-like.

The obtained patterns of two-electron energy levels for the Harper model as a function of λ for more values of the interaction strength, such as $U = 0$ (non-interacting), 0.1 and 10, are shown in Fig. 3. It is only when U takes intermediate values such as $U = 1$ (see also Fig. 1b) where the chaotic level-repulsion features become more obvious. This is expected for two electrons in disordered chains since critical chaos sets in for sizes equal to the one-particle localization length only as long as U is intermediate [13]. A complementary plot is shown in Fig. 4 where the critical spectra of the Harper model are displayed as a function of the interaction strength U instead, focusing near the origin $E = 0$. For intermediate U we also see some chaotic signs of level-repulsion in the two-electron energy levels which can appear in a gap region.

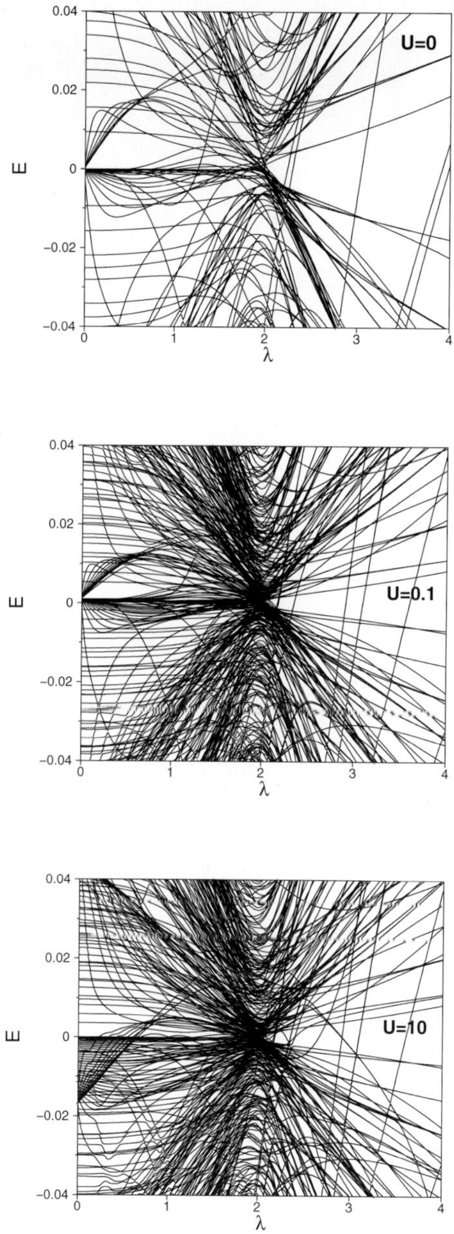

Fig. 3. The two-electron energy levels close to $E = 0$ as a function of λ for interaction strength **a** $U = 0$, **b** $U = 0.1$ and **c** $U = 10$. The chain size is $N = 95$

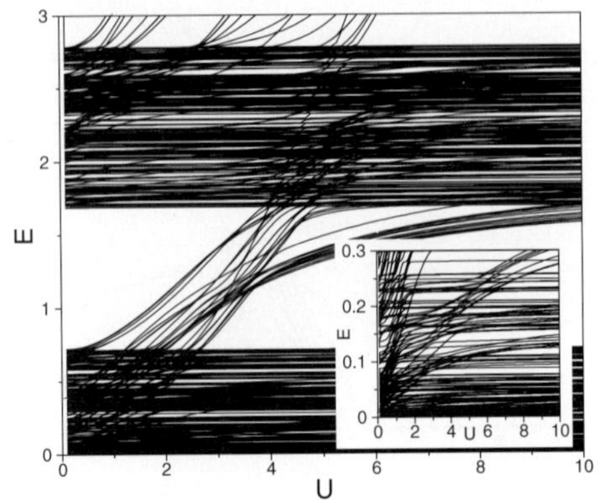

Fig. 4. The same as in Fig. 3 but now the energy levels as a function of the interaction strength U at the critical point $\lambda_c = 2$. In the inset the energy region near $E = 0$ is enlarged

2.3 The Density of States

The densities of states for one and two interacting electrons of the Harper model can be obtained from the computed energy levels. In our calculations we have considered sufficiently large quasiperiodic systems for various quasiperiodicity strengths λ and interaction U. It must be understood that the quasiperiodicity does not involve any ensemble averaging as disordered systems do. In Fig. 5 typical pictures of the critical integrated density of states ($\lambda = 2$) for one and two interacting electrons display the expected "devil's staircase" type characteristics (also for two interacting electrons) with many bands separated by gaps. The difficulty obtaining the averaged behavior needed (see Sect. 3.4) for the integrated density in order to compute the level statistics is obvious.

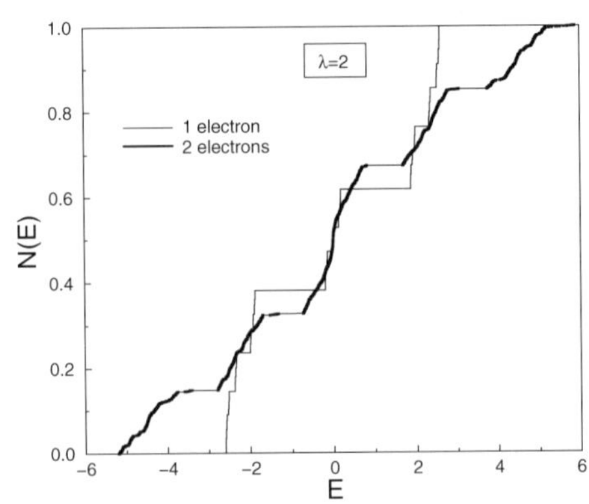

Fig. 5. The integrated "devil's staircase" type density of states at the critical point $\lambda_c = 2$ for both one ($N = 1597$) and two interacting electrons with $U = 1$ ($N = 233$)

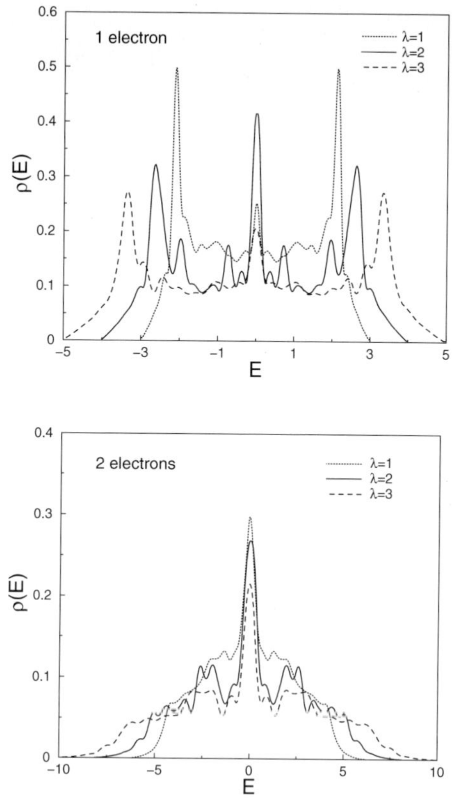

Fig. 6. a The density of states of one-electron spectra averaged over many σ's, for the delocalized $\lambda = 1$ (dotted line), the critical $\lambda_c = 2$ (*solid line*) and the localized $\lambda = 3$ (*dashed line*) regions. The chain size is $N = 1000$, 10000 while 250 values of σ are used so that $2\pi\sigma$ covers the region $[0, \pi]$. **b** The density of states for two-electrons with $U = 1$ averaged over σ's for the delocalized $\lambda = 1$ (*dotted line*), the critical $\lambda_c = 2$ (solid line) and the localized $\lambda = 3$ (*dashed line*) regions. The chain size is $N = 45$ and 1000 values of σ are used so that $2\pi\sigma$ covers the region $[0, \pi]$

In order to produce some kind of "averaged" density of states also for the Harper model we have examined how its spectra change by considering all possible values for the incommensurability parameter σ from 0 to 0.5. The corresponding density of states "averaged over σ" for strengths λ below, above and at the critical point $\lambda = 2$ is displayed in Fig. 6a for one electron and Fig. 6b for two interacting electrons. We observe that these densities of states are different for periodic one- and two-dimensional lattices which have two edge van-Hove like singularities (one-dimensional) and one midband logarithmic singularity (two-dimensional), respectively. For the one electron case in Fig. 6a some one-dimensional characteristics can be still seen, while for two interacting electrons two-dimensional features are also displayed in Fig. 6b. Another interesting point may be the fact that at the origin $E = 0$ both the one and two-electron spectra display a peak, which becomes higher when $\lambda_c = 2$ only for one electron. In the delocalized ($\lambda < 2$) and the localized ($\lambda > 2$) regimes the one-electron central peaks are similar to each other being smaller in comparison to the critical case. This is a consequence of duality of the Harper model which seems to be violated for two interacting electrons.

2.4 The Nearest Level $P(S)$ Distribution

We have also investigated the level-statistics for one- and two-electron Harper model by "unfolding" the energy levels via the removal of the density of states averaged over σ, shown in Fig. 6. This is a simple rescaling procedure of the eigenvalues according to the above "averaged" density of states and it is equivalent to a simple division of the raw energy spacings $E_{i+1} - E_i$ by a mean level-spacing Δ obtained from many levels for various σ inside a local energy window around E_i. Since for the "unfolded" levels $\mathcal{E}_i = N_{av}(E_i)$, where N_{av} is the averaged integrated density around E_i, the mean level-spacing becomes exactly equal to one so that we can study the corresponding fluctuations of the density of states, such as the nearest-level distribution function $P(S)$ with $S = \mathcal{E}_{i+1} - \mathcal{E}_i$, free from the averaged behavior. The $P(S)$ distributions are obtained for finite size systems with the required "averaged" density obtained in the previous Sect. 3.3 by using all the energy levels for various σ. The computed $P(S)$ after such "unfolding" procedure over σ can be seen in Fig. 7 when all energy levels in the band are kept. The peculiar $P(S)$ obtained in this case for one and two electrons has power-law tails which resemble the well-known inverse power-law gap-width distributions [19]. This is due to the dominant role of the widely varying gaps in the spectrum when all energy levels are kept. Instead, the $P(S)$ obtained when levels only near $E \approx 0$ are kept is shown in Fig. 8. In this energy window $P(S)$ is also compatible with the critical semi-Poisson distribution function.

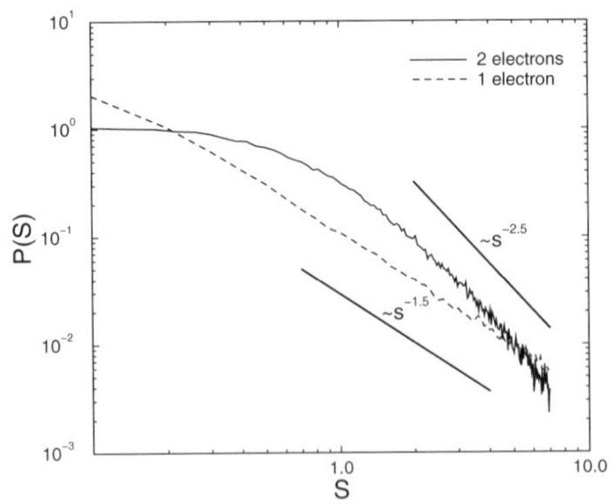

Fig. 7. Log-log plot of the $P(S)$ distribution function for one and two interacting electrons averaged over many σ's by keeping the majority of the energy levels in the band. The total number of eigenvalues was about one million in each case. In this plot the power-law tails are dominated by the gapwidth distribution

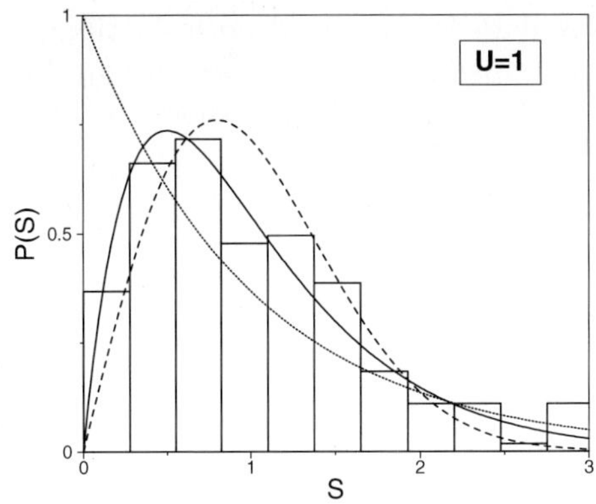

Fig. 8. The $P(S)$ distribution for two electrons of strength U using energy data restricted in a small energy window around $E \approx 0$

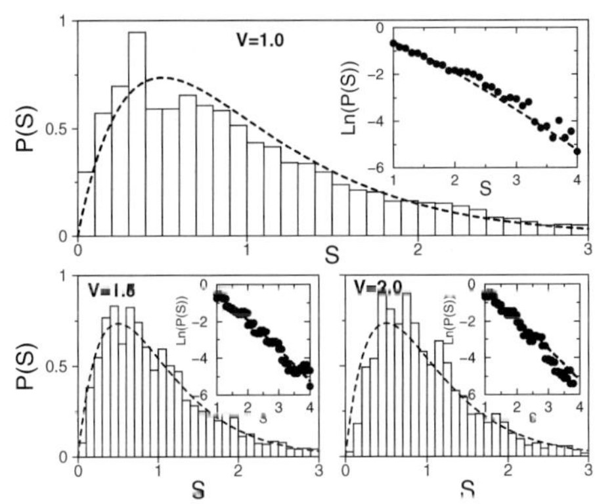

Fig. 9. The $P(S)$ distribution function (obtained from the bandwidths which is equivalent to "unfolding" over fluxes) for the diagonal Fibonacci model of potential strength V when averages over ten V's around $V = 1$, $V = 1.5$ and $V = 2.0$ are taken. In the insets the tails of the distributions are also shown in semi-log plots together with the semi-Poisson curve $P(S) = 4S \exp(-2S)$ (*broken lines*)

We have also examined the bandwidth and the gapwidth statistics. For the Harper model the bandwidths obey the semi-Poisson distribution [12] and the gapwidths a fractal inverse power-law behavior [19]. Our results verify both the semi-Poisson and the fractal power-law behavior. The integrated spacing density of the gapwidths for one electron obeys the power-law $\propto S^{-3/2}$ and for two electrons $\propto S^{-5/2}$, which after differentiation give the gapwidth distributions which obey the power-laws $\propto S^{-1/2}$ and $\propto S^{-3/2}$, respectively. We find that the bandwidth distribution for the one-electron spectra of the Fibonacci model shown in Fig. 9 also follows the semi-Poisson, similarly to the Harper model in [12]. This is equivalent to $P(S)$ after "unfolding" over various flux values since the bands are determined for all fluxes from 0 to π which specify levels inside the bands. The data for the bandwidth distribution obtained from the extreme periodic and antiperiodic boundary conditions, by taking also averages over various values of the strength V, are shown in Fig. 9 together with the semi-Poisson curve. They are in satisfactory agreement so that we can conclude that the semi-Poisson curve seems to be the appropriate distribution to describe the level-statistics not only for the Harper model at criticality but also for the Fibonacci model.

3 The Multifractal Q-Matrix

Finally, we examine the structure of the corresponding Hilbert space network which is appropriate for the description of N-interacting particles in a one-dimensional quasiperiodic potential [13]. This network is made of "sites" which consist of non-interacting states and "hoppings" which are induced by the many-body interaction. Since the Hubbard interaction is local the many-electron Hamiltonian matrix of (1) can be written in a basis build from the one-particle states Ψ_α, $\alpha = 1, 2, ..., N$ with corresponding energy eigenvalues ϵ_α. In such Hilbert space matrix the "sites" of the equivalent lattice network have energy $\sum_\alpha^N \epsilon_\alpha$ and the "hoppings" are defined from the interaction matrix elements

$$Q_{\alpha,\beta}^{\gamma,\delta} = \sum_n \psi_\alpha^*(n)\psi_\beta^*(n)\psi_\gamma(n)\psi_\delta(n), \tag{2}$$

which also includes the interaction U as a proportionality factor and involves the amplitudes $\psi_\alpha(n) =< n|\Psi_\alpha >$ of the single particle states $\Psi_\alpha = \sum_{n=1}^N \psi_\alpha |n\rangle$. The matrix Q since it consists of single particle states can be viewed as a special single particle correlation function.

We have studied numerically the Q-matrix structure in the four-dimensional space of $\alpha, \beta, \gamma, \delta$ for the critical one-particle states at λ_c. The results are presented in Fig. 10a where slopes D_q are determined from scaling the moments which involve the quadruple summations $\frac{1}{q-1} \ln \sum_{\alpha,\beta,\gamma,\delta} |Q_{\alpha,\beta}^{\gamma,\delta}|^q$ versus $\ln N$ and the measure is normalized with $\sum_{\alpha,\beta,\gamma,\delta} |Q_{\alpha,\beta}^{\gamma,\delta}| = 1$. This problem becomes a lot easier if we fix two of the parameters, e.g. γ_0, δ_0 to vary within ten sites around the origin, so that we consider only the remaining two-dimensional (α, β) problem. In this fashion we can consider larger sizes for computing the moments. The results are shown in Fig. 10b and the slopes from both procedures define approximate fractal dimensions D_q shown in Fig. 11. For example, the definition

$$D_q = \frac{1}{q-1} \frac{\ln \sum_{\alpha,\beta} |Q_{\alpha,\beta}^{\gamma_0,\delta_0}|^q}{\ln N}, \tag{3}$$

gives D_q for the normalized measure $\sum_{\alpha,\beta} |Q_{\alpha,\beta}^{\gamma_0,\delta_0}| = 1$ for fixed γ_0 and δ_0.

In the dual $\lambda < 2$ or $\lambda > 2$ cases the computed D_q in the restricted two-parameter space tend towards the value 1 if $q \neq 0$ with $D_0 = 2$, as it was already predicted for periodic systems in [13]. At the critical point $\lambda_c = 2$ we find that

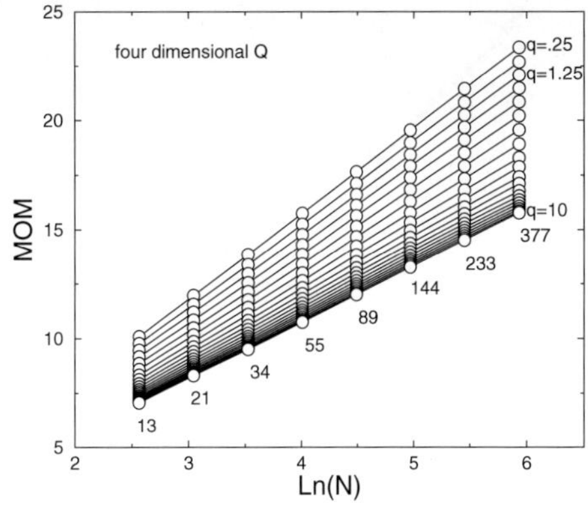

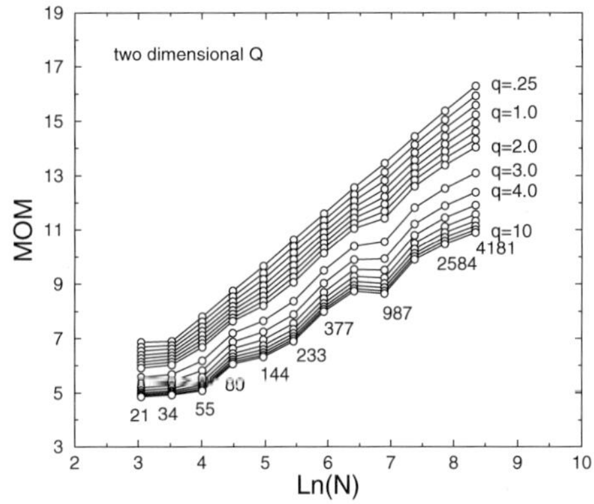

Fig. 10. a The log-log plot of the scaling of the moments (Eq. (2)) versus size in the four-dimensional Q-space at $\lambda_c = 2$. **b** The log-log plot of the scaling of moments versus size when averages over ten sites around the centre are taken for two from the four variables of Q at $\lambda_c = 2$. This allows much larger chains to be considered. The periodicity between the n-th and $n \pm 6$ Fibonacci generations is also seen

the fractal dimensions obey a universal behavior with the approximate linear equation

$$D_q = 2 - \Lambda q, \quad \Lambda \approx 0.125, \tag{4}$$

which is valid for small-q. For localized one-body states the predicted enhancement of the localization length [13] depends on $f(\alpha = 2)$

$$\xi_2/\xi_1 \propto U^2 \xi_1^{2f(\alpha(2))}/\xi_1^3 \tag{5}$$

where $\xi_{1,2}$ are the one, two particle localization lengths with $f(\alpha(q)) \approx 2 - \Lambda q^2$, where $f(\alpha)$ is the multifractal spectral density. For localized states $f(\alpha(2)) \approx 1.75$ was found in [13], which gave the enhancement factor $\xi_2 \propto \xi_1^{1.5}$ of the one-particle localization length ξ_1 due to the interactions. For the Harper model

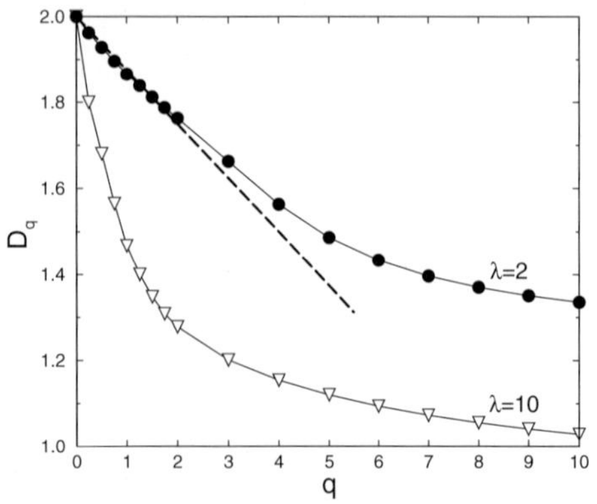

Fig. 11. The computed fractal dimensions of the Hilbert space matrix structure at $\lambda_c = 2$. The broken line is the linear small-q approximation of (4)

at criticality for $\Lambda \approx 0.125$ we obtain $f(\alpha(2)) \approx 1.5$ which from (5) gives no enhancement of the one-particle localization length ($\xi_2 \propto \xi_1$), as expected at $\lambda = 2$ where the one-particle localization length ξ_1 has already reached the system size N.

4 Discussion

In summary, we have investigated the one dimensional quasiperiodic Harper and Fibonacci models for one and two interacting electrons. The quasiperiodic models are like quantum chaotic models in the sense that they allow no additional ensemble-averaging as in disordered systems but only energy-averaging. However, in order to study the level-statistics one can introduce averages over applied flux values (by varying the boundary conditions) or over the incommensurability parameter (by varying σ). Fortunately, the average over various fluxes in one-dimension was shown to be equivalent to the statistics of bandwidths and led to a semi-Poisson $P(S)$ which can be obtained solely from the bandwidth distribution [12]. In this article we verify that the same bahavior holds for another quasiperiodic model. We also obtain interesting results by examining a second alternative, which is averaging over the quasiperiodicity σ, instead of the flux. They are summarized in the following: Verification that the one-electron Harper model has critical spectral fluctuation characterized by a semi-Poisson $P(S)$ while the same distribution also describes the spectral fluctuations of the Fibonacci model. Moreover, we have examined the fate of the critical singular continuous one-electron spectra when two interacting electrons are considered. We find chaotic features for two-electrons in the presence of interactions. Under appropriate conditions (for example, when all the spectrum is considered) we show that the statistics is dominated by the gapwidth distributions with inverse power-law tails, while for levels only near the band center for two interacting electrons the critical semi-Poisson law is also obtained.

We have also answered some questions related to the many-body Hilbert space structure of the spatial overlap Q which consists of one-electron states and turns out to be multifractal. The related exponents D_q are computed and at the critical point show no localization length enhancement, justifying previous calculations which predict the localization length enhancement $\xi_2 \propto \xi_1^{1.5}$ for two electrons in disordered systems. The above results indicate that many-body effects viewed in simple two-electron models at criticality imply fractal characteristics of the many-body Q-space structure.

Acknowledgements

This is a work done under an RTN network collaboration on "Nanoscale dynamics, coherence and computation". Spiros Evangelou is grateful for the great hospitality at Hamburg University, Physics Department, during a short visit where most of this work was done.

References

1. D.L. Shepelyansky, Phys. Rev. Lett. **73**, 2607 (1994)
2. P. Jacquod and D.L. Shepelyansky, Phys. Rev. Lett. **75**, 3501 (1995)
3. X. Waintal, D. Weinmann and J.-L. Pichard, Eur. Phys. J. B **7**, 451 (1999)
4. O. Halfpap, I.Kh. Zharekeshev, A. MacKinnon and B. Kramer, Ann. Phys. (Leipzig) **7**, 503, (1998)
5. O. Halfpap, T. Kawarabayashi and B. Kramer, Ann. Phys. (Leipzig) **7**, 483, (1998)
6. B.I. Shklovskii et al., Phys. Rev. B **47**, 11487 (1993)
7. I. Zharekeshev and B. Kramer, Phys. Rev. Lett. **79**, 717 (1997)
8. S.N. Evangelou, J. Phys. A **23**, L317 (1990); M. Schreiber and H. Grussbach, Phys. Rev. Lett. **67**, 607, (1991)
9. E.B. Bogomolny, U. Gerland, C. Schmidt, Phys. Rev. E **59**, R1315 (1999)
10. D. Braun, G. Montambaux and M. Pascaud, Phys. Rev. Lett. **81**, 1062 (1998)
11. G. Katomeris and S.N. Evangelou, Eur. Phys. J. B **16**, 133 (2000)
12. S.N. Evangelou and J.-L. Pichard, Phys. Rev. Lett. **84**, 1643 (2000)
13. X. Waintal and J.-L. Pichard, Eur. Phys. J. B **6**, 117 (1998)
14. F. Evers and A.D. Mirlin, Phys. Rev. Lett. **84**, 3690 (2000)
15. S. Aubry and G. Andre, Ann. Isr. Phys. Soc. **3**, 133 (1980); P.G. Harper, Proc. Roy. Soc. London Sect. A **68**, 874, 879 (1955); D.R. Hofstadter, Phys. Rev. B**14**, 2239 (1976)
16. H. Hiramoto and M. Kohmoto, Int. J. Mod. Phys. B**6**, 281 (1992) and references therein
17. J.X. Zhong and R. Mosseri, J. Phys. C **7**, 8383 (1995)
18. F. Piechon, Phys. Rev. Lett. **76**, 4372 (1996)
19. T. Geisel, R. Ketzmerick, and G. Petschel, Phys. Rev. Lett. **66**, 1651 (1991)
20. A. Barelli, J. Bellisard, P. Jacquod and D.L. Shepelyansky, Phys. Rev. Lett. **77**, 4752 (1996)
21. S.N. Evangelou and D. Katsanos, PRB **56**, 12797 (1997)
22. K. von Klitzing, G. Dorda and M. Pepper, Phys. Rev. Lett. **45**, 494 (1980)
23. D. Shechtman, I. Blech, D. Gratias and J.W. Cahn, Phys. Rev. Lett. **53**, 1951 (1984)
24. M. Inui, S.A. Trugman and E. Abrahams, Phys. Rev. B **49**, 3910 (1994)

Superconductivity from the Repulsive Electron Interaction – from 1D to 3D

Hideo Aoki

Department of Physics, University of Tokyo, Hongo, Tokyo 113-0033, Japan

1 Introduction

There is a growing realisation that the high-Tc superconductivity found in the cuprates in the 1980's has an electronic mechanism – namely, anisotropic pairing from the repulsive electron-electron interaction. Superconductivity from electron repulsion is conceptually interesting in its own right, and has indeed a long history of discussion. In fact, in the field of electron gas, i.e., electron system with the Coulombic electron-electron interaction, Kohn and Luttinger [1] pointed out, as early as in the 1960's, that the electron gas should become superconducting with anisotropic pairing (having nonzero relative angular momenta) at sufficiently low temperatures in a perturbation theory. This becomes an exact statement for dilute enough electron gas, where p-wave pairing (with the relative angular momentum = 1) should arise, as far as the static interaction is concerned [2,3].

While these have to do with the long-range Coulomb interaction where the dominant fluctuation is charge fluctuation, the problem we would like to address here is the opposite limit of short-range repulsion, as appropriate for strongly-correlated systems such as transition metal oxides. There, the dominant fluctuation is the spin fluctuation. The most widely used model is the Hubbard model having the on-site repulsion, U. If the one-band Hubbard model, the simplest possible model for repulsively correlated electron systems, superconducts, the interest is not only generic but may be practical as well, which has indeed been a challenge in the physics of high T_C superconductivity.

To develop a theory for that, it is instructive to start with one-dimensional (1D) systems. When the system is purely 1D, we have an exact effective theory, which is the Tomonaga-Luttinger theory that is exactly solvable in terms of the bosonisation and renormalisation. So we start with this, where no superconducting phase is shown to exist when the interaction is repulsive. When there are more than one chains, which is called ladders, a superconducting phase appears. If one closely looks at the pairing wavefunction, this is a pairing having opposite signs across two bands where the key process is the interband pair hopping.

We then show that this physics has a very natural extension to two-dimensional (2D) systems. There, anisotropic (usually d having the relative angular momentum of 2) pairing superconductivity can arise. If one looks at the pairing wavefunction, this is a pairing having opposite signs across the key interband pair-hopping processes. The key process is dictated by the peak in the spin structure (usually antiferromagnetic).

We finally look at how this kind of anisotropic pairing superconductivity can be "optimised", namely how we can make T_C higher. We first note that "T_C is very low in the electron mechanism" in that T_C is usually two orders of magnitude lower than the electronic energy. The main reason is that the node in the gap function, which has to exist for the anisotropic pairing, intersects the Fermi surface. So we can propose, and show, that systems that have a disconnected Fermi surface has much higher T_C.

2 1D – Tomonaga-Luttinger Theory and the Physics of Ladders

2.1 Tomonaga-Luttinger Theory

It was Tomonaga who pioneered the many-body physics in 1D. In his 1950 paper [4] the essence of the whole idea is already there, although the theory is now often called Tomonaga-Luttinger. When the system is 1D, the Fermi energy, E_F, intersects the band at two points, left-moving branch (L) and the right-moving one (R; Fig. 1a). The dispersion around these points may be approximated as linear functions of the wavenumber, k. When we do this, every electron-hole excitation across E_F becomes a creation operator of a sound wave (which is a boson).

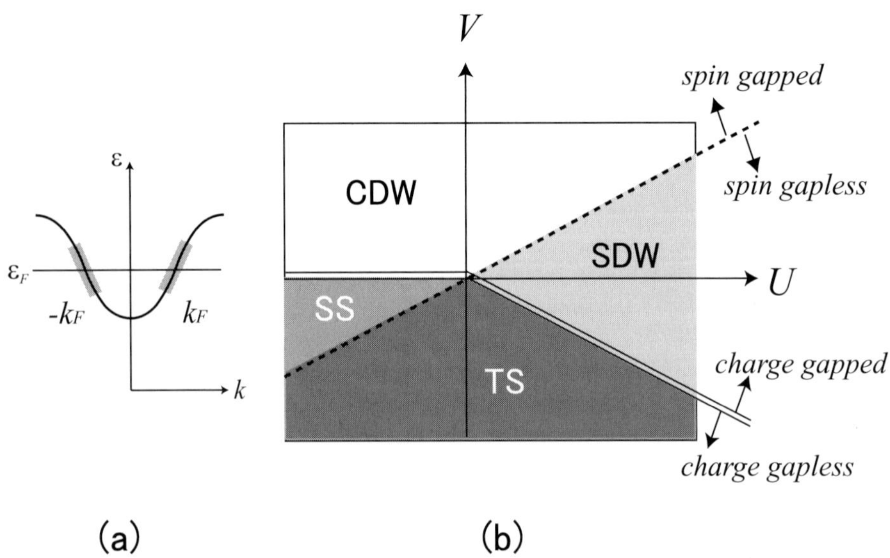

Fig. 1. a Tomonaga-Luttinger model, in which the low-energy excitations around the Fermi energy (shaded) at $k = \pm k_F$ are considered for 1D systems. **b** Weak-coupling result for the phase diagram against the on-site repulsion, U, and the off-site repulsion, V, for the 1D extended Hubbard model at half filling. SS: spin-singlet superconductivity, TS: triplet superconductivity

As for the electron-electron interaction, the matrix elements may be classified into four categories: (i) backward scattering, where one electron at R jumps to L while another from L to R, (ii) forward scattering, where R jumps to R, L jumps to L, (iii) umklapp scattering, where ($[R, R]$ jumps to $[L, L]$ or vice versa), (iv) forward scattering within each branch ($[R, R]$ to $[R, R]$ or $[L, L]$ to $[L, L]$). Tomonaga-Luttinger theory [5,7–11] is a weak-coupling theory (i.e., theory for the case when the electron-electron interaction is weak enough) that deals with low-energy processes. For that we can integrate out the higher-energy processes in the perturbational renormalisation-group sense. We can then look at the flow of the renormalisation equation, and its end point called the fixed point. To discuss the nature of the fixed-point Hamiltonian, it is convenient to bosonise (i.e., to write everything in terms of boson operators). The final result for the effective Hamiltonian is written in terms of two boson fields, spin phase (ϕ) and charge phase (θ), whose stiffness (coefficients of $(\partial\phi)^2, (\partial\theta)^2$) is given in terms of only two quantities, K_σ, K_ρ, which determine everything, including whether the ground state is superconducting. To be more precise, in 1D even a "long-range" order can only have a two-point correlation that decays with a power law ($\propto 1/r^\alpha$) where the exponent α is dictated, for each of the order parameters considered, by K_σ, K_ρ.

For every Hamiltonian originally given, we can calculate the four scattering parameters, and then renormalise them. If we look at the phase diagram (Fig. 1b) for the extended Hubbard model (where we have an off-site interaction, V, on top of the on-site one, U), there is no superconducting phase when all the interactions U and V are repulsive (> 0).

Incidentally, there is no magnetism, either, for a single chain. This is due to the well-known Lieb-Mattis theorem, which dictates that electrons in 1D are entirely non-ferromagnetic. The proof makes use of the fermion statistics of electrons, where a key factor is that two electrons cannot pass each other in 1D, or, in the words in Mattis's textbook [6], "neighbours remain neighbours till death did them part".

2.2 Pairing in Ladders, or $1 + 1 \neq 2$

When there is more than one chain with inter-chain hopping and/or interaction, the physics can be, and is indeed, entirely different. The model, then, becomes multi-band (i.e., n-band system for n-leg ladder). The Fermi energy can intersect the dispersion at $2n$ points (Fig. 2), so the model is what can be called *multiband Tomonaga-Luttinger model*. The multiband Tomonaga-Luttinger model has been studied in various context, including the excitonic phase in electron-hole systems [12], transport properties [13] and interband excitations in quantum wires as detected by Raman spectroscopy by Sassetti et al. [14]

In the context of the high T_C, the idea of superconductivity in multi-chain (or "ladder") systems was kicked off theoretically in 1986, when Schulz [15] proposed a possible relation between ladders at half-filling and Haldane's conjecture for spin chains. He made the following reasoning: If we consider repulsively interacting electrons on a ladder, the undoped system will be a Mott insulator, so that

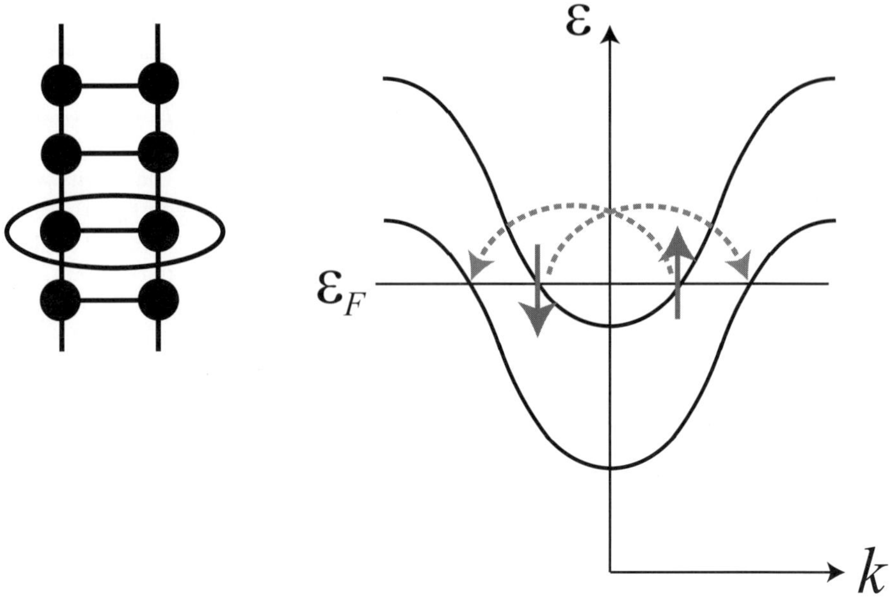

Fig. 2. Multi-band 1D model (*right panel*) for the Hubbard model on a ladder (*left*). An oval on the ladder represents the inter-chain pairing

we may consider the system as an $S = 1/2$ antiferromagnetic (AF) Heisenberg magnet on a ladder for large Hubbard repulsion U. Schulz's analysis [15] is that an AF $S = n/2$ single chain, which is exactly the Haldane's system [16,17], is similar to an $S = 1/2$ AF ladder with n-legs. For the spin chains, Haldane [16] has conjectured that the spin excitation should be gapless for half-odd-integer spins (n: odd) or gapful for integer spins (n: even). If the situation is similar in ladders, a ladder having an even number of legs will have a spin gap, associated with a 'spin-liquid' ground state where the quantum fluctuation is so large that the AF correlation decays exponentially.

Dagotto et al. [18] and Rice et al. [19] then suggested the possibility of superconductivity associated with the spin gap. The presence of a spin gap, i.e., a gap in the spin excitation which is indicative of a quantum spin liquid, in the two-leg ladder (or, more generally, in even legs) is a good news for superconductivity, since an idea proposed by Anderson [20] in the context of the high-T_C superconductivity suggests that a way to obtain superconductivity is to carrier-dope spin-gapped systems. The superconductivity in even-leg ladders is in accord with this. Subsequently superconductivity has been reported for a cuprate with a ladder structure [21], although it later turned out that this material has a rather strong two-dimensionality that may dominate the superconductivity.

For doped systems, the conjecture for superconductivity [19] is partly based on an exact diagonalisation study for finite t-J models on a two-leg ladder [30].

This was then followed by analytical [31] and numerical [32–35] works on the doped t-J ladder, for which the region for the dominant pairing correlation appears at a lower side of the exchange coupling J than in the case of a single chain.

On the other hand, the Hubbard model on a ladder is of general interest [36]. Although the Hubbard crosses over to the t-J model for $U \to \infty$, we have only an infinitesimal J there, so the result for the t-J model does not directly answers this. Since there is no exact solution for the Hubbard ladder, we can proceed in two ways: for small U we can adopt an analytic method, which is the weak-coupling renormalisation-group theory, where the band structure around the Fermi points is linearised in the continuum limit to treat the interaction with a perturbative renormalisation group. The weak-coupling theory with the bosonisation and renormalisation-group techniques has been applied to the two-leg Hubbard ladder [22–27].

The Hamiltonian of the two-leg Hubbard ladder is given in standard notations as

$$\begin{aligned}\mathcal{H} = -t \sum_{\alpha\langle i\rangle\sigma} (c_{i\sigma}^{\alpha\dagger} c_{i+1\sigma}^{\alpha} + \text{h.c.}) - t_{\perp} \sum_{i\sigma} (c_{i\sigma}^{1\dagger} c_{i\sigma}^{2} + \text{h.c.}) \\ + U \sum_{\alpha i} n_{i\uparrow}^{\alpha} n_{i\downarrow}^{\alpha},\end{aligned} \tag{1}$$

where $\alpha (= 1, 2)$ specifies the chains, or in the momentum space as

$$\begin{aligned}\mathcal{H} = -2t \sum_{\mu k\sigma} \cos(k) c_{k\sigma}^{\mu\dagger} c_{k\sigma}^{\mu} - 2t_{\perp} \sum_{k\sigma} c_{k\sigma}^{0\dagger} c_{k\sigma}^{0} \\ + U \sum (\text{interaction of the form } c^{\dagger} c^{\dagger} cc),\end{aligned} \tag{2}$$

where μ specifies the bonding ($\mu = 0$) and anti-bonding ($\mu = \pi$) bands, so labelled since $k_y = 0, \pi$, respectively.

The part of the Hamiltonian, H_{d}, that can be diagonalised in the bosonisation includes only intra- and inter-band forward-scattering processes arising from the intrachain forward-scattering terms. We can then define bosonic operators as in the single-chain case. If we introduce the phase variables as in the single-chain case, H_{d}, written in terms of them, is separated into the spin-part H_{spin} and the charge-part H_{charge}. While H_{spin} is already diagonalised, H_{charge} can be made so with a linear transformation, and the diagonalised H_{charge} is written in terms of the correlation exponent $K_{\rho i}, i = 1, 2$. So we end up with the total Hamiltonian that reads

$$H = \text{kinetic energy} + H_{\text{d}} + \text{pair-hopping terms}. \tag{3}$$

Here, the pair-hopping (or pair-scattering) term represents those part of the interaction Hamiltonian, in which a pair of electrons is scattered via the interaction to another pair of electrons.

At half-filling, the system reduces to a spin-liquid insulator having both charge and spin gaps [23]. When carriers are doped to the two-leg Hubbard ladder, on the other hand, the relevant scattering processes at the fixed point in the renormalisation-group flow are the pair hopping across the bonding and anti-bonding bands (Fig. 2), $c_{\uparrow}^{\pi\dagger} c_{\downarrow}^{\pi\dagger} c_{\downarrow}^{0} c_{\uparrow}^{0}$ + h.c. in k space, and the backward-scattering process within each band. The importance of the pair-hopping across the two bands for the dominance of pairing correlation in the two-leg Hubbard ladder is reminiscent of the Suhl-Kondo mechanism, which was proposed back in the 1950's for superconductivity in a quite different context of the s-d model for the transition metals [37,38].

The renormalisation results in a formation of gaps in both of the two spin modes and a gap in one of the charge modes. This leaves one charge mode massless, where the mode is characterised by a critical exponent K_ρ. Then the correlation of the intraband singlet pairing,

$$\sum_{\sigma} \sigma (c_{k\sigma}^{0} c_{-k,-\sigma}^{0} - c_{k\sigma}^{\pi} c_{-k,-\sigma}^{\pi}), \tag{4}$$

decays like $1/r^{1/(2K_\rho)}$, where K_ρ should be close to unity in the weak-coupling regime. So this should be the dominant phase, which is, expressed in real space as $c_{i\sigma}^{1} c_{i,-\sigma}^{2} - c_{i,-\sigma}^{1} c_{i\sigma}^{2}$, an *interchain* singlet pairing.

2.3 How to Detect Pairing in Quantum Monte Carlo Studies?

The perturbational renormalisation group is in principle guaranteed to be valid only for sufficiently small interaction strengths ($U \ll t$), so that its validity for finite $U(\sim t)$ has to be checked. To be more precise, the renormalisation approach can tell whether the interaction flows into weak coupling (with the relevant mode gapless) or into strong-coupling regime (gapful) for small enough interactions, but the framework itself (i.e., the perturbational expansion) might fail for stronger interactions.

This is where numerical studies come in. Numerical calculations for finite U have been performed with the exact diagonalisation, DMRG or quantum Monte Carlo (QMC) methods [39–44], but in an earlier stage the results are scattered, where some of the results seemed inconsistent with the weak-coupling prediction: a DMRG study by Noack et al. for the doped Hubbard ladder shows the enhancement of the pairing correlation over the $U = 0$ result strongly depends on the inter-chain hopping, $t_\perp$ [28,45]. Quantum Monte Carlo (QMC) results also exhibit an absence [46] or presence [47] of the enhancement depending on the hopping parameters and/or band filling.

Recently, however, a QMC study by Kuroki et al. [48] has resolved the puzzle, and has clearly detected an enhanced pairing correlation. A key factor found there in detecting superconductivity in any numerical calculation, which also resolves the origin of the former discrepancies, is: we have to question a very tiny energy scale ($\ll$ starting electronic energy scale, t, U) in detecting the pairing. This immediately implies that the discreteness of energy levels in finite systems examined in QMC studies enormously affects the pairing correlation – If the level separation is greater than the energy scale we want to look at, any feature in the

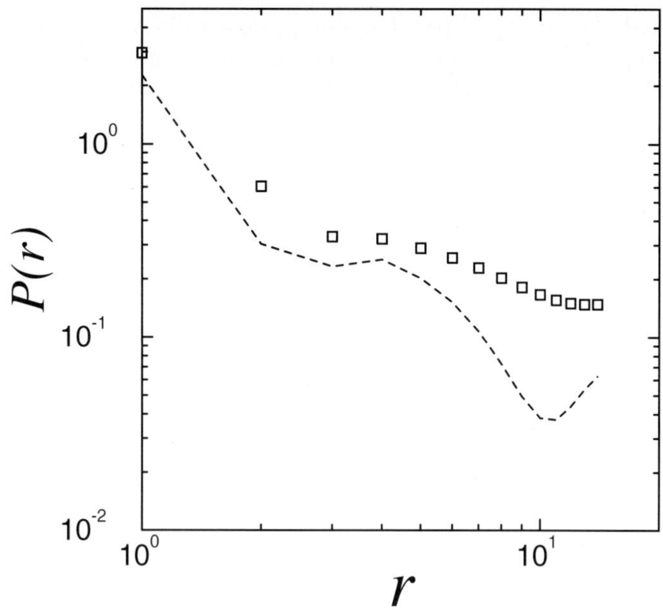

Fig. 3. The pairing correlation function plotted against the real space distance r in a 30-rung Hubbard ladder having 56 electrons for $U = 1$ with $t_\perp = 1.975$ (*square*) [48]. The *dashed line* is the non-interacting result

correlation function will be easily washed out. This can be circumvented if we tune the parameters so as to make the separation between the levels just below and above E_F tiny (i.e., to make the LUMO-HOMO nearly degenerate in the quantum chemical language), which should be a reasonable way to approach the bulk limit where the levels are dense. The importance of small offsets between the highest occupied and lowest unoccupied levels has also been stressed by Yamaji et al. for small systems [47].

So we have applied the (projector) Monte Carlo method [49] to look into the ground state correlation function $P(r) \equiv \langle O_{i+r}^\dagger O_i \rangle$ of this pairing for finite values of $U(\sim t)$. We show in Fig. 3 the result for $P(r)$ for $t_\perp = 0.98$, $U = 1$ and the band filling $n = 0.867 = 52$ electrons/ (30 rungs $\times$ 2 sites). The $U = 0$ result (dashed line) for these two values of $t_\perp$ are identical because the Fermi sea remains unchanged. If we turn on U, we can see that a large enhancement over the $U = 0$ result emerges at large distances.

We have deliberately chosen the value of $t_\perp = 0.98$ to make the one-electron energy levels of the bonding and anti-bonding bands lie close to each other around the Fermi level within $0.004t$. This is much smaller than the energy scale we question (which is the spin gap in the present 1D case). In fact, a 5% change in $t_\perp = 0.98 \to 1.03$, for which the LUMO-HOMO separation blows up to $\sim 0.1t$, washes out the enhancement in the correlation function. In the latter case the renormalisation of higher energy modes has to stop at this energy scale, so that the interband pair hopping process will not be renormalised into a strong coupling, while in the weak-coupling theory the renormalisation all the way down to the Fermi level is assumed.

While it is difficult to determine the decay exponent of the pairing correlation $P(r)$, we can fit the data by assuming a trial function expected from the weak-coupling theory, $P(r) \propto c/\sqrt{r} + (1-c)/r^2 + [\cos(2k_F^0 r) + \cos(2k_F^\pi r)]/r^2$, where the overall decay at large distances is assumed to be $\propto 1/\sqrt{r}$ as dictated in the weak-coupling theory. This form reproduces the result surprisingly accurately.

3 Three-leg Ladder and 1D-2D Crossover

Now, the physics of ladders can provide quite an instructive line of approach for understanding the physics in two-dimensional systems via the crossover from 1D to 2D (two-dimensions). So let us first look at the three-leg ladder.

3.1 Three-Leg Ladder

If we return to ladders, one can naively expect that ladders with odd-number (e.g., 3) of legs will have no spin gap, which would then signify an absence of dominating pairing correlation ('even-odd conjecture for superconductivity'). As far as the spin gap in undoped ladders is concerned, experiments on a class of cuprates, $Sr_{n-1}Cu_nO_{2n-1}$ having n-leg ladders, have supported the conjecture [51–55]. So it was believed that odd-numbered legs only have the usual $2k_F$ spin-density wave (SDW) rather than superconductivity.

Kimura et al. [56,57], however, showed that that is too simplistic a view, and that, while the even-odd conjecture for the spin gap is certainly correct, an odd-number of legs does indeed superconduct by exploiting the spin-gapped mode. In that work the pairing correlation in the three-leg Hubbard ladder has been examined [50].

We start with the weak-coupling theory for correlation functions for the three-leg Hubbard ladder (Fig. 4). Arrigoni has looked into a three-leg ladder with weak Hubbard-type interactions with the perturbational renormalisation-group technique to conclude that gapless and gapful spin excitations coexist in three legs [58].

He has actually enumerated the numbers of gapless charge and spin modes on the phase diagram spanned by the doping level and the interchain hopping, $t_\perp$. He found that, at half-filling, one gapless spin mode exists. For general band filling, one gapless spin mode remains in the region where the Fermi level intersects all the three bands in the noninteracting case. From this, Arrigoni argues that the $2k_F$ SDW correlation should decay as a power law as expected from experiments. Arrigoni's result indicates that two gapful spin modes exist in addition. The charge modes, on the other hand, consists of two gapless modes and one gapful mode.

The question we address then is what happens when gapless and gapful spin modes *coexist*. This is an intriguing problem, since it may well be possible that

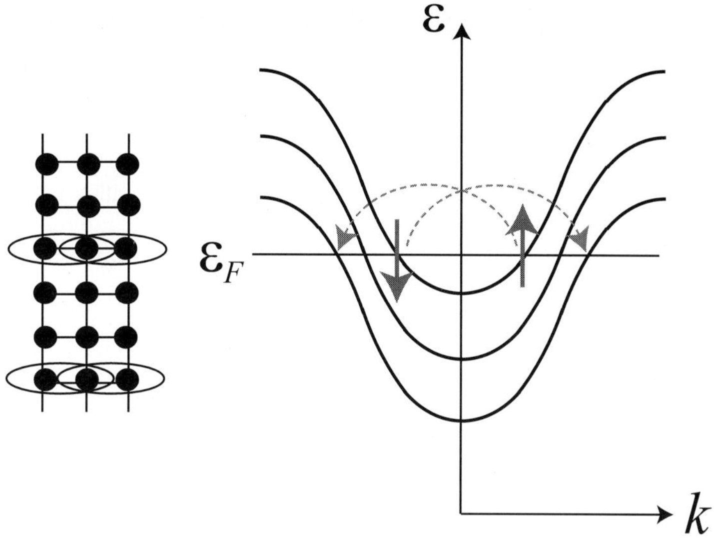

Fig. 4. Three-leg ladder (*inset*) and its dispersion. The *dashed arrows* represent the pair-hopping process, and *ovals* represent the inter-chain pairing

the presence of gap(s) in *some* out of multiple spin modes may be sufficient for the dominance of a pairing correlation. Schulz [59] has independently shown similar results for a subdominant $2k_F$ SDW and the interchain pairing correlations.

Physically, the picture that emerges as we shall describe below, is that the two spin gaps, which are relevant to the pairing, arise as an effect of the pair-hopping process that is the many-body matrix element transferring two electrons simultaneously across the outermost bands (i.e., the top and bottom bands for a three-leg ladder) (Fig. 4). In this sense the mechanism is reminiscent of the situation in the two-leg case or the Suhl-Kondo mechanism [37,38].

The correlation functions can be calculated with the bosonisation method [5] for the three-leg Hubbard model.

We can define three bosonic operators, where we diagonalise H_{charge} in terms of the three correlation exponents $K_{\rho i}, i = 1, 2, 3$. As Arrigoni pointed out [58] the pair-hopping processes across the top and bottom bands become relevant as the renormalisation is performed. In order to actually calculate the correlation functions, we have to express the relevant scattering processes in terms of the phase variables. The fixed-point Hamiltonian density, H^*, takes the form, in terms of the phase variables,

$$\begin{aligned} H^* \propto & - g_{\text{back}}(1)\cos[2\phi^{1+}(x)] - g_{\text{back}}(3)\cos[2\phi^{3+}(x)] \\ & + 2g_{\text{pairhopping}}(1,3)\cos[\sqrt{2}\chi^{1-}(x)]\sin\phi^{1+}(x)\sin\phi^{3+}(x), \end{aligned} \tag{5}$$

where $g_{\text{back}}(1), g_{\text{back}}(3)$ are negative large quantities, and $g_{\text{ph}}(1,3)$ is a positive large quantity. This indicates the following. Two spin phases, ϕ^{1+}, ϕ^{3+}, become long-range ordered and fixed, respectively, while ϕ^{2+} is not fixed to give a gapless spin mode. Similarly, the difference in the charge phases, θ^i, for the outermost bands,

$$\chi^{1-} \equiv \frac{1}{\sqrt{2}}(\theta^{1-} - \theta^{3-}), \tag{6}$$

is ordered and fixed, and the charge gap opens for this particular mode.

Now we can calculate the correlation functions, since the gapless fields have already been diagonalised, while the remaining gapful fields have the respective expectation values.

Among various order parameters, the dominant one (with the longest tail in the correlation) is the singlet pairing across the central and edge chains (Fig. 4), which is, in the band picture,

$$O_d \sim \sum_\sigma \sigma(\psi_\sigma^{1+}\psi_{-\sigma}^{1-} - \psi_\sigma^{3+}\psi_{-\sigma}^{3-}). \tag{7}$$

We call this "d" for the following reason. Since we have taken a continuum limit along the chain, it is not straightforward to name the symmetry of a pairing. However, we could call the above pair as d-wave-like in that the pairing, in addition to being off-site on the rung, is a linear combination of a bonding band and an anti-bonding band with opposite signs. Since the relevant pair-hopping is across these bands, we can say that there is a node in the pair wavefunction along the line that bisects the relevant pair-hopping.

Calculation of the correlation function of O_d gives

$$\langle O_d(x)^\dagger O_d(0)\rangle \sim x^{-\frac{1}{3}\left(\frac{1}{K^*_{\rho 2}}+\frac{1}{2K^*_{\rho 3}}\right)}, \tag{8}$$

which is the dominant ordering. In the weak-interaction limit ($U \to +0$), where all the K^*'s tend to unity, the "d"-pairing correlation decays as slowly as $x^{-1/2}$, while the other correlations decay like x^{-2}. We can see that the interchain pairing *exploits* the charge gap and the spin gaps to reduce the exponent of the correlation function, in contrast to the intrachain pairing. Namely, we would have (-1) in the exponent if the spin were gapless. This alone would only result in a $1/r$ decay, but the charge mode χ_{1-} is further locked, which further reduces the exponent (down to $1/\sqrt{r}$ in the limit $U \to 0$).

Now, how does the pairing correlation in the three-leg Hubbard ladder look like when U is finite? Our QMC result for the three-leg Hubbard ladder exhibits an enhancement of the pairing correlation even for finite coupling constants, $U/t = 1 \sim 2$ [57]. As in the two-leg case with a finite U, we have taken care that levels below and above the Fermi level are close.

3.2 1D-2D Crossover

We have seen that the weak-coupling theory (perturbational renormalisation + bosonisation) predicts that the interband pair hopping between the innermost

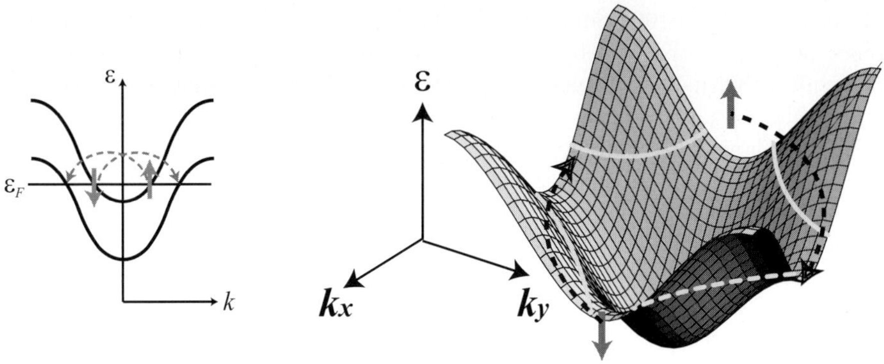

Fig. 5. Band dispersion of the square lattice, on which the Fermi surface for E_F close to half filling and a typical pair-hopping process (*dashed arrows*) are shown. Similar plot for a ladder is attached for comparison

and outermost Fermi points $(k_x, k_y) \approx (\pm k_F^0, 0), (\pm k_F^\pi, \pi)$ becomes relevant (i.e., increases with the renormalisation). This concomitantly makes the two-point correlation of the interchain singlet decay slowly with distance. In k-space the dominant component of this pair reads

$$\sum_\sigma \sigma\ (c^0_{k_F^0,\sigma} c^0_{-k_F^0,-\sigma} - c^\pi_{k_F^\pi,\sigma} c^\pi_{-k_F^\pi,-\sigma}). \tag{9}$$

Now, when E_F intersects the outermost-band top and the innermost-band bottom with $k_F^0 \simeq \pi, k_F^\pi \simeq 0$ (Fig. 5 left), *intrachain nearest-neighbor* singlet pair also has a dominant Fourier-component equal to eqn.(9) with a phase shift π relative to the interchain pairing. Thus, a linear combination which amounts to the $d_{x^2-y^2}$ pairing should become dominant. We shall see this is exactly what happens in the 2D square lattice around the half filling, Fig. 5 right.

4 Superconductivity from the Repulsive Interaction in 2D

As we have seen, the two-leg and three-leg Hubbard ladders do superconduct, so what will happen if we consider n-leg ladders for $n = 3, 4, ..., \infty$ to reach the 2D square lattice. This view enables us to have a fresh look at the 2D Hubbard model.

So we move on to the repulsive Hubbard model on the square lattice. The seminal notion that the high-T_C conductivity in cuprates should be related to strong electron correlation was first put forward by Anderson [61]. There the superconductivity is expected to arise from the pairing interaction mediated by spin fluctuations (usually antiferromagnetic). A phenomenology along this line such as the self-consistent renormalisation [62–65] has succeeded in reproducing anisotropic d-wave superconductivity. Microscopically, the repulsive Hubbard model, a simplest possible model for correlated electrons, should capture the physics in cuprates [60]. Some analytical calculations have suggested the occurrence of $d_{x^2-y^2}$-wave superconductivity in the 2D Hubbard model [66–70]. In particular, fluctuation exchange approximation (FLEX), developed by Bickers et al. [71], has also been applied to the Hubbard model on the square lattice [72,73] to show the occurrence of the superconductivity.

Numerical calculations have also been performed extensively. Finite binding energy [74,75] and pairing interaction vertex [76–78] were found in those calculations. Variational Monte Carlo calculations show that a superconducting order lowers the variational energy [79,80]. Nevertheless, there had been a reservation against the occurrence of superconductivity in the Hubbard model because the pairing correlation functions do not show any symptom of long-range behavior in some of the works [78,81,82].

Again, Kuroki et al. [48] showed for the first time that QMC does indeed exhibit symptoms of superconductivity if we take proper care of a small energy scale involved, i.e., the d-wave pairing correlation becomes long-tailed when the Fermi level lies between a narrowly separated levels residing on the k-points across which the dominant pair hopping occurs. An enhancement of the pairing correlation has in fact been found by exact diagonalisation [47] and by density matrix renormalisation group [45] when E_F lies close to the k-points $(0, \pi)$ and $(\pi, 0)$. Although the $d_{x^2-y^2}$-like nature of the pairing was suggested [45], $d_{x^2-y^2}$ pairing correlation itself has not been calculated. So, in our quest for 2D, we first calculate the correlation function with QMC. Here we employ the ground-state, canonical-ensemble QMC [83], where we take the free Fermi sea as the trial state.

4.1 Anisotropic Pairing in 2D

First, let us look at why the attractive interaction is by no means a necessary condition for superconductivity, which can quite generally arise from repulsive electron-electron interactions, which seems to be still not realised well enough. If we look at the BCS gap equation, we can immediately see that superconductivity can readily arise from repulsive interactions. The gap equation reads

$$1 = -V_\phi \sum_{\vec{k}'} \frac{1}{2\varepsilon(\vec{k}')} \tanh\left[\frac{1}{2}\beta\varepsilon(\vec{k}')\right],$$

$$V_\phi = \frac{\langle V(\vec{k,k'})\Delta(\vec{k})\Delta(\vec{k}')\rangle_{\mathrm{FS}}}{\langle \Delta^2(\vec{k})\rangle_{\mathrm{FS}}}, \tag{10}$$

where $\varepsilon(\vec{k})$ is the band energy measured from the chemical potential, $\beta = 1/(k_B T_C)$, $V(\vec{k,k'})$ the pair-hopping matrix element, $\Delta(\vec{k})$ the BCS gap function, and $\langle ... \rangle_{\mathrm{FS}}$ is the average over the Fermi surface. So, if $\Delta(\vec{k})$ has nodes across $\vec{k} \leftrightarrow \vec{k}'$ (i.e., changes sign before and after the pair hopping), the originally repulsive $V > 0$ acts effectively as an attraction,

$$\langle V(\vec{k,k'})\Delta(\vec{k})\Delta(\vec{k}')\rangle_{\mathrm{FS}} < 0.$$

This most typically happens for the $d_{x^2-y^2}$ pairing with $\Delta(\vec{k}) \propto \cos(k_x) - \cos(k_y)$ when the dominant pair hopping occurs across $\vec{k} \sim (0, \pi) \leftrightarrow \vec{k}' \sim (\pi, 0)$.

When the spin fluctuation is antiferromagnetic, most typically in bipartite lattices such as a square lattice, the importance of the interactions around $(0, \pi)$ and $(\pi, 0)$ in the 2D Hubbard model has been suggested by various authors [67–70,77,80,85–88].

Group theoretically, the square lattice has a tetragonal symmetry, so that everything, including the gap function, should be an irreducible representation of the tetragonal group. The $d_{x^2-y^2}$ pairing indeed belongs to the B_{1g} representation of this group.

4.2 Quantum Monte Carlo Study for the 2D Hubbard Model

In the context of our QMC study for the 2D Hubbard model, we have to take finite systems that have the k-points around $(0, \pi)$ and $(\pi, 0)$ close in energy. Namely, our expectation from the study on ladders is that the pair hopping processes across around $(0, \pi)$ and $(\pi, 0)$ may result in $d_{x^2-y^2}$ pairing, $\sum_{\vec{k}}[\cos(k_x) - \cos(k_y)]c_{\vec{k}\uparrow}c_{-\vec{k}\downarrow}$ in 2D, but an enhanced pairing correlation should be detected only when the level offset between the discrete levels around those points is small.

We take 78 electrons in 10×10 sites ($n = 0.78$) with $t_y = 0.999$ with periodic boundary condition in both directions. We have taken $t_y = 0.999$, because the number of electrons considered here would have an open shell (with a degeneracy in the free-electron Fermi sea) for $t_y = 1$, which will destabilise QMC convergence. Taking $t_y = 0.999$ lifts the degeneracy to give a tiny (< 0.01) but finite $\Delta\varepsilon^0$. In Fig. 6 we plot the $d_{x^2-y^2}$ pairing correlation, defined as

$$
\begin{aligned}
P(r) &= \sum_{|\Delta x|+|\Delta y|=r} \langle O^\dagger(x+\Delta x, y+\Delta y) O(x,y) \rangle, \\
O(x,y) &= \sum_{\delta=\pm 1,\sigma} \sigma (c_{x,y,\sigma} c_{x+\delta,y,-\sigma} - c_{x,y,\sigma} c_{x,y+\delta,-\sigma}),
\end{aligned} \tag{11}
$$

where the correlation for $U = 1$ is clearly seen to be enhanced over that for $U = 0$ especially at large distances.

We can readily show that when the level spacing becomes too large (e.g., $\Delta\varepsilon^0 \sim 0.1$), the enhancement is washed out. In the present choice the energy levels around $(0, \pi), (\pi, 0)$ are close ($< 0.01t$), while the other levels lie more than $\sim 0.1t$ away from E_F^0. One might thus raise a criticism that the scattering processes involving the states away from E_F^0 are unduly neglected. We can however show (not displayed here) that when other levels exist around E_F an enhanced $d_{x^2-y^2}$ correlation is obtained as well.

How about the band-filling (n) dependence? We have calculated the long-range part of the correlation, $S \equiv \sum_{r \geq 3} P(r)$, for various values of n keeping $\Delta\varepsilon^0 < 0.01t$ throughout. The result, displayed in Fig. 7, shows that the enhancement in S for $U = 1$ has a maximum around a finite doping. Thus the message

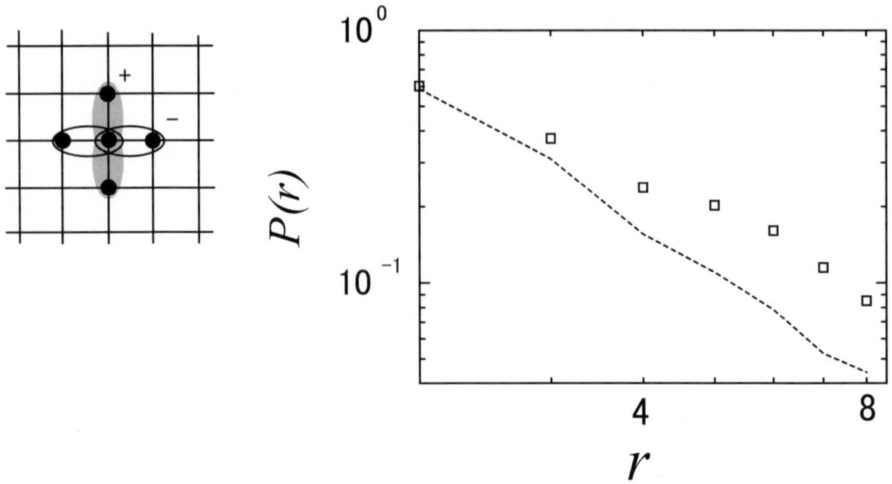

Fig. 6. QMC result for the $d_{x^2-y^2}$ pairing correlation for a 10×10 square lattice with 78 electrons for $U = 1$ and $t_y = 0.999$ (*square*) [48]. The *dashed line* represents the noninteracting case. Inset depicts the square lattice with the ovals representing a d-wave pair

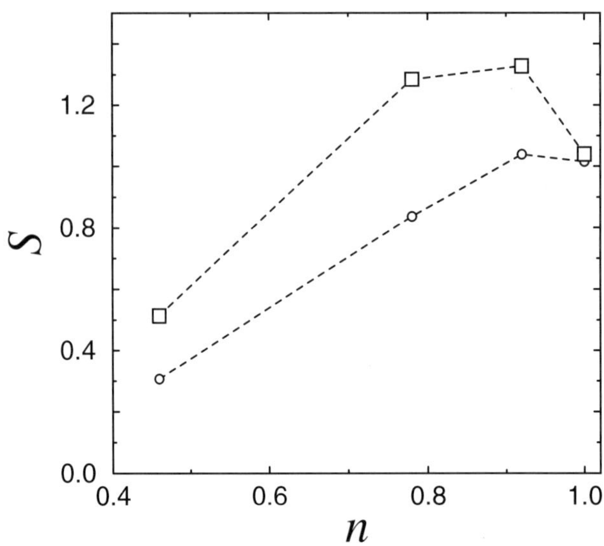

Fig. 7. The integrated pairing correlation plotted against the band filling n for a 10×10 lattice with $U = 0$ (○) or $U = 1$ (*square*) [48]

here is that the $d_{x^2-y^2}$ pairing is favoured *near, but not exactly at*, half-filling. The fact that the better nesting does not necessarily imply the more enhanced pairing correlation has also been shown in another numerical work in the context of organic superconductivity [89].

5 Which is More Favourable for Superconductivity, 2D or 3D?

The theoretical results described so far indicate that the superconductivity near the AF instability in 2D has a 'low T_C' $\sim O(0.01t)$ (t: transfer integral), i.e., two orders of magnitude smaller than the original electronic energy, but still 'high T_C' $\sim O(100\text{ K})$ for $t \sim O(1\text{ eV})$. So, identifying the conditions for higher T_C in the repulsive Hubbard model is one of the most fascinating goals of theoretical studies. For instance, while the high-T_C cuprates are layer-type materials with Cu_2 planes in which the supercurrent flows, the question is whether the two-dimensionality is promoting or degrading the superconductivity.

Hence Arita et al. [90] have questioned: (i) Is a 2D system more favourable for spin-fluctuation mediated superconductivity than in three dimensions (3D)? (ii) Can other pairing, such as a triplet p-pairing in the presence of ferromagnetic spin fluctuations, become competitive? We take the single-band, repulsive Hubbard model as a simplest possible model, and look into the pairing with the FLEX method in ordinary (i.e., square, trianglar, fcc, bcc, etc) lattices in 2D and 3D. The FLEX method has an advantage that systems having large spin fluctuations can be handled.

As for 3D systems, Scalapino et al. [91] showed for the Hubbard model that paramagnon exchange near a spin-density wave instability gives rise to a strong singlet d-wave pairing interaction, but T_C was not discussed there. Nakamura et al. [92] extended Moriya's spin fluctuation theory of superconductivity [63] to 3D systems, and concluded that T_C is similar between the 2D and 3D cases provided that common parameter values (scaled by the band width) are taken. However, the parameters there are phenomelogical ones, so we wish to see whether the result remains valid for microscopic models.

As for the triplet pairing, the possibility of triplet pairing mediated by ferromagnetic fluctuations has been investigated for superfluid ^{3}He [93], a heavy fermion system UPt_3 [94], and most recently, an oxide Sr_2RuO_4 [95]. It was shown that ferromagnetic fluctuations favour triplet pairing first by Layzer and Fay [2] before the experimental observation of p-wave pairing in ^{3}He. For the electron gas model, Fay and Layzer [2] and later Chubukov [96] have extended the Kohn-Luttinger theorem [1] to p-pairing for 2D and 3D electron gas in the dilute limit. Takada [97] discussed the possibility of p-wave superconductivity in the dilute electron gas with the Kukkonen-Overhauser model [98]. As for lattice systems, the 2D Hubbard model with large enough next-nearest-neighbor hopping (t') has been shown to exhibit p-pairing for small band fillings [99]. Hlubina [100] reached a similar conclusion by evaluating the superconducting vertex in a perturbative way [101]. However, the energy scale of the p-pairing in the Hubbard model, i.e., T_C, has not been evaluated so far.

Here we show that (i) d-wave instability mediated by AF spin fluctuation in 2D square lattice is much stronger than those in 3D, while (ii) p-wave instability mediated by ferromagnetic spin fluctuations in 2D are much weaker than the d-instability. These results, which cannot be predicted a priori, suggest that for the Hubbard model the 'best' situation for the pairing instability is the 2D case with dominant AF fluctuations.

We consider the single-band Hubbard model with the transfer energy $t_{ij} = t(= 1$ hereafter) for nearest neighbors along with $t_{ij} = t'$ for second-nearest neighbors, which is included to incorporate the band structure dependence. The FLEX starts from a set of skeleton diagrams for the Luttinger-Ward functional to generate a (k-dependent) self energy based on the idea of Baym and Kadanoff [102]. Hence the FLEX approximation is a self-consistent perturbation approximation with respect to on-site interaction U.

To obtain T_C, we solve the eigenvalue (Éliashberg) equation,

$$\lambda \Sigma^{(2)}(k) = \frac{T}{N} \sum_{k'} \Sigma^{(2)}(k') |G(k')|^2 V^{(2)}(k-k'), \tag{12}$$

where $\Sigma^{(2)}(k)$ is the anomalous self energy, $k \equiv (\vec{k}, i\omega_n)$ with $\omega_n = (2n-1)\pi T$ being Matsubara frequencies, and the pairing interaction, $V^{(2)}$, comprises contributions from the transverse spin fluctuations, longitudinal spin fluctuations and charge fluctuations, namely,

$$\begin{aligned} V^{(2)}(k,k') &= -U^2 \left[\frac{1}{2}\chi_{\mathrm{ch}}(k-k') \right. \\ &\quad \left. -\frac{1}{2}\chi^{zz}(k-k') + \chi^{\pm}(k+k') \right] \\ &= -\left[\frac{U^3\chi_{\mathrm{irr}}^2(k-k')}{1-U^2\chi_{\mathrm{irr}}^2(k-k')} \right] - \left[\frac{U^2\chi_{\mathrm{irr}}(k+k')}{1-U\chi_{\mathrm{irr}}(k+k')} \right] \end{aligned}$$

Here χ_{ch} is the charge susceptibility, $\chi^{zz}(\chi^{\pm})$ the longitudinal (transverse) spin susceptibility, and

$$\chi_{\mathrm{irr}}(q) \equiv -(T/N) \sum_k G(k)G(k+q)$$

the irreducible susceptibility constructed from the dressed Green's function. The dressed Green's function, $G(k)$, obeys the Dyson equation,

$$G(k)^{-1} = G^0(k)^{-1} - \Sigma(k), \tag{13}$$

where G^0 is the bare Green's function, and Σ the self energy with

$$\Sigma(k) = \frac{1}{N} \sum_q G(k-q) V^{(1)}(q). \tag{14}$$

If we take RPA-type bubble and ladder diagrams for the interaction $V^{(1)}$, we have

$$\begin{aligned} V^{(1)}(q) &= \frac{1}{2}U^2\chi_{\mathrm{irr}}(q) \left[\frac{1}{1+U\chi_{\mathrm{irr}}(q)} \right] \\ &\quad + \frac{3}{2}U^2\chi_{\mathrm{irr}}(q) \left[\frac{1}{1-U\chi_{\mathrm{irr}}(q)} \right] - U^2\chi_{\mathrm{irr}}(q), \end{aligned}$$

which completes the set of equations.

Since we have $\Sigma^{(2)}(k) = \Sigma^{(2)}(-k)$ for the spin-singlet pairing whereas $\Sigma^{(2)}(k) = -\Sigma^{(2)}(-k)$ for the spin-triplet pairing, $V^{(2)}(k,k')$ becomes a function of $k - k' = q$ with

$$V^{(2)}(q) = -\frac{3}{2}\left[\frac{U^2\chi_{\rm irr}(q)}{1-U\chi_{\rm irr}(q)}\right] + \frac{1}{2}\left[\frac{U^2\chi_{\rm irr}(q)}{1+U\chi_{\rm irr}(q)}\right] \tag{15}$$

for the singlet pairing, and

$$V^{(2)}(q) = \frac{1}{2}\left[\frac{U^2\chi_{\rm irr}(q)}{1-U\chi_{\rm irr}(q)}\right] + \frac{1}{2}\left[\frac{U^2\chi_{\rm irr}(q)}{1+U\chi_{\rm irr}(q)}\right] \tag{16}$$

for the triplet pairing. $T = T_C$ is identified as the temperature at which the maximum eigenvalue $\lambda_{\rm Max}$ reaches unity.

Let us start with the 2D case having strong AF fluctuations. We have first obtained $\chi_{\rm RPA}(q) = \chi_0/(1 - U\chi_0)$ as a function of the momentum for the Hubbard model on a nearly half-filled ($n = 0.85$) square lattice, where a dominant AF spin fluctuation is seen as $\chi_{\rm RPA}$ peaked around (π, π). We can then plug this into the Éliashberg equation (12) to plot in Fig. 8a $\lambda_{\rm Max}$ as a function of temperature T.

T_C is identified as the temperature at which $\lambda_{\rm Max}$ becomes unity, which occurs at $T \sim 0.02$ for the square lattice, in accord with previous results [72,103].

If we move on to the case with ferromagnetic spin fluctuations where triplet pairing is expected, this situation can be realised for relatively large $t'(\simeq 0.5)$ and away from half-filling in the 2D Hubbard model. Physically, the van Hove singularity shifts toward the band bottom with t', and the large density of states at the Fermi level for the dilute case favours the ferromagnetism. We have found that $\lambda_{\rm Max}$ becomes largest for $n = 0.3$, $t' = 0.5$.

$\chi_{\rm RPA}$ is indeed peaked at Γ ($\vec{k} = (0,0)$). The question then is the behavior of $\lambda_{\rm Max}$ as a function of T, which shows that $\lambda_{\rm Max}$ is much smaller than that in the AF case.

A low T_C for the ferromagnetic case contrasts with a naive expectation from the BCS picture, in which the Fermi level located around a peak in the density of states favours superconductivity. We may trace back two-fold reasons why this does not apply. First, if we look at the dominant ($\propto 1/[1 - U\chi_0(q)]$) term of the pairing potential $V^{(2)}$ itself in eqs. (15) and (16), the triplet pairing interaction is only one-third of that for singlet pairing. Second, the factor $|G|^2$ for the ferromagnetic case is smaller than that in the AF case, which implies that the self-energy correction is larger in the former. Larger self-energy (smaller $|G|^2$) works unfavourably for superconductivity as seen in the Éliashberg equation (12). When we take a larger repulsion U to increase the triplet pairing attraction (susceptibility), this makes the self-energy correction even stronger.

Let us now move on to the case of d-wave pairing in the 3D Hubbard model, for which FLEX was first applied by Arita et al. [90]. In simple-cubic systems, we find that the Γ_3^+ representation of O_h group [31] has the largest $\lambda_{\rm Max}$. We have found that $\lambda_{\rm Max}$ for this symmetry becomes largest for $n = 0.8$, $t' = -0.2 \sim -0.3$

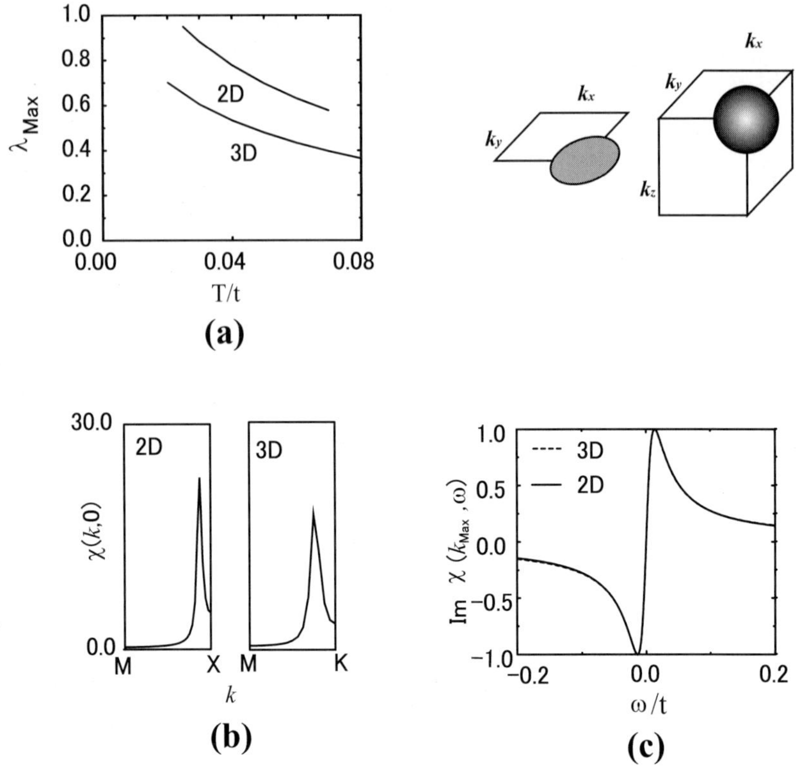

Fig. 8. **a** The maximum eigenvalue of the Éliashberg equation against temperature for the Hubbard model on a square lattice with $n = 0.85$ and $U = 4$ and on a cubic lattice with $n = 0.8$, the second-neighbour hopping $t' = -0.2$ and $U = 8$. $\chi(\boldsymbol{k}, 0)$ **b** as a function of wavenumber and $\mathrm{Im}\chi(\boldsymbol{k}_{\mathrm{Max}}, \omega)$ **c** (normalised by its maximum) as a function of ω/t at $T = 0.03t$ are also shown for the two lattices. Top right panel schematically shows the regions, in respective dimensions, that contribute to the pairing

and $U = 8 \sim 10$. In Fig. 8a, we superpose λ_{Max} as a function of T, where we can immediately see that the pairing tendency in 3D is much *weaker* than that in 2D.

Why is the d-superconductivity much stronger in 2D than in 3D? We can pinpoint the origin by looking at the various factors involved in the Éliashberg equation. Namely we question the height of $V^{(2)}$ and $|G|^2$ along with the width of the region, both in the momentum sector and in the frequency sector, over which $V^{(2)}(k)$ contributes to the summation over $k \equiv (\vec{k}, i\omega_n)$. We found that the maximum of $|G|^2$ is in fact larger in 3D than in 2D. The width of the peak in χ_{RPA} on the frequency and momentum axes is surprisingly similar between 2D and 3D as displayed in Fig. 8b,c. Note that if the frequency spread of the susceptibility scaled not with t but with the *band width*, as Nakamura et al. [92] have assumed, λ_{Max} would have become larger. Now, λ in the Éliashberg equation (12) is $\propto (a/L)^D$, where L is the linear dimension of the system and

a the width in the momentum space for the effective attraction, this factor is much smaller in 3D than in 2D as far as the main contribution of $V^{(2)}$ to the pairing occurs through special points in the k-space (e.g., (π, π) or (π, π, π) for the antiferromagnetic spin fluctuation exchange pairing). So we can conclude that this is the main reason why 2D is more favourable than 3D.

To summarise this section, d-pairing in 2D is the best situation for the repulsion originated (i.e., spin fluctuation mediated) superconductivity in the Hubbard model. Monthoux and Lonzarich [104] have also concluded for 2D systems, by making use of a phenomenological approach, that the d-wave pairing is much stronger than p-wave pairing, which is consistent with the present result. In this sense, the layer-type cuprates do seem to hit upon the right situation.

This is as far as one-band models having simple Fermi surfaces are concerned. Indeed, if we turn to heavy fermion superconductors, for instance, in which the pairing is thought to be meditated by spin fluctuations, the T_C, when normalized by the band width W, is known to be of the order of $0.001W$. Since the present result indicates that T_C, normalized by W, is $\sim 0.0001W$ at best in the 3D Hubbard model, we may envisage that the heavy fermion system must exploit other factors such as the multiband. Neverthless, a recent experimental finding [105] that a heavy-fermion compound Ce(Rh,Ir,Co)In_5 has the higher T_c for the more two-dimensional lattice (with larger c/a) is consistent with our prediction.

6 How to Realise Higher T_C in Anisotropic Pairing – Disconnected Fermi Surfaces

Ironically, the main question about the superconductivity from the electronic mechanism is "why is T_C so low?", which has been repeatedly raised in literatures. Namely, one remarkable point is $T_c \sim O(0.01t)$, estimated for the repulsive Hubbard model in the two-dimensional (2D) square lattice, is *two orders of magnitudes smaller than* the starting electronic energy (i.e., the hopping integral t), although this gives the right order for the cuprates' T_c. We have seen that even the best case, as far as these ordinary lattices are concerned, has $T_c \sim O(0.01t)$. As discussed in [106], there are good reasons why T_c is so low: One reason is the effective attraction mediated by spin fluctuations is much weaker than the original electron-electron interaction, U. Another important reason is the presence of nodes in the superconducting gap function greatly reduces T_c: While the main pair-scattering, across which the gap function has opposite signs to make the effective interaction attractive, there are other pair scatterings around the nodes that have negative contributions to the effective attraction by connecting k-points on which the gap has the same sign.

So a next important avenue to explore is: can we improve the situation by going over to multiband systems. Kuroki and Arita [106] have shown that this is indeed the case if we have *disconnected Fermi surfaces*. In this case T_c is dramatically enhanced, because the sign change in the gap function can avoid the Fermi pockets, where all the pair-scattering processes contribute positively [106,107]. This has been numerically shown to be the case for the triangular

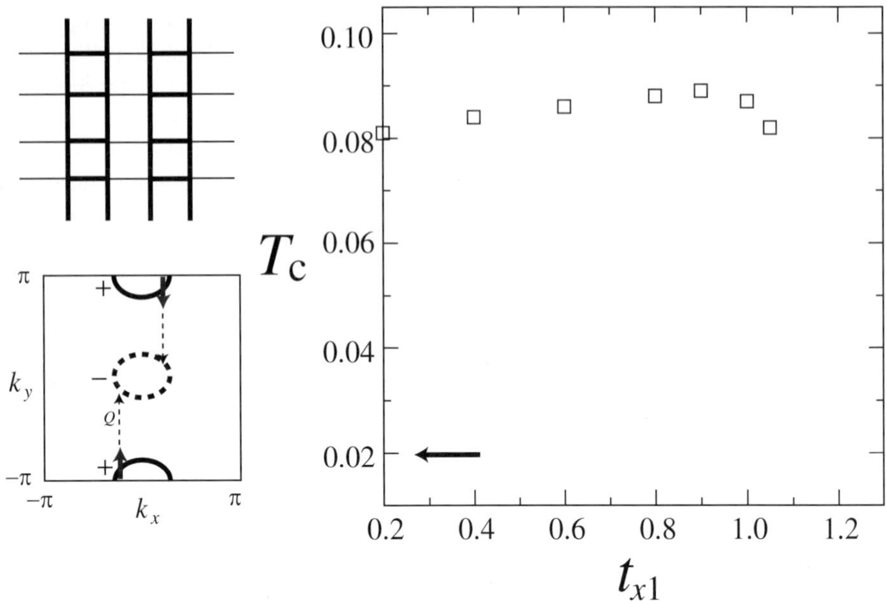

Fig. 9. The T_C estimated for the lattice depicted in the top left panel having disconnected Fermi surface (*bottom left*) as a function of the weak transfer (*thin lines in the top left panel*; t_{x1}) [106]. The *arrow* indicates typical T_C for ordinary (e.g., *square*) lattices

lattice (for spin-triplet pairing) [108] and a square lattice with a period-doubling [106], where T_c as estimated with FLEX is as high as $O(0.1t)$.

To be more precise, the key ingredients are: (a) when the Fermi surface is nested, the spin susceptibility $\chi(\vec{q}, \omega)$ has a peak. (b) When a multiband system with a disconnected Fermi surface has an inter-pocket nesting (i.e., strong inter-pocket pair scattering and weak intra-pocket one) the gap function has the same sign (s-wave symmetry) within each pocket, and the nodal lines can happily run in between the pockets. The estimated T_c for the two-dimensional (2D) Hubbard model on such lattices is indeed almost an order of magnitude higher, $T_c \sim 0.1t$, as displayed in Fig. 9 along with the lattice structure.

As for the dimensionality of the system, we have shown above that 2D systems are generally more favourable than 3D systems as far as the spin-fluctuation-mediated superconductivity in ordinary lattices (square, triangular, fcc, bcc, etc) are concerned. Now, if one puts the idea for the disconnected Fermi surface on the above observation on the dimensionality, a natural question is: can we conceive 3D lattices having disconnected Fermi surfaces that have high T_c's. More specifically, can the disconnected Fermi surface overcome the disadvantage of 3D? If we express our idea more explicitly, what we have in mind is the *interband nesting* (or Suhl-Kondo process in its broader context) in the 3D disconnected Fermi surface as depicted in Fig. 10.

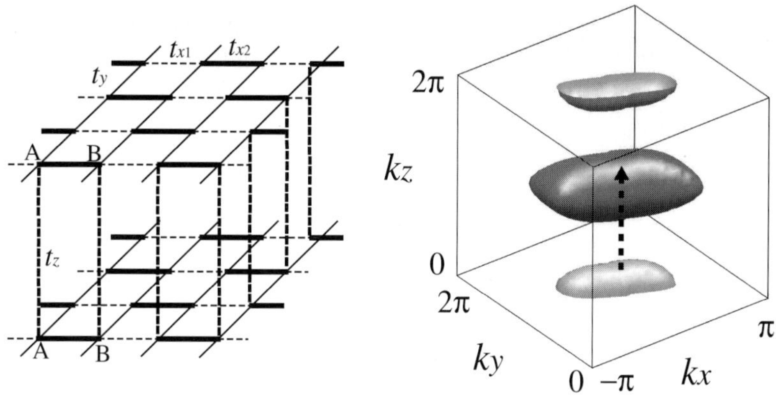

Fig. 10. Interband nesting (*arrow in the right panel*) in 3D on a disconnected Fermi surface, which is exemplified here for the stacked bond-alternating lattice (*left*)

There, the nesting vector runs across the two bands, and this is envisaged to give the attractive pair-scattering interaction. So the gap function should be nodeless within each band, while the gap has opposite signs between the two bands.

In our most recent study [109] we have found that a stacked bond-alternating lattice (Fig. 10) has a compact and disconnected (i.e., a pair of ball-like) Fermi pockets. We have shown that T_c is $O(0.01t)$, which is the same order of that for the square lattice, and remarkably high for a 3D system. We have further found that the T_c can be made even higher ($\sim O(0.1t)$) in a model in which the original Kuroki-Arita 2D system having disconnected Fermi surface is stacked. So the final message obtained here, starting from 1D and ending up with 3D, is that 3D material with considerably high T_c can be expected if we consider appropriate lattice structures.

7 Closing Remarks

So we have seen the electronic properties of electron systems with short-range repulsive interactions, starting from 1D up to 3D systems. While the quasi-1D ladders already contain seeds for the d-wave pairing, anisotropic pairing has more degrees of freedom in 2D and 3D where the topology (e.g., disconnected Fermi surfaces) of the Fermi surface can greatly favour higher Tc. Finally it would be needless to stress that the electron correlation is such a fascinating subject that there are many open questions to be explored. Among them, a question we can ask is what would happen to the superconductivity from the repulsive electron interaction when disorder is introduced in the system. Then we have a problem of dirty superconductors, i.e., an interplay of interaction and disorder. For ladders there are some discussions on this. For instance, Kimura et al. [110]

have looked at the dirty double wire (i.e., two-band Tomonaga-Luttinger system with impurities), and noted that in the phases where the pairing correlation is dominant, the Anderson localisation is absent despite the system being quasi-1D. How this would be extended to higher dimensions is an interesting issue.

Acknowledgements

First and foremost, I wish to thank Professor Bernard Kramer for many years of interactions and discussions, dating back to the advent period when the Anderson localisation just began to form a main stream in the condensed matter physics. For the physics on Tomonaga-Luttinger and electron-mechanism superconductivity I would like to thank Kazuhiko Kuroki for collaboration. I also wish to acknowledge Ryotaro Arita, Takashi Kimura, Miko Eto, Michele Fabrizio and Seiichiro Onari for collaborations. Some of the works described here have been supported by Grants-in-aids from the Japanese Ministry of Education, and I also thank Yasumasa Kanada for a support in 'Project for Vectorised Supercomputing' in the numerical works.

References

1. W. Kohn and J.M. Luttinger, Phys. Rev. Lett. **15**, 524 (1965)
2. D. Fay and A. Layzer: Phys. Rev. Lett. **20**, 187 (1968); A. Layzer and D. Fay: Int. J. Magn. **1**, 135 (1971)
3. The pairing for more dilute electron gas is shown to be s-wave, where the dynamical interaction becomes negative (see, [97])
4. S. Tomonaga: Prog. Theoret. Phys. **5**, 544 (1950)
5. For reviews of the Tomonaga-Luttinger theory, see J. Sólyom: Adv. Phys. **28**, 201 (1979); V.J. Emery in *Highly Conducting One-Dimensional Solids*, ed. by J.T. Devreese et al.(Plenum, New York, 1979), p. 247
6. D.C. Mattis: *The Theory of Magnetism I* (Springer-Verlag, Berlin, 1988)
7. A. Luther and V.J. Emery: Phys. Rev. Lett. **33**, 589 (1974)
8. P.A. Lee: Phys. Rev. Lett. **34**, 1247 (1975)
9. C.M. Varma and A. Zawadowski: Phys. Rev. B **32**, 7399 (1985)
10. K. Penc and J. Sólyom: Phys. Rev. B **41**, 704 (1990)
11. A.M. Finkel'stein and A.I. Larkin: Phys. Rev. B **47**, 10461 (1993)
12. N. Nagaosa and T. Ogawa: Solid State Commun. **88**, 295 (1993)
13. T. Kimura, K. Kuroki, H. Aoki and M. Eto: Phys. Rev. B **49**, 16852 (1994); T. Kimura, K. Kuroki, and H. Aoki: *ibid* **53**, 9572 (1996)
14. M. Sassetti, F. Napoli and B. Kramer: Phys. Rev. B **59**, 7297 (1999)
15. H.J. Schulz: Phys. Rev. B **34**, 6372 (1986)
16. F.D.M. Haldane: Phys. Lett. A **93**, 464 (1993)
17. Y. Nishiyama, N. Hatano and M. Suzuki: J. Phys. Soc. Jpn **64**, 1967 (1994)
18. E. Dagotto, J. Riera, and D.J. Scalapino: Phys. Rev. B **45**, 5744 (1992)
19. T.M. Rice, S. Gopalan and M. Sigrist: Europhys. Lett. **23**, 445 (1993); Physica B **199 & 200**, 378 (1994)
20. P.W. Anderson: Science **235**, 1196 (1987)
21. M. Uehara, T. Nagata, J. Akimitsu, H. Takahashi, N. Môri and K. Kinoshita: J. Phys. Soc. Jpn. **65**, 2764 (1996)

22. A.M. Finkel'stein and A.I. Larkin: Phys. Rev. B **47**, 10461 (1993)
23. L. Balents and M.P.A. Fisher: Phys. Rev. B **53**, 12133 (1996)
24. M. Fabrizio: Phys. Rev. B **48**, 15838 (1993)
25. M. Fabrizio, A. Parola and E. Tosatti: Phys. Rev. B **46**, 3159 (1992).
26. N. Nagaosa and M. Oshikawa: J. Phys. Soc. Jpn. **65**, 2241 (1996)
27. H.J. Schulz: Phys. Rev. B **53**, R2959 (1996)
28. R.M. Noack, S.R. White and D.J. Scalapino: Phys. Rev. Lett. **73**, 882 (1994)
29. R.M. Noack, S.R. White and D.J. Scalapino: Physica C **270**, 281 (1996)
30. E. Dagotto, J. Riera and D.J. Scalapino: Phys. Rev. B **45**, 5744 (1992)
31. M. Sigrist and K. Ueda: Rev. Mod. Phys. **63**, 239 (1991)
32. D. Poilblanc, H. Tsunetsugu and T.M. Rice: Phys. Rev. B **50**, 6511 (1994)
33. H. Tsunetsugu, M. Troyer and T.M. Rice: Phys. Rev. B **49**, 16078 (1994); *ibid.* **51**, 16456 (1995)
34. C.A. Hayward et al.: Phys. Rev. Lett. **75**, 926 (1995); C.A. Hayward and D. Poilblanc: Phys. Rev. B **53**, 11721 (1996)
35. K. Sano: J. Phys. Soc. Jpn. **65**, 1146 (1996)
36. Multiband Tomonaga-Luttinger model that includes pair transfer between two bands [K.A. Muttalib and V.J. Emery, Phys. Rev. Lett. **57**, 1370 (1986)] or charge-charge coupling between the chains [K. Kuroki and H. Aoki, Phys. Rev. Lett. **72**, 2947 (1994)] has been studied
37. H. Suhl, B.T. Mattis and L.R. Walker: Phys. Rev. Lett. **3**, 552 (1959)
38. J. Kondo: Prog. Theor. Phys. **29**, 1 (1963)
39. J.E. Hirsch: Phys. Rev. B **31**, 4403 (1985)
40. M. Imada and Y. Hatsugai: J. Phys. Soc. Jpn. **58**, 3572 (1989)
41. N. Furukawa and M. Imada: J. Phys. Soc. Jpn. **61**, 3331 (1992)
42. S. Sorella, E. Tosatti, S. Baroni, R. Car and M. Parrinello: Int. J. Mod. Phys. B **1**, 993 (1988)
43. S.R. White, D.J. Scalapino, R.L. Sugar, E.Y. Loh, J.E. Gubernatis and R.T. Scalettar: Phys. Rev. B **40**, 506 (1991)
44. W. von der Linden, I. Morgenstern and H. de Raedt: Phys. Rev. B **41**, 4669 (1990)
45. R.M. Noack, S.R. White and D.J. Scalapino: Europhys. Lett. **30**, 163 (1995); Physica C **270**, 281 (1996)
46. Y. Asai: Phys. Rev. B **52**, 10390 (1995)
47. K. Yamaji and Y. Shimoi: Physica C **222**, 349 (1994); K. Yamaji, Y. Shimoi, and T. Yanagisawa: *ibid* **235–240**, 2221 (1994)
48. K. Kuroki and H. Aoki: Phys. Rev. B **56**, R14287 (1997); J. Phys. Soc. Jpn **67**, 1533 (1998)
49. S. Sorella et al.: Int. J. Mod. Phys. B **1**, 993 (1988); S.R. White et al.: Phys. Rev. B **40**, 506 (1989); M. Imada and Y. Hatsugai: J. Phys. Soc. Jpn. **58**, 3752 (1989)
50. Three-leg t-J ladder has also been examined by T.M. Rice, S. Haas, M. Sigrist and F.C. Zhang: Phys. Rev. B **56**, 14655 (1997) with exact diagonalisation and mean-field approximation, and by S.R. White and D.J. Scalapino: Phys. Rev. B **57**, 3031 (1998) with DMRG
51. M. Azuma, Z. Hiroi, M. Takano, K. Ishida and Y. Kitaoka: Phys. Rev. Lett. **73**, 3463 (1994)
52. K. Ishida, Y. Kitaoka, Y. Tokunaga, S. Matsumoto, K. Asayama, M. Azuma, Z. Hiroi and M. Takano: Phys. Rev. B **53**, 2827 (1996)
53. K. Kojima, A. Karen, G.M. Luke, B. Nachumi, W.D. Wu, Y.J. Uemura, M. Azuma and M. Takano: Phys. Rev. Lett. **74**, 2812 (1995)
54. Z. Hiroi and M. Takano: Nature **377**, 41 (1995)

55. H. Mayaffre, P. Auban-Senzier, D. Jérome, D. Poilblanc, C. Bourbonnais, U. Ammerahl, G. Dhalenne, and A. Revcolevschi: Science **279**, 345 (1998)
56. T. Kimura, K. Kuroki and H. Aoki: Phys. Rev. B **54**, R9608 (1996)
57. T. Kimura, K. Kuroki and H. Aoki: J. Phys. Soc. Jpn. **66**, 1599 (1997); **67**, 1377 (1998)
58. E. Arrigoni: Phys. Lett. A **215**, 91 (1996); Phys. Rev. Lett. **83** 128 (1999); Phys. Rev. B **61** 7909 (2000). See also, for the even-odd conjecture for the spin gap, U. Ledermann, K. Le Hur, and T.M. Rice, Phys. Rev. B **62**, 16383 (2000); U. Ledermann, Phys. Rev. B **64**, 235102 (2001)
59. H.J. Schulz in Correlated Fermions and Transport in Mesoscopic Systems, ed. by T. Martin, G. Montambaux and J.T.T. Van (Editions Frontieres, Gif-sur-Yvette, 1996), p. 81
60. To be precise the CuO_2 plane consists of a square array of Cu d orbitals and O p orbitals, which are usually modelled by the "d-p" model. Quantum Monte Carlo has indeed detected a long-ranged pairing correlation [K. Kuroki and H. Aoki: Phys. Rev. Lett. **76**, 4400 (1996)]. The d-p model can be mapped to a one-band Hubbard model [see M.S. Hybertsen, E.B. Stechel, M. Schlüter, and D.R. Jennison: Phys. Rev. B **41**, 11068 (1990); M.S. Hybertsen and M. Schlüter in *New Horizons in Low-Dimensional Electron Systems* ed. by H. Aoki et al. (Kluwer, Dordrecht, 1992) p. 229, and refs therein]
61. P.W. Anderson: *A Carrier in Theoretical Physics* (World Scientific, Singapore, 1994) and refs therein
62. T. Moriya, Y. Takahashi, and K. Ueda: J. Phys. Soc. Jpn, **59**, 2905 (1990); Physica C **185–189**, 114 (1991)
63. T. Ueda, T. Moriya, and Y. Takahashi in *Electronic Properties and Mechanisms of High-T_C Superconductors* ed. T. Oguchi *et al.* (North Holland, Amsterdam, 1992), p. 145; J. Phys. Chem. Solids **53**, 1515 (1992)
64. T. Moriya and K. Ueda: J. Phys. Soc. Jpn. **63**, 1871, (1994)
65. P. Monthoux, A.V. Balatsky, and D. Pines: Phys. Rev. B **46**, 14803 (1992)
66. N.E. Bickers, D.J. Scalapino, and S.R. White: Phys. Rev. Lett. **62**, 961 (1989)
67. I.E. Dzyaloshinskii: Zh.Eksp. Teor. Fiz. **93**, 2267 (1987)
68. H.J. Schulz: Europhys. Lett. **4**, 609 (1987)
69. D.J. Scalapino: Phys. Rep. **250**, 329 (1995)
70. J.V. Alvarez, J. Gonzalez, F. Guinea and M.A.H. Vozmediano: J Phys. Soc. Jpn **67**, 1868 (1998)
71. N.E. Bickers, D.J. Scalapino, and S.R. White: Phys. Rev. Lett. **62**, 961 (1989); N.E. Bickers and D.J. Scalapino: Ann. Phys. (N.Y.) **193**, 206 (1989)
72. T. Dahm and L. Tewordt: Phys. Rev. B **52**, 1297 (1995)
73. J.J. Deisz, D.W. Hess, and J.W. Serene: Phys. Rev. Lett. **76**, 1312 (1996)
74. E. Dagotto et al.: Phys. Rev. B **41**, 811 (1990)
75. G. Fano, F. Ortolani, and A. Parola: Phys. Rev. B **42**, 6877 (1990)
76. S.R. White et al.: Phys. Rev. B **39**, 839 (1989); Phys. Rev. B **40**, 506 (1989)
77. T. Husslein et al.: Phys. Rev. B **54**, 16179 (1996)
78. S. Zhang, J. Carlson, and J.E. Gubernatis: Phys. Rev. Lett. **78**, 4486 (1997)
79. T. Giamarchi and C. Lhuillier: Phys. Rev. B **43**, 12943 (1991)
80. T. Nakanishi, K. Yamaji, and T. Yanagisawa: J. Phys. Soc. Jpn. **66** 294 (1997)
81. N. Furukawa and M. Imada: J. Phys. Soc. Jpn. **61**, 3331 (1992)
82. A. Moreo: Phys. Rev. B **45**, 5059 (1992)
83. G. Sugiyama and S.E. Koonin: Ann. Phys. **168**, 1 (1986); S. Sorella et al.: Int. J. Mod. Phys. B **1**, 993 (1988); S.R. White et al.: Phys. Rev. B **40**, 506 (1989); M. Imada and Y. Hatsugai: J. Phys. Soc. Jpn., **58**, 3752 (1989)

84. K. Kuroki, T. Kimura, and H. Aoki: Phys. Rev. B **54**, R15641 (1996)
85. P.A. Lee and N. Read: Phys. Rev. Lett. **58**, 2691 (1987)
86. D.M. Newns et al.: Phys. Rev. Lett. **69**, 1264 (1992)
87. R.S. Markiewicz et al.: Physica C **217**, 381 (1993) and references therein
88. J. Gonzalez and J.V. Alvarez: Phys. Rev. B **56**, 367 (1997)
89. K. Kuroki and H. Aoki: Phys. Rev. B **60**, 3060 (1999)
90. R. Arita, K. Kuroki and H. Aoki: Phys. Rev. B **60**, 14585 (1999); J. Phys. Soc. Jpn. **69**, 1181 (2000)
91. D.J. Scalapino, E. Loh, Jr., and J.E. Hirsh: Phys. Rev. B **34**, 8190 (1986)
92. S. Nakamura, T. Moriya and K. Ueda: J. Phys. Soc. Jpn **65**, 4026 (1996)
93. A.J. Leggett: Rev. Mod. Phys. **47**, 331 (1975)
94. H. Tou et al.: Phys. Rev. Lett. **77**, 1374 (1996); *ibid*, **80**, 3129 (1998)
95. Y. Maeno et al.: Nature **372**, 532 (1994); T.M. Rice, M. Sigrist: J. Phys. Condens. Matter **7**, L643 (1995)
96. M.Y. Kagan and A.V. Chubukov: JETP Lett. **47**, 614 (1988); A.V. Chubukov: Phys. Rev. B **48**, 1097 (1993)
97. Y. Takada: Phys. Rev. B **47**, 5202 (1993)
98. C. A. Kukkonen and A. W. Overhauser: Phys. Rev. B **20**, 550 (1979)
99. A.V. Chubukov and J.P. Lu: Phys. Rev. B **46**, 11163 (1992)
100. R. Hlubina: Phys. Rev. B **59**, 9600 (1999)
101. H. Takahashi [J. Phys. Soc. Jpn, **68**, 194 (1999)] on the other hand concludes that *p*-wave channel is most attractive for dilute ($n \sim 0.1$) 2D Hubbard model on the $t' = 0$ square lattice
102. G. Baym and L.P. Kadanoff: Phys. Rev. **124**, 287 (1961); G. Baym: Phys. Rev. **127**, 1391 (1962)
103. Here a finite T_C in 2D systems is thought of as a measure of T_C when the layers are stacked to Josephson-couple
104. P. Monthoux and G.G. Lonzarich: Phys. Rev. B **59**, 14598 (1999)
105. P.G. Pagliuso, R. Movshovich, A.D. Bianchi, M. Nicklas, N.O. Moreno, J.D. Thompson, M.F. Hundley, J.L. Sarrao, Z. Fisk: Physica B **312**, 129 (2002)
106. K. Kuroki, and R. Arita: Phys. Rev. B **64**, 024501 (2001)
107. T. Kimura, H. Tamura, K. Kuroki, K. Shiraishi, H. Takayanagi, and R. Arita: Phys. Rev. B **68**, 132508 (2002)
108. K. Kuroki and R. Arita: Phys. Rev. B **63**, 174507 (2001)
109. S. Onari, K. Kuroki, R. Arita and H. Aoki: Phys. Rev. B **65**, 184525 (2002); in preparation
110. T. Kimura, K. Kuroki, and H. Aoki: Phys. Rev. B **51**, 13860 (1995)

Quantum Coherence in Low-Dimensional Interacting Fermi Systems

Ulrich Eckern and Cosima Schuster

Institute of Physics, University of Augsburg, 86135 Augsburg, Germany

1 Introduction

Disorder versus interaction induced metal-insulator transitions are still a central problem in solid state physics. For non-interacting electrons in disordered systems [1] the scaling hypothesis of localization [2] successfully predicts many of the universal features of the transition from metallic to insulating behavior. However, the influence of the interaction on the transition is not equally well understood [3]; recent investigations of an apparent metal-insulator transition in two-dimensional systems even question the main assumptions of the scaling hypothesis [4]. Hereby, attention is also directed to randomly and periodically distorted one-dimensional systems. In one dimension, an infinitesimal amount of disorder (with backscattering contributions) leads to a localized ground state for non-interacting electrons, but with interaction [5] or for special realizations of the disorder [6,7] this result may change.

In contrast to higher dimensional systems, one-dimensional models are often accessible to a detailed theoretical (analytical and numerical) treatment. Using the density matrix renormalization group (DMRG) algorithm, we investigate a lattice model for spinless fermions as well as the Hubbard model in the presence of various potentials. We determine the ground state energy and the phase sensitivity [8] of a ring of interacting spinless fermions as well as the local density, from which the decay of the Friedel oscillations is obtained. For the Hubbard model, we concentrate on the Friedel oscillations.

We start with a brief review of the effects of a single impurity in an interacting one-dimensional Fermi system. Generalizing these results we discuss how the interplay of disorder and interaction can lead to an extended ground state. For example, spinless fermions and electrons in a special quasi-periodic potential [9] show a metal-insulator transition at a finite potential strength, independent of interaction [10,11], whereas the phase diagrams in the presence of random [12] and periodic potentials [13] show a delocalized phase for strong attractive interactions only. The investigation of interacting-electron models is more difficult: The phase sensitivity, a very practical observable in the spinless-fermion case, cannot be used for the Hubbard model [14]. For this case, we hence characterize our system through the Friedel oscillations; in addition, these should be experimentally accessible.

2 Methods and Models

2.1 Methods

The DMRG is a quasi-exact numerical method to determine the ground state properties, i.e. the ground state and the ground state energy, of long one-dimensional (non-integrable) systems with reasonable accuracy [15]. It is useful to implement the spinless-fermion model in terms of the equivalent spin chain, and the Hubbard model as two coupled spin chains. In particular, we calculate the phase sensitivity, i.e. the reaction of the system to a change in the boundary condition. This quantity is very useful to determine numerically the localization-delocalization transition for systems with finite size. We model the boundary condition via a magnetic flux, which results in an additional phase in the hopping terms. The energy levels depend on the total flux, Φ, only. Thus we determine in particular the energy difference between periodic ($c_N = c_0$, $\Phi = 0$) and anti-periodic ($c_N = -c_0$, $\Phi = \pi$) boundary conditions, $\Delta E = (-)^{N_f}[E(0) - E(\pi)]$. We denote by L the length of the chain, by N the number of sites, and by N_f the number of fermions. The factor $(-1)^{N_f}$ cancels the odd-even effects resulting from the change in the ground state for odd compared to even particle numbers. The phase sensitivity, $N\Delta E$, is independent of N for the metallic state. In an insulator, on the other hand, the system does not feel a twist in the boundary condition, i.e. $N\Delta E$ is expected to decrease with system size. Using the DMRG, it is possible to extend the tractable system lengths for the spin chain to about $N \approx 100-200$. In our simulations we perform five lattice sweeps and keep 300 to 500 states per block. Local quantities as e. g. the density can be obtained within an error of 10^{-6} in the Hubbard model when using open boundary conditions, with about 300 states taken into account for chains with about 80 sites. This requires a memory of about 700 MB.

Note that in one dimension, a useful – for the interpretation of the numerical data – formulation on the basis of the bosonization technique [16] is available: The low lying excitations of the non-interacting as well as the interacting fermions are sound waves, i.e. the Fermi system can be described as a non-interacting Bose system, called a Luttinger liquid [17]. The main advantage of the bosonization is that the kinetic and interaction terms are described at the same level, i.e. the Hamiltonian of the interacting Fermi system is diagonalized. Starting from the linearized energy dispersion of non-interacting fermions,

$$H^0 = \sum_{k>0} v_F(k - k_F)c_k^+ c_k + \sum_{k<0} v_F(-k - k_F)c_k^+ c_k \, , \tag{1}$$

it was shown [18] that the bosonic Hamiltonian

$$H_B^0 = \frac{2\pi v_F}{L} \sum_{q>0} \left[\rho_R(q)\rho_R(-q) + \rho_L(-q)\rho_L(q)\right] \tag{2}$$

is equivalent to (1). The corresponding (bosonic) phase variables can be introduced in an intuitive way, writing the particle density of the Fermi system as

$$\rho(x) = \sum_i \delta(x - x_i) \to \partial_x\phi \sum_n \delta(\phi(x) - n\pi) , \tag{3}$$

and assuming that ϕ increases monotonically by π each time x passes the location of a particle, $x = x_i$, i.e. the particles are thought to be located at the points where $\phi(x) = n\pi$. Creating a particle means introducing a kink of height π into $\phi(x)$. The field operator, ψ^+, is thus a displacement operator, $\exp[i\pi\int_0^x \mathrm{d}x' \Pi(x')]$, see e. g. [19]. Then, by defining $\Pi(x) = \partial_x\theta(x)/\pi$, we obtain

$$\psi^+ = [n_0 + \partial_x\varphi(x)/\pi]^{1/2} \mathrm{e}^{i\theta(x)} \mathrm{e}^{i[k_F x + \varphi(x)]} . \tag{4}$$

Here we set $\partial_x\phi(x) = n_0 + \partial_x\varphi(x)/\pi$. The exponential φ-term has to be introduced to achieve the anti-commutation relations of the fermionic operators. With the above definitions, $\varphi(x)$ and $\partial_x\theta(x)/\pi$ are conjugate variables.

2.2 Models

As a first example, we consider a generalized anisotropic Heisenberg (XXZ) model:

$$\begin{aligned} H_{\text{spin}} = &-\sum_{n=1}^{N} J_n(u) \left(\sigma_n^x\sigma_{n+1}^x + \sigma_n^y\sigma_{n+1}^y + \Delta\sigma_n^z\sigma_{n+1}^z\right) \\ &-\sum_n^N h_n\sigma_n^z + N\frac{K}{2}u^2 , \end{aligned} \tag{5}$$

where we include an alternating coupling, $J_n(u) = J[1 + (-1)^n u]$; h_n denotes a random magnetic field. The loss in lattice energy due to the dimerization is taken into account within the harmonic approximation with K as stiffness constant. For the clean XXZ model, i.e. for $u = 0$, and for zero total magnetization, $M = \sum_n\langle\sigma_n^z\rangle = 0$, one finds three phases [20]: a ferromagnetic phase for $\Delta \geq 1$, separated by a first-order transition from a gapless phase for $-1 \leq \Delta < 1$ (whose low lying excitations are given by those of a Luttinger liquid); and an antiferromagnetic phase for $\Delta < -1$. The transition from the Luttinger to the antiferromagnetic phase is of Berezinskii-Kosterlitz-Thouless type. The corresponding fermionic model is obtained via the Jordan-Wigner transformation. Changing the notation, $J \to t$, $J\Delta \to -V/2$, and $h_i \to -\epsilon_i/2$, and neglecting constant energy shifts like $\sum_i h_i$ and $\Delta(2N_f - N/2)$, we obtain

$$H_{\text{fermion}} = -\sum_i t_i \left(c_i^+ c_{i+1} + \text{h. c.}\right) + \sum_i V_i n_i n_{i+1} + \sum \epsilon_i n_i , \tag{6}$$

where $t_i(u) = t[1 + (-1)^i u_t]$, and $V_i(u) = V[1 + (-1)^i u_V]$. The random on-site energies ϵ_i can be considered as due to local impurity potentials. In the bosonized form, the Hamiltonian can be written as follows:

$$H_{\rm B} = \int \frac{dx}{2\pi} \Big\{ \frac{v}{g} \left[\partial_x \varphi(x)\right]^2 + vg \left[\pi \Pi(x)\right]^2 + B_u \sin[2\varphi(x)] + B_\epsilon(x) \cos[2\varphi(x)] \Big\} . \tag{7}$$

The velocity v of the bosonic excitations is given by $v = [\pi t \sin(2\eta)]/(\pi - 2\eta)$, and $g = \pi/4\eta$, where η parameterizes the interaction according to $V = -2t\cos(2\eta)$. The terms containing B_u and B_ϵ are proportional to the dimerization and the disorder, respectively.

Generally speaking, the Hubbard model is thought to be the prototypical model to describe the interplay between kinetic energy ($\rightarrow$ delocalization) and local interaction ($\rightarrow$ localization) for electronic systems. In particular, the Hubbard chain is exactly solvable by means of the Bethe ansatz, which in fact is very useful to determine some parameters [21,22] of the Luttinger description, especially the parameters g_c and g_s. However, no closed expression can be given for these quantities in the Hubbard model (but see also next paragraph). An interaction with longer range, the more generic case, leads to even more phases in the ground state phase diagram [23]. The Hamiltonian of the Hubbard model is given by

$$H_{\rm Hubb} = -t \sum_{i,\sigma}^{N} \left(c^{+}_{i,\sigma} c_{i+1,\sigma} + \text{h. c.}\right) + U \sum_{i}^{N} n_{i,\uparrow} n_{i,\downarrow} + \sum_{i,\sigma} \epsilon_i n_{i,\sigma} . \tag{8}$$

In the clean case, it shows three phases. Phase one occurs for $U < 0$, where the spin excitation spectrum has a gap – thus $g_s = 0$ – and the low-lying charge excitations can be described by those of a Luttinger liquid with $1 < g_c < 2$. Phase two arises for $U \geq 0$ and away from half filling, where spin and charge excitations are those of a Luttinger liquid with $g_s = 1$ and $1/2 \leq g_c \leq 1$. For small U, g_c is given by $g_c = 1 - U/(\pi v_F)$ in phase one and two. The last phase occurs for $U > 0$ and half filling, where the charge excitations have a gap – $g_c = 0$ – and the spin excitations are of Luttinger type with $g_s = 1$. The crucial point here is that the non-interacting electron system is unstable with respect to an attractive interaction. Moreover, non-interacting electrons are a singular point in the phase diagram for half filling. In the bosonized form, (8) can be rewritten as follows:

$$H_{\rm B} = \frac{1}{4\pi} \sum_{\mu={\rm c,s}} \int dx \Big\{ v_\mu \frac{\left[\partial_x \varphi_\mu(x)\right]^2}{g_\mu} + v_\mu g_\mu \left[2\pi \Pi_\mu\right]^2 + B_s \cos[2\varphi_s(x)] + B_c \cos[(4k_F - G)x + 2\varphi_c(x)] + B_\epsilon(x) \cos[\varphi_s(x)] \cos[\varphi_c(x)] \Big\} , \tag{9}$$

where G is a reciprocal lattice vector. In the clean case, H_B shows the spin-charge separation which is characteristic for Luttinger liquids. The B_s-term with

$B_s \propto U$ arises from backscattering ($q \approx 2k_F$) of two electrons with opposite spin and is responsible for the spin gap whenever $U < 0$. The B_c-term with $B_c \propto U$ arises from Umklapp scattering ($q \approx 4k_F$) of two electrons with opposite spin and generates the charge gap for $U > 0$ and half filling. In addition, the B_s-term is marginal for a repulsive interaction, thus leading to logarithmic corrections for zero magnetization. The particle-density, $n(x)$, and the magnetization, $m(x)$, of course, describe real electrons and therefore couple spin and charge,

$$n(x) = n_0 + \frac{\partial_x \varphi_c}{\pi} + \frac{k_F}{\pi} \cos[2k_F x + \varphi_c] \cos[\varphi_s] \,, \tag{10}$$

$$m(x) = \frac{\partial_x \varphi_s}{2\pi} + \frac{k_F}{2\pi} \sin[2k_F x + \varphi_c] \sin[\varphi_s] \,, \tag{11}$$

where $n_0 = N_e/N$ is the electron density; N_e is the number of electrons. The ground state magnetization, $M = \sum_i \langle s_i^z \rangle = (N_\uparrow - N_\downarrow)/2$, vanishes when no external field is present.

3 Quantum Coherence of Spinless Fermions

Using the phase sensitivity we study in the following the ground state phase diagrams with respect to the parameters interaction and potential strength.

3.1 A Single Impurity

The behavior of an interacting Fermi system in the presence of a local potential scatterer is the most studied example of an impurity effect. In this case, the free motion of the fermions inside the ring is influenced mainly by the backscattering at the impurity ($\pm k_F \to \mp k_F$). As discussed by Kane and Fisher [24] the impurity strength scales to zero (i.e. the defect becomes transparent) for an attractive interaction, and scales to infinity (i.e. it becomes completely reflective) for a repulsive interaction, according to the renormalization group equations

$$\frac{\mathrm{d}\epsilon_0}{\mathrm{d}\ln L} = (1-g)\epsilon_0 \,, \qquad \frac{\mathrm{d}}{\mathrm{d}\ln L}\frac{4t^2}{\epsilon_0} = (1-1/g)\frac{4t^2}{\epsilon_0} \,, \tag{12}$$

where L is the length of the chain. Here we consider, for example, an impurity at the end of the chain, $x = 0$. For strong ϵ_0 the transition rate – the effective hopping matrix element between the two sites next to the barrier – is given by $4t^2/\epsilon_0$. Accordingly, the phase sensitivity is given by

$$N\Delta E = \frac{\pi v g}{2} - \epsilon_0 \left(\frac{N}{N_0}\right)^{1-g} \tag{13}$$

in the case of a weak potential, and by

$$N\Delta E = \frac{4t^2}{|\epsilon_0|} \left(\frac{N}{N_0}\right)^{1-1/g} \tag{14}$$

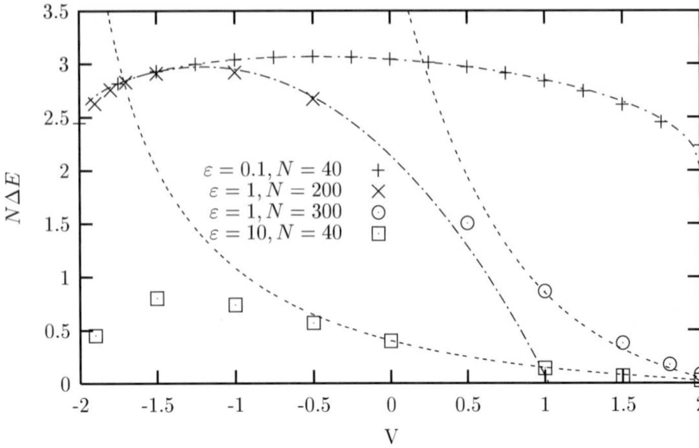

Fig. 1. Phase sensitivity as a function of interaction for a system with one impurity [7]. The lines indicate the behavior appropriate for a weak (*dash-dotted lines*) and a strong scatterer (*dashed lines*), respectively, see (13) and (14)

in the case of a strong barrier, where N_0 is a cut-off ($N_0 \approx 2$). A detailed analytical and numerical study of the phase sensitivity in the presence of a single defect can be found in [7] and [12]. For illustration we show the results for a scatterer of intermediate strength in Fig. 1, where the crossover from the weak potential to the weak link is seen for long systems. Considering a random potential or a global $2k_F$ lattice distortion instead of a local defect, the effective scaling of the perturbations qualitatively is given as follows:

$$\boxed{\begin{array}{llll} \text{local} & \langle \epsilon_0 \cos[2\varphi(0)] \rangle & \propto \epsilon_0 N^{1-g} & \to V_c = 0 \\[2ex] \text{random} & \langle \int \mathrm{d}x \, \epsilon(x) \cos[2\varphi(x)] \rangle & \propto W N^{3/2-g} & \to V_c = -1 \\ & \{\text{using } [\int \mathrm{d}x \, \epsilon(x)]^2 \sim W^2 N\} & & \\[2ex] \text{global} & \langle \int \mathrm{d}x \, u \sin[2\varphi(x)] \rangle & \propto u N^{2-g} & \to V_c = -\sqrt{2} \end{array}} \tag{15}$$

The last two cases are considered below; see also [12] and [13], respectively, for details.

3.2 Random Potential

Our main results for a random potential are a universal behavior of the rms-value of the logarithmic phase sensitivity and the zero-temperature phase diagram: We find that the rms-value of the logarithmic phase sensitivity increases with system size $\propto N^{2/3}$ in the localized region; and we determine the delocalized phase which appears for an attractive interaction [12].

In the first step, generalizing the single impurity result to the case of a weak random potential, we obtain

$$\langle N\Delta E\rangle = \frac{\pi v g}{2} - \frac{W\sqrt{N_0}}{\sqrt{6\pi}}\left(\frac{N}{N_0}\right)^{(3-2g)/2} , \tag{16}$$

$$\sigma^2_{N\Delta E} = \frac{W^2 N_0}{12}\left(1-\frac{2}{\pi}\right)\left(\frac{N}{N_0}\right)^{3-2g} , \tag{17}$$

where we introduce disorder by taking the $\{\epsilon_n\}$, see (6), uniformly distributed over the interval $[-W/2, W/2]$. Again, a repulsive interaction tends to enhance the effective strength of the defects and an attractive interaction reduces it. Especially, for $g > 3/2$, i.e. $V < -1$, the strength of each defect vanishes so fast that disorder becomes an irrelevant perturbation: there is no localization. In the second step, we concentrate our discussion on the localized phase, $V > -1$. Assuming that only one relevant length scale, the localization length ξ, exists, we see that $\xi \propto W^{2/(2g-3)}$ for weak disorder. The rms-value, $\sigma_{\ln(N\Delta E)}$, for small systems is proportional to $N^{(3-2g)/2}$, see (17). A crossover is apparent in the numerical data for $N \approx \xi$, when the fluctuations of $N\Delta E$ are comparable to its average. For large systems we find the fluctuations to be proportional to $N^{2/3}$, as in the non-interacting case. Explicitly, we find from our numerical data

$$\sigma_{\ln(N\Delta E)} \approx 0.027\left(NW^{2/(3-2g)}\right)^{2/3} , \tag{18}$$

where the prefactor applies for $V = 1.2$ ($g \approx 0.71$). In summary, in the weak disorder limit, we verified quantitatively several predictions for disordered Luttinger liquids. In the localized region, we determined the localization length and the distribution of the phase sensitivity.

3.3 Periodic Potential

Including the dimerization, u, see (5), the clean spin system becomes localized by forming spin singlets on neighboring sites for $\Delta < 0$, i.e. for antiferromagnetic coupling. An excitation gap opens due to the usual Peierls mechanism [25]. Leaving aside the question whether a finite u can be stabilized, we note that the dimerization is already relevant for $\Delta < \sqrt{2}/2$ [26,27], and the ground state wave function is localized. The interaction-dimerization phase diagram was determined in [13]. The schematic phase diagram in terms of the parameters of the spinless-fermion model is shown in Fig. 2. We find – similar to the case of the random potential – for an attractive interaction a region where the distortion is irrelevant in the sense of the RG scaling. A stable dimerization, u_0, however, is only established for a repulsive interaction where the fermionic energy gain overcomes the energy loss of the lattice itself. Thereby u_0 grows with increasing interaction until the competing order of the charge density wave (CDW) – indicated by $\Delta(V)$ – sets in. Adding an impurity (dashed lines in Fig. 2) to

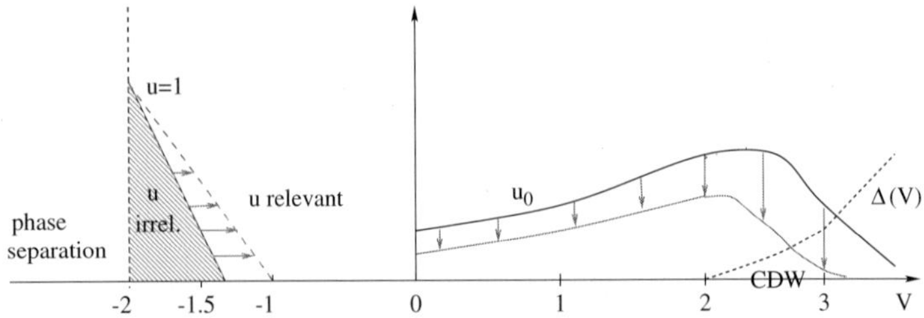

Fig. 2. Schematic ground state phase diagram of the dimerized model. The shaded region indicates a delocalized phase, where the dimerization is irrelevant. In the range $V \geq 0$ we also sketch the equilibrium dimerization u_0 and the correlation gap, $\Delta(V)$

the dimerized system the delocalized region grows [28]. On the other hand the stable dimerization and the excitation gap are reduced. In summary, the numerical results confirm the prediction of the renormalization group treatment concerning the transition point for weak distortion. In addition, it was possible to quantitatively determine the boundary of the delocalized phase for strong distortions.

3.4 Quasi-periodic Potential

Let us now consider interacting spinless fermions on a ring in the Aubry-André potential such that

$$H_{\mathrm{qp}} = -t \sum_m \left(c^+_{m+1} c_m + \mathrm{h.\,c.} \right) + V \sum_m n_{m+1} n_m + 2\mu \sum_m n_m \cos(\alpha m) \, . \quad (19)$$

The quasi-periodic potential shows features of both the random and the periodic potential. In the numerical analysis the value of $\alpha/2\pi$ is approximated by the ratio of successive Fibonacci numbers – $F_n = F_{n-2} + F_{n-1} = 0, 1, 2, 3, 5, 8, 13, \ldots$ – as is customary in the context of quasi-periodic systems [29]. Choosing $N = F_n$, we retain the periodicity of the quasi-periodic potential on the ring. In the non-interacting case, the one-dimensional quasi-periodic Aubry-André model is rigorously known to exhibit a metal-insulator transition for all states in the spectrum as a function of the strength of the quasi-periodic potential [9]. The ground state wave function is extended for small, and localized for large μ. At the critical value the wave functions decrease algebraically. In the case of incommensurate densities the metal-insulator transition is found at $\mu_c \approx 1$, and $N \Delta E \propto (\mu_c - \mu)^\nu$ with $\nu \approx 1$. This indicates that the influence of the interaction at this particular type of metal-insulator transition is not strong enough to change the universality class of the model. For commensurate densities, on the other hand, we find a Peierls-like behavior, similar to the situation with a periodic potential, with a metal-insulator transition at a certain value of the attractive interaction, provided μ is small. Thus, the physics of the model at commensurate densities is dominated by the Peierls resonance condition which becomes irrelevant only for a strong attractive interaction; for details see [11].

4 Local Distortions

Returning to single impurities, we discuss now Friedel oscillations in spin chains and metals. Note that the exponent of the decay of the oscillations is related to the scaling of the impurity, see (13): In case the impurity is reflecting (and thus localizing) the decay of the oscillations is expected to be slower than the x^{-1}-behavior of non-interacting particles. Otherwise, the impurity disturbs the system only locally and is thus an irrelevant perturbation.

4.1 Friedel Oscillations

Oscillations of the local density – and similarly of the magnetization – arise because the density response $n(\mathbf{q})$, given by the Lindhard function $\chi(\mathbf{q})$, is not analytic at $2k_F$. For non-interacting electrons the response function is given by [30]

$$\begin{aligned} \chi(\mathbf{q}) &= -2e^2 \int \frac{\mathrm{d}^D k}{(2\pi)^D} \frac{f(\mathbf{k}-\mathbf{q}/2) - f(\mathbf{k}+\mathbf{q}/2)}{\varepsilon(\mathbf{k}-\mathbf{q}/2) - \varepsilon(\mathbf{k}+\mathbf{q}/2)} \\ &\overset{3D}{=} -e^2 \left(\frac{mk_F}{\pi^2}\right) \left[\frac{1}{2} + \frac{1-x^2}{4x} \ln\left|\frac{1+x}{1-x}\right|\right] \qquad (20) \\ &\overset{1D}{=} -e^2 \left(\frac{2m}{\pi k_F}\right) \frac{1}{2x} \ln\left|\frac{1+x}{1-x}\right| . \qquad (21) \end{aligned}$$

Here, $x = q/2k_F$, and $f(\mathbf{k})$, $\varepsilon_{\mathbf{k}}$ denote the Fermi function and the energy of a free electron with momentum $\mathbf{k}$, respectively. The transformation to space coordinates then leads to

$$n(\mathbf{r}) \sim \cos(2\mathbf{k_F r})/r^D . \qquad (22)$$

We expect that in one dimension correlation effects change the exponent of the decay,

$$n(x) = a\epsilon \frac{\cos(2k_F x)}{x^\delta} , \qquad (23)$$

where a denotes a prefactor independent of the impurity strength ϵ. In the non-interacting case it is well known that $\delta = 1 = D$. Experimentally the Friedel oscillations in a spin system can be measured by nuclear magnetic resonance spectroscopy [31]. The density oscillations of metals are more difficult to determine, but it has been reported that they can directly be observed with a scanning tunneling microscope [32]. Furthermore, oscillations in the local density of states are a signature of the Friedel oscillations too [33].

4.2 Spinless Fermions

The Friedel oscillations in a spinless-fermion system have been calculated before, analytically for repulsive interaction [34] using the bosonization technique, and

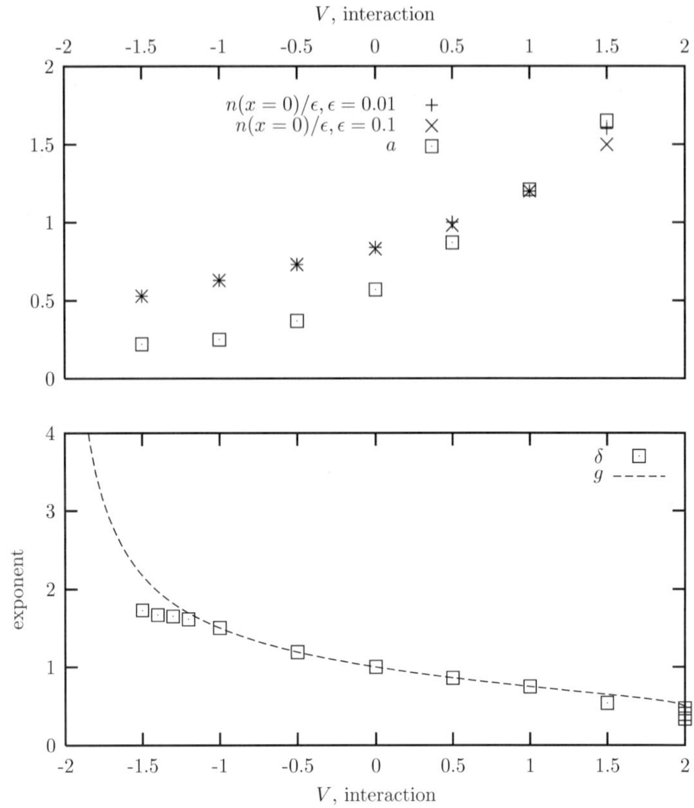

Fig. 3. The *lower panel* shows the exponent δ, obtained numerically, for half filling ($n_0 = 1/2$) versus interaction. The results are compared with the prediction $\delta = g$ from bosonization. The *upper panel* shows the local density at $x = 0$ and the prefactor a, determined for $x = 5 \dots 20$, versus interaction

numerically for weak and intermediate interaction [35] using the DMRG. The results are confirmed by boundary conformal field theory [36]. The exponent $\delta = g$ is intimately related to the scaling relation of a single impurity, (12), mentioned above. In the case of a repulsive interaction ($g < 1$), where the impurity is relevant, the decay, x^{-g}, is slow in comparison to the increase of the system with N. In the other case ($g > 1$) where the impurity is irrelevant, the oscillations decay fast and the impurity remains local. We have determined the exponent of the decay and the prefactor of the Friedel oscillations in the Luttinger-liquid phase for a boundary defect with $\epsilon_1 = -\epsilon_N$, see [37]. In Fig. 3 (lower panel) we compare the exponent obtained numerically with the predictions from bosonization and conformal field theory. For small couplings, $-1 < V < 1$, the results are in good agreement. The deviations near $V = -2$ are related to the divergence of the Luttinger parameter g. As the impurity becomes more and more irrelevant, the oscillations die out and we find an exponential decay only.

On the other hand, at $V = 2$ the Umklapp scattering becomes marginal; therefore a logarithmic correction has to be taken into account which alters the exponent in the weak impurity case. Thus, this deviation near $V = 2$ is absent for other fillings and in the strong impurity limit. In addition, the $4k_F$-oscillations become stronger near $V = 2$. The scaling of the impurity strength according to the renormalization group equation of [24] is mainly absorbed in the exponent of the decay, i.e. the decay is slow for $V > 0$ ($g < 1$), and fast for $V < 0$ ($g > 1$). For weak impurities, a is independent of the strength of the defect while $n(x = 0)$ depends linearly on it, compare Fig. 3 (upper panel). Nevertheless, both quantities depend on interaction. They grow (due to a larger influence of the impurity) with increasing interaction. The strong impurity results, where $n(x = 0)$ saturates, are not included in Fig. 3.

4.3 Hubbard Model

We also investigate the Friedel oscillations in the one-dimensional Hubbard chain induced by boundaries and by defects, for the cases of half-filled and third-filled bands. Due to the rich phase diagram with spin-gap and charge-gap phases we compare and discuss three representative cases. Using bosonization the Friedel oscillations are given by [34]

$$\langle n(x)\rangle - n_0 \propto \frac{\cos\left(2k_F x + \psi_c\right)}{x^{(g_s+g_c)/2}}, \tag{24}$$

$$\langle m(x)\rangle \propto \frac{\sin\left(2k_F x + \psi_s\right)}{x^{(g_s+g_c)/2}}, \tag{25}$$

where ψ_c and ψ_s denote arbitrary phase shifts. In the Luttinger-liquid phase a marginal operator from backscattering is always present and leads, for vanishing magnetization $M = 0$, to logarithmic corrections [38]. The Friedel oscillations in the Hubbard chain with $M \neq 0$ are considered in detail in [39].

In the first step, we analyze the numerical data assuming an algebraic decay and performing a linear regression of the log-log representation as before. We find that the exponent obtained for a defect in the middle of a chain, δ_{m}, is clearly different from the exponent obtained for a boundary defect, δ_{b}, see Fig. 4. Furthermore, both deviate from the value predicted by bosonization (≈ 0.95, see [22]). Taking into account the analysis of the marginal operator in [38], where the two relevant exponents of the logarithmic corrections were determined, we fit our data again including these corrections. Thus we determine in the second step the corrected exponents (≈ 0.95 and ≈ 0.9, see Fig. 4), which agree reasonably well with the prediction of bosonization. Of course, the logarithmic corrections of the asymptotic regime cannot be determined from our numerical results but the related deviations of the exponents on a short length scale. Increasing the impurity strength the data do not converge to the prediction from bosonization as they do in the spinless case. However, applying the logarithmic corrections we find that the oscillations decay always slower than x^{-1} in the Luttinger region, in agreement with the prediction.

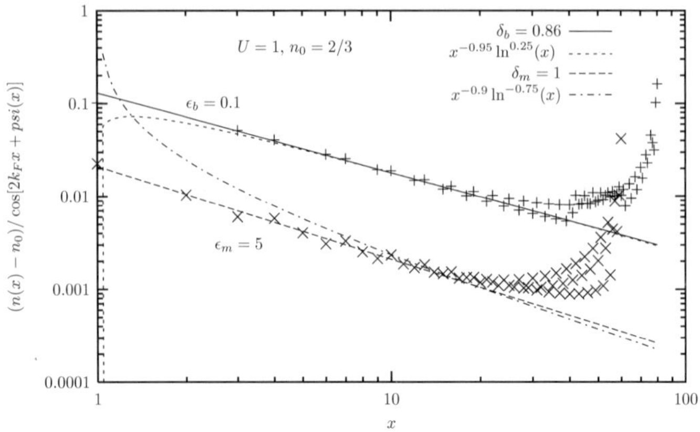

Fig. 4. Comparison of the Friedel oscillations in the Hubbard model. We set $U = 1$, $n_0 = 2/3$; the number of lattice sites is $N = 81$. The *upper curve* shows density oscillations induced by a boundary potential with $\epsilon_b = 0.1$, the *lower curve* shows the oscillations around a defect in the middle of the chain with $\epsilon_m = 5$. The data were evaluated with $\psi_c^b(x) = \pi/2$ for the boundary impurity, and with $\psi_c^m(x) = 0$ for the defect in the middle

In the phases with gap we can distinguish two cases. First we consider density (magnetization) oscillation in the spin-gap (charge-gap) phase. As seen in (10) and (11) the spin (charge) degrees couple *only* in the cosine term to the density (magnetization). Therefore we can set $g_s = 0$ ($g_c = 0$) when evaluating $\langle n(x) \rangle$ ($\langle m(x) \rangle$) and hence obtain a rather slow decay. Especially, we find $\langle m(x) \rangle \sim \cos(\pi x) x^{-0.5}$ for a strong repulsive interaction and at half filling, i.e. in the limit of the isotropic Heisenberg model. In the second case, as already stated by Luther and Emery [40] in the seventies, the decay of the density oscillations is exponential in the charge-gap phase; similarly the magnetization decreases exponentially in the spin-gap phase.

5 Summary

This article summarizes our recent theoretical studies of one-dimensional lattice models (spinless fermions, Hubbard model), which aim at a better understanding of the interplay of interaction and distortions on ground state and low-temperature transport properties. Most of our results are based on the density matrix renormalization group (DMRG) algorithm, supplemented by analytical considerations (bosonization, conformal field theory). Using the DMRG, we considered the ground state energy and the phase sensitivity of interacting spinless fermions on a ring. We found, for an attractive interaction in the presence of either a random, periodic, or quasi-periodic potential, a delocalized phase of finite extension. The zero-temperature phase diagrams have been determined.

In addition, we investigated in detail the Friedel oscillations induced by single defects. We confirmed the predictions of conformal field theory for weak interaction, but near phase transitions deviations are found in the form of vanishing or additional oscillations. For the Hubbard chain, we studied these oscillations – in the density and the magnetization – in the spin-gap, charge-gap, and Luttinger-liquid phase. We found an exponential or a very slow algebraic decay of the oscillations in the gapped phases. In the Luttinger-liquid phase, we concentrated on the question of logarithmic corrections. Differences in the behavior near a boundary compared to an impurity in the bulk have been pointed out.

Acknowledgement

We thank P. Schwab for helpful discussions, and P. Brune and P. Schmitteckert for providing us with their DMRG algorithms. Financial support by the Deutsche Forschungsgemeinschaft (SPP 1073, SFB 484) is gratefully acknowledged.

References

1. B. Kramer and A. MacKinnon, Rep. Prog. Phys. **56**, 1469 (1993)
2. E. Abrahams, P.W. Anderson, D.C. Licciardello, and T.V. Ramakrishnan, Phys. Rev. Lett. **42**, 673 (1979)
3. D. Belitz and T.R. Kirkpatrick, Rev. Mod. Phys. **66**, 261 (1994)
4. S.V. Kravchenko, D. Simonian, M.P. Sarachik, W. Mason, and J.E. Furneaux, Phys. Rev. Lett. **77**, 4938 (1996); D. Simonian, S.V. Kravchenko, M.P. Sarachik, and V.M. Pudalov, Phys. Rev. B **57**, R9420 (1998)
5. C.A. Doty and D.S. Fisher, Phys. Rev. B **45**, 2167 (1992)
6. S. Eggert and I. Affleck, Phys. Rev. B **46**, 10866 (1992)
7. C. Schuster and U. Eckern, Ann. Phys. (Leipzig) **11**, 901 (2002)
8. B.S. Shastry and B. Sutherland, Phys. Rev. Lett. **65**, 243 (1990)
9. S. Aubry and G. André, Ann. Israel Phys. Soc. **3**, 133 (1980)
10. A. Eilmes, U. Grimm, R.A. Römer, and M. Schreiber, Eur. Phys. J. B **8**, 547 (1999)
11. C. Schuster, R.A. Römer, and M. Schreiber, Phys. Rev. B **65**, 115114 (2002)
12. P. Schmitteckert, T. Schulze, C. Schuster, P. Schwab, and U. Eckern, Phys. Rev. Lett. **80**, 560 (1998)
13. C. Schuster and U. Eckern, Eur. Phys. J. B **5**, 395 (1998)
14. C. Schuster, P. Brune, and U. Eckern: 'Impurities in a Hubbard-chain'. In: *High Performance Computing in Science and Engineering '01*, edited by E. Krause and W. Jäger (Springer-Verlag, Heidelberg, 2002), pp. 157–166
15. S.R.White, Phys. Rev. Lett. **69**, 2863 (1992); I. Peschel, X. Wang, M. Kaulke, and K. Hallberg, *Density-Matrix Renormalization: A New Numerical Method in Physics* (Lecture Notes in Physics, vol. 528, Springer-Verlag, Heidelberg, 1999)
16. F.D.M. Haldane, Phys. Rev. Lett. **47**, 1840 (1981)
17. J.M. Luttinger, J. Math. Phys. **4**, 1154 (1963); D.C. Mattis and E.H. Lieb, J. Math. Phys. **6**, 304 (1965)
18. A. Luther and I. Peschel, Phys. Rev. B **9**, 2911 (1974)
19. H.J. Schulz, Lecture notes from the Jerusalem Winter School on Theoretical Physics: *Correlated Electron Systems*, Dec. 1991–Jan. 1992, arXiv:cond-mat/9302006
20. V.E. Korepin, N.M. Bogoliubov, and A.G. Izergin, *Quantum inverse scattering method and correlation functions* (Cambridge University Press, Cambridge, 1993)
21. E.H. Lieb and F.Y. Wu, Phys. Rev. Lett. **20**, 2435 (1968)

22. H.J. Schulz, Phys. Rev. Lett. **64**, 2831 (1990)
23. J. Voit, Phys. Rev. B **45**, 4027 (1992)
24. C.L. Kane and M.P.A. Fisher, Phys. Rev. Lett. **68**, 1220 (1992)
25. R.E. Peierls, *Quantum Theory of Solids* (Oxford University Press, Oxford, 1955)
26. M. Kohmoto, M. den Nijs, and L.P. Kadanoff, Phys. Rev. B **24**, 5229 (1981)
27. F.C. Alcaraz and A.L. Malvezzi, J. Phys. A: Math. Gen. **28**, 1521 (1995)
28. C. Schuster and U. Eckern, Ann. Phys. (Leipzig) **8**, 585 (1999)
29. U. Grimm, Habilitationsschrift, Technische Universität Chemnitz (1999)
30. N.W. Ashcroft and N.D. Mermin, *Solid state physics*, chap. 17 (W.B. Saunders Company, Philadelphia, 1976)
31. J. Bobroff et al., Phys. Rev. Lett. **86**, 4116 (2001)
32. A. Depuydt et al., Phys. Rev. B **60**, 2619 (1999)
33. S. Eggert, Phys. Rev. Lett. **84**, 4413 (2000)
34. R. Egger and H. Grabert, Phys. Rev. Lett. **75**, 3505 (1995)
35. P. Schmitteckert, P. Schwab, and U. Eckern, Europhys. Lett. **30**, 543 (1995)
36. Y. Wang, J. Voit, and Fu-Cho Pu, Phys. Rev. B **54**, 8491 (1996)
37. C. Schuster, *Local distortions in spin chains*, to be published
38. T. Giamarchi and H.J. Schulz, Phys. Rev. B **39**, 4620 (1989)
39. G. Bedürftig, B. Brendel, H. Frahm, and R.M. Noack, Phys. Rev. B **58**, 10225 (1998)
40. A. Luther and V.J. Emery, Phys. Rev. Lett. **33**, 589 (1974)

Spin Transport Properties of a Quantum Dot

Maura Sassetti[1], Alessandro Braggio[1], and Fabio Cavaliere[1]

INFM-Lamia, Dipartimento di Fisica, Università di Genova, Via Dodecaneso 33, 16146 Genova, Italy

1 Introduction

During the last decade, spin phenomena in the transport properties of low-dimensional quantum systems have become a subject of increasing interest [1,2]. Several fundamental effects have been found when controlling transport of electrons one by one in quantum dots. For instance, parity effects in the Coulomb blockade of quantum dots with very small numbers of electrons have been detected experimentally [3]. In addition, the spin blockade effect, especially in the non-linear current-voltage characteristic of one-dimensional (1D) quantum dots has been predicted [4].

The recent experimental realization of semiconductor-based 1D quantum wires [5–7] has opened new perspectives to systematically investigate the influence of interactions, spins and impurities on electron transport properties. Also the electronic transport in Carbon nanotubes has been the subject of continuous experimental and theoretical efforts [8]. Recently, even single organic molecules have been successfully attached to metallic contacts such that Coulomb blockade and non-linear transport spectra could be detected, including characteristic spin features when incorporating a magnetic atom into the molecule [9,10].

There are perspectives of applications of quantum nano-structures in spin-electronics, quantum computing and communication [11]. Previous works have been focusing on spin transport in two dimensional (2D) quantum dots connected to non-interacting leads and in the presence of a magnetic field [12], including an oscillating magnetic electron spin resonance component [13]. Spin transport in circuits with ferromagnetic elements and in the presence of a Luttinger-liquid interaction [14–16] have been considered. In view of the applications envisaged, the theory of the spin control of electron transport in the presence of correlations is of fundamental interest since in nanoscale devices the latter can be very important.

Many of the fundamental transport properties of quantum dots are well understood. Nevertheless, important aspects, as the temperature behaviors of the Coulomb peaks, and the spin properties in the regions of the excited states are not completely worked out. It is therefore useful to study the combined effect of the electron interaction and additional scattering potentials in 1D quantum systems on a microscopic level. The Luttinger liquid model for fermions in 1D allows for taking into account exactly the interaction between the fermions [17–19]. For this model, in the presence of repulsive interaction, the linear conductance of a clean, infinitely long, spin degenerate Luttinger liquid has been predicted to be $G_\rho = 2e^2 g_\rho/h$, the universal conductance renormalized by the interaction constant g_ρ [20]. However, DC-transport experiments on quasi-1D semiconductor quantum wires at relatively high temperatures showed only quantization in terms of the universal quantum conductance $G_0 = 2e^2/h$. This has been interpreted either in terms of a compensating, self-consistent renormalization of the external electric driving field [21–23] or by attaching to the quantum wire non-interacting leads [24–26] which eventually dominate the conductance. At lower temperatures, deviations from the universal conductance steps have been found [5–7]. This has been attributed to weak scattering at the random impurity potential in a Luttinger liquid [27].

When the electron density in a GaAs/AlGaAs-quantum wire fabricated by using the cleaved-edge overgrowth technique is decreased by applying a voltage to an external gate, eventually, even the lowest electronic subband can be depopulated [28] such that Coulomb blockade is achieved. Here, it is assumed that the electron density is so low that a few maxima of the random potential of the impurities are higher than the Fermi level. Then, a *quasi-1D* quantum island can be formed between two successive potential maxima. Electron transport becomes dominated by charging effects [29–31]. The linear conductance shows discrete peaks that correspond to transferring exactly one additional electron through the quantum island. In this region, it has been detected that the temperature dependence of the area $\mathcal{A}$ of the conductance peaks is modified by the correlations. Instead of being independent of T, as predicted by the so-called orthodox theory, it shows non-analytical power-law behaviors $\mathcal{A}(T) \propto T^\lambda$ where the exponent $\lambda > 0$ depends on the interaction parameter. Formerly, such a behavior has been predicted in the sequential tunneling regime for resonant transport through a quantum dot coupled to a Luttinger liquid [32] and for a double potential barrier in a Luttinger liquid [33]. However, the interaction parameter deduced from the exponent λ has been found to be *different* from the one obtained from the charging energy [34,28].

In non-linear transport, collective excited states of the electrons in the island can contribute to the current transport [35,36]. The low-temperature current-voltage characteristics shows fine structures within the Coulomb steps which reflect the excitation spectra of the electrons confined within the quantum dot. In view of the vast amount of theoretical understanding of 1D non-Fermi liquids [37], it is desirable to provide quantitative results for charging and correlation phenomena in these systems by using realistic and experimentally accessible models. This should provide unique experimental information on the physics of the Luttinger liquid, one of the most important non-Fermi liquid paradigms in modern theoretical physics.

Previously, efforts in this direction have been made by studying transport through a single barrier embedded in a Luttinger liquid [38,39]. The effect of long range interactions was investigated [40,41]. Charging in the presence of two impurities has been studied for spinless electrons with an interaction range much smaller than the distance between the two impurities [42,43]. The crossover from a Luttinger liquid with long-range interaction to a Wigner crystal in the presence of two symmetrically arranged potential barriers has been investigated [44]. Charging effects for two strong impurities in a single-channel quantum wire with long-range electron interaction in the presence of the electron spin have been studied [34]. The effect of a magnetic field on the spin polarization of the current has been investigated in detail [45].

In this review, we present a theory for the sequential spin and charge transport through a quantum dot within a Luttinger liquid. In the linear region, we confirm the above mentioned parity effect of the Coulomb blockade found in experiments. In addition, we predict a temperature dependence of the position of the conductance peaks induced by the spin. In the non-linear region, we find that spin-charge separation effects strongly affect the current-voltage characteristics. The spin leads to rich structure in the non-linear differential conductance that reflects both the collective spin density excitations and the orientation of the total spin in the quantum dot. The paper is organized as follows. Section 2 briefly describes the model. In Sect. 3, the effective action is provided and the corresponding physics is described. In Sect. 4, the conditions for transport are discussed. Some quantitative results of spin effects on the linear and non-linear differential conductance are summarized in Sect. 5.

2 The Tomonaga-Luttinger Model

2.1 The Tomonaga-Luttinger Hamiltonian

Interacting fermion systems with spin in 1D can be diagonalized exactly by using the bosonization technique [17–19,37]. The Hamiltonian is a quadratic form ($\hbar = 1$),

$$H_0 \equiv H_\rho + H_\sigma = \sum_{\nu=\rho,\sigma} \frac{v_\nu g_\nu}{2} \int \mathrm{d}x \left\{ \Pi_\nu^2(x) + \frac{1}{g_\nu^2} [\partial_x \vartheta_\nu(x)]^2 \right\} . \qquad (1)$$

The electrons are represented by conjugate bosonic fields Π_ρ, ϑ_ρ, and $\Pi_\sigma, \vartheta_\sigma$ which are associated with collective charge and spin density excitations, respectively. The correlations are visible in the parameters g_ρ and g_σ that represent the strengths of the charge and spin interactions

$$g_\rho = \frac{1 + V_{\mathrm{ex}}}{[1 - V_{\mathrm{ex}}^2 + 4(1 + V_{\mathrm{ex}})V_{\mathrm{f}}]^{1/2}} , \qquad g_\sigma = \left(\frac{1 + V_{\mathrm{ex}}}{1 - V_{\mathrm{ex}}} \right)^{1/2} . \qquad (2)$$

They depend on the electron screened Coulomb interaction $V(x - x')$, with $V_{\mathrm{f}} \equiv \hat{V}(q \to 0)/2\pi v_{\mathrm{F}}$ the forward scattering contribution of the Fourier transform $\hat{V}(q)$, and $V_{\mathrm{ex}} \equiv \hat{V}(2k_{\mathrm{F}})/2\pi v_{\mathrm{F}}$ the exchange interaction related to the $2k_{\mathrm{F}}$ component of $\hat{V}(q)$. The latter is generally smaller as compared with V_{f}. Note that the finite value of the exchange interaction breaks the SU2 invariance of the spin components. For repulsive interaction one has $g_\rho < 1$ and $g_\sigma > 1$. The

Fermi velocity is v_{F}. The dispersions of the charge and spin density excitations are [46,47]

$$\omega_\rho(q) = \frac{v_{\mathrm{F}}(1+V_{\mathrm{ex}})}{g_\rho}|q| \equiv v_\rho|q| , \qquad \omega_\sigma(q) = \frac{v_{\mathrm{F}}(1+V_{\mathrm{ex}})}{g_\sigma}|q| \equiv v_\sigma|q| . \tag{3}$$

2.2 Potential Barriers

In this model, a quantum dot can be described by a double barrier consisting of two delta-function potentials $V_{\mathrm{t}}\,\delta(x-x_\pm)$ at the positions $x_\pm = \pm a/2$. The corresponding backward scattering contribution to the energy can be written as

$$H_{\mathrm{t}} = U_{\mathrm{t}} \sum_{r=\pm} \cos\left[2k_{\mathrm{F}}x_r + \sqrt{2\pi}\vartheta_\rho(x_r)\right] \cos\left[\sqrt{2\pi}\vartheta_\sigma(x_r)\right] , \tag{4}$$

where $U_{\mathrm{t}} = \rho_0 V_{\mathrm{t}}$ with the electron density $\rho_0 = 2k_{\mathrm{F}}/\pi$. Equation (4) is the potential energy of the impurities seen by the charges represented by the density $\rho(x) \equiv \rho_\uparrow(x) + \rho_\downarrow(x)$

$$\rho(x) = \rho_0 + \sqrt{\frac{2}{\pi}}\,\partial_x\vartheta_\rho(x) + \rho_0 \cos\left[2k_{\mathrm{F}}x + \sqrt{2\pi}\vartheta_\rho(x)\right] \cos\left[\sqrt{2\pi}\vartheta_\sigma(x)\right] .$$

The second term in (2.2) accounts for the slowly varying part of the charge fluctuations. The third term represents the charge density excitations involved in the $2k_{\mathrm{F}}$ backscattering interference between left- and right-moving electrons. It also couples the charge with the long wave length part of the spin density $\rho_\uparrow(x) - \rho_\downarrow(x) \approx \sqrt{2}\partial_x\vartheta_\sigma(x)/\sqrt{\pi}$, which is considered here with respect to a zero mean value of the spin.

2.3 Charge and Spin Currents

The presence of the two localized impurities separates the charge and spin degrees of freedom at bulk positions $x \neq x_\pm$ from those at the barriers. It is useful to introduce symmetric and antisymmetric variables for particle and spin densities ($\nu = \rho, \sigma$)

$$N_\nu^\pm = \sqrt{\frac{2}{\pi}}\left[\vartheta_\nu(x_+) \pm \vartheta_\nu(x_-)\right] . \tag{5}$$

The quantity N_ρ^- is associated with the fluctuations of the particle number within the dot as compared to the mean particle number $n_0 = a\rho_0$. The quantity N_σ^- represents the z-component of the total spin in the dot. Correspondingly, the numbers of particles and spins transferred across the dot are represented by N_ρ^+ and N_σ^+. In terms of these new variables, the coupling to the external bias voltages can be written as

$$H_V = -e\left(\frac{V}{2}N_\rho^+ + V_{\mathrm{g}}N_\rho^-\delta\right) . \tag{6}$$

Here we have assumed a source-drain bias voltage V which drops symmetrically at the barriers and a gate voltage V_g inside the quantum dot. The parameter δ contains information about the experimental setup. It is essentially the ratio between gate and total capacitances.

The tunneling Hamiltonian in the new variables can be cast into the form

$$H_\mathrm{t} = U_\mathrm{t} \sum_{\alpha=0,1} \prod_{\nu=\rho,\sigma} \cos\left[\frac{\pi}{2}\left(N_\nu^- + \nu_0 a - \alpha\right)\right] \cos\left[\frac{\pi}{2}\left(N_\nu^+ - \alpha\right)\right]. \tag{7}$$

The DC charge and spin currents are defined as $I_\rho = I_\uparrow + I_\downarrow$ and $I_\sigma = I_\uparrow - I_\downarrow$, respectively. They can be evaluated by considering the stationary limits of the charge and the spin transfers through the dot in the presence of the bias voltage

$$I_\rho = \frac{e}{2}\langle \dot{N}_\rho^+(t\to\infty)\rangle\,, \qquad I_\sigma = \frac{e}{2}\langle \dot{N}_\sigma^+(t\to\infty)\rangle\,. \tag{8}$$

The brackets $\langle\cdots\rangle$ include a thermal average over the collective excitations at $x \neq x_\pm$ and a statistical average performed with the reduced density matrix for the degrees of freedom at $x = x_\pm$.

3 The Effective Action

In order to evaluate the current-voltage characteristic, we perform a thermal average over the "bulk modes" at $x \neq x_\pm$. This is a particular choice that implies the assumption of a thermal distribution for the spin and the charge density excitations both in the leads and in the dot. This can be done by using the imaginary-time path integral method [48]. The result of the integration is an effective action S_eff which depends only on the four variables N_ν^r, with $r = \pm; \nu = \rho,\sigma; \beta = 1/k_\mathrm{B}T$

$$\begin{aligned} S_\mathrm{eff}[N_\rho^\pm, N_\sigma^\pm] = \int_0^\beta \mathrm{d}\tau \Big\{ & H_\mathrm{t}[N_\rho^+, N_\sigma^\pm] + H_V[N_\rho^\pm] \\ & + \sum_{r=\pm,\nu=\rho,\sigma} \int_0^\beta \mathrm{d}\tau'\, N_\nu^r(\tau) K_\nu^r(\tau-\tau') N_\nu^r(\tau') \Big\}\,. \end{aligned} \tag{9}$$

The Fourier transforms, at Matsubara frequencies $\omega_n = 2\pi n/\beta$ (n integer), of the dissipative kernels $K_\nu^r(\tau)$ are

$$K_\nu^r(\omega_n) = \frac{\pi}{8}\frac{1}{\mathcal{G}_\nu(\omega_n;0,0) + r\mathcal{G}_\nu(\omega_n;-a/2,a/2)}\,. \tag{10}$$

These kernels depend on the two-point time ordered propagators $\mathcal{G}_\nu(\tau;x,x') = \langle \mathbf{T}_\tau \vartheta_\nu(x,\tau)\vartheta_\nu(x',0)\rangle$. The average is here defined with respect to H_0 in (1) with the solution

$$\mathcal{G}_\nu(\omega_n;x,x') = \frac{g_\nu}{2|\omega_n|}\exp\left[-\frac{|\omega_n(x-x')|}{v_\nu}\right]. \tag{11}$$

The kernels $K_\nu^r(\omega_n)$ contain the collective bulk modes which introduce the interaction effects described below. It is easy to see that $K_\nu^+(\omega_n \to 0) = 0$ but $K_\nu^-(\omega_n \to 0) \neq 0$ [34,48]. The latter describe the costs in energy for changing the numbers of charges and/or spins on the island between the potential barriers. The corresponding action is

$$S_0[N_\rho^-, N_\sigma^-] = \sum_{\nu=\rho,\sigma} \frac{E_\nu}{2} \int_0^\beta \mathrm{d}\tau \, [N_\nu^-(\tau)]^2 \,, \tag{12}$$

with the energies $E_\nu = 2K_\nu^-(\omega_n \to 0)$. For $\nu = \rho$, this corresponds to the charge addition energy that has to be supplied/is gained in order to transfer/remove one charge to/from the island as compared with the neutral configuration. Correspondingly, for $\nu = \sigma$, the spin addition energy E_σ is needed/gained in order to change the z-component of the spin in the dot by exactly $\pm 1/2$. For the above model, that contains a zero range interaction, these energies can be easily calculated

$$E_\rho = \frac{\pi}{2a} \frac{v_\rho}{g_\rho} \,, \qquad E_\sigma = \frac{\pi}{2a} \frac{v_\sigma}{g_\sigma} \,. \tag{13}$$

Without interactions, $g_\nu = 1$, the addition energies are equal and still finite, $E_\rho = E_\sigma$. They represent the energy that has to be paid because of the Pauli principle with the discreteness of the dot levels. For strong Coulomb interaction we have $E_\rho \gg E_\sigma$.

The frequency dependent parts of the kernels describe the dynamical effects of the "external leads" and of the excited states in the quantum dot via the spectral densities [33,34]

$$J_\nu(\omega) = \frac{2}{\pi} \mathrm{Im} \left[K_\nu^+(\omega_n \to i\omega) + K_\nu^-(\omega_n \to i\omega) \right] \,. \tag{14}$$

They can be decomposed into two parts that describe the influences of the leads and dot,

$$J_\nu(\omega) = \frac{\omega}{2g_\nu} \left[1 + \varepsilon_\nu \sum_{m=1}^{\infty} \delta(\omega - m\varepsilon_\nu) \right] \,. \tag{15}$$

The energies

$$\varepsilon_\rho = \frac{\pi v_\rho}{a} \,, \qquad \varepsilon_\sigma = \frac{\pi v_\sigma}{a} \,, \tag{16}$$

correspond to the quantization energies of the collective charge and spin waves in the quantum dot. Differently from the charge and spin addition energies they do not change neither the spin polarization nor the total number of electrons in the dot.

Although we have obtained an explicit expression of the four energy scales in eqs. (13) and (16), as a function of the interaction strength, we expect that the coupling to gates and the presence of long range electron interactions and inhomogeneity could renormalize the charge sector [34,49]. For this reason we will treat the two charge energy scales E_ρ and ε_ρ as independent parameters, with values much larger than the discretization energies of the spin degrees of freedom.

4 Current Transport

4.1 Conditions for Linear and Nonlinear Transport

To calculate explicitly the electrical current one has to solve the equations of motion for $N_\nu^\pm$. For barriers much higher than the charge and spin addition energies, $U_\mathrm{t} \gg E_\rho, E_\sigma$, the dynamics is dominated by tunneling events that connect the minima of H_t in the 4D space of the variables, $N_\rho^+, N_\rho^-, N_\sigma^+, N_\sigma^-$ [45]. The transitions between the minima of H_t correspond to different physical processes of transferring electrons from one side to the other of the quantum dot. If the temperature is higher than the tunneling rate through the single barrier, the dominant processes are sequential tunnelings [32]. The transfer of charge occurs via uncorrelated single-electron hops into and out of the island, associated with corresponding changes in the total spin. In this region, the minima correspond to pairs of even and odd integer N_ρ^- and N_σ^-, respectively, and vice versa. The dominant transport processes are those which connect minima via transitions $N_\rho^- \to N_\rho^- \pm 1$ associated with changes of the spin $N_\sigma^- \to N_\sigma^- \pm 1$. For each of these also the external charge and the spin change by $N_\nu^+ \to N_\nu^+ \pm 1$. The degeneracy of these minima is lifted by the charge and spin addition energies which force the system to select favorable charge and spin states of the electron island. These selection rules become essential at low temperatures, $k_\mathrm{B}T < E_\rho \pm E_\sigma$, when the current can flow through the dot only under particular circumstances. The latter can be achieved in experiment by tuning external parameters, like source-drain or gate voltages, in order to create degenerate charge states in the island.

4.2 Quantum Rates

In sequential tunneling the transport dynamics can be described by a master equation for charges and spins in terms of backward and forward tunneling rates through the left and right barriers with spin up and down. In terms of the total spectral density $J_\mathrm{tot}(\omega) = J_\rho(\omega) + J_\sigma(\omega)$, (15), one has

$$\Xi(E,T) = \frac{\Delta^2}{4} \int_{-\infty}^{\infty} \mathrm{d}t \mathrm{e}^{iEt} \mathrm{e}^{-W(t)} , \tag{17}$$

with

$$W(t) = \int_0^\infty \mathrm{d}\omega \frac{J_\mathrm{tot}(\omega)}{\omega^2} \left\{ [1-\cos(\omega t)] \coth\frac{\beta\omega}{2} + i\sin(\omega t) \right\} e^{-\omega/\omega_\mathrm{c}} , \tag{18}$$

where Δ is the tunnel matrix element, related to $U_{\rm t}$ via the WKB approximation [50], and $\omega_{\rm c}$ is the frequency cutoff of the Luttinger model. From the expressions of the spectral densities in (15) one can separate the contributions of the modes outside and inside the quantum dot [45]

$$\Xi(E,T) = \sum_{l,m=-\infty}^{\infty} W_l^{\rho}(\varepsilon_\rho) W_m^{\sigma}(\varepsilon_\sigma)\gamma(E - l\varepsilon_\rho - m\varepsilon_\sigma, T)\,. \tag{19}$$

The rate $\gamma(E,T)$ is

$$\gamma(E,T) = \frac{\Delta^2}{4\omega_{\rm c}}\left(\frac{\beta\omega_{\rm c}}{2\pi}\right)^{1-1/g_{\rm eff}} \left|\Gamma\left(\frac{1}{2g_{\rm eff}} + \frac{i\beta E}{2\pi}\right)\right|^2 \frac{e^{-|E|/\omega_{\rm c}}\, e^{\beta E/2}}{\Gamma(1/g_{\rm eff})}\,, \tag{20}$$

where

$$\frac{1}{g_{\rm eff}} = \frac{1}{2}\left(\frac{1}{g_\rho} + \frac{1}{g_\sigma}\right)\,, \tag{21}$$

is the effective interaction of the continuum of excitations present in the leads. At $E = 0$, the rate shows the typical power law $\gamma(0,T) \propto T^{-1+1/g_{\rm eff}}$ driven by $g_{\rm eff}$.

The charge and spin density excitations are represented by the weights $W_m^{\nu}(\varepsilon_\nu)$. A corresponding analytic expression is available only for $k_{\rm B}T \ll \varepsilon_\nu$, $(\nu = \rho, \sigma)$ [45]

$$W_m^{\nu}(\varepsilon_\nu) \approx \left(\frac{\varepsilon_\nu}{\omega_{\rm c}}\right)^{1/2g_\nu} \frac{\Gamma(1/2g_\nu + m)}{m!\,\Gamma(1/2g_\nu)} e^{-m\varepsilon_\nu/\omega_{\rm c}}\theta(m)\,. \tag{22}$$

As already discussed, these expressions are obtained by assuming a thermal distribution for the charge and spin modes in the leads and in the dot. At very low temperatures this implies that any excited collective charge and spin state, occupied during transport, will decay into the ground state long time before the next electron can tunnel into or out from the dot at very low temperatures. This is a way to describe phenomenologically the presence of relaxation phenomena that are faster than the corresponding tunneling events. Microscopic processes responsible for the relaxation of the plasmons are for example induced by the coupling between the electron and the phonons [51]. In the case of the spin density, which are not associated with spin flip processes, there are several different relaxation mechanisms. These range from spin orbit coupling to a direct coupling with phonons [52].

5 Results

5.1 Linear Conductance

In the linear regime, $V \to 0$, and at low temperatures, $k_{\rm B}T \ll E_\rho \pm E_\sigma$, starting with the island containing n electrons, an additional electron can enter only if the difference between the ground state energies $E_0(n+1)$ of $n+1$ electrons and $E_0(n)$ of n electrons is aligned with the chemical potentials of the external semi-infinite leads. The ground state energy of n electrons with the z-component of the spin $s_n/2$ (s_n integer) is determined by the charge and spin addition energies

$$E_0(n, s_n) = \frac{1}{2}[E_\rho(n - n_{\mathrm{g}})^2 + E_\sigma s_n^2] , \tag{23}$$

where the particle number $n_{\mathrm{g}} \equiv eV_{\mathrm{g}}\delta/E_\rho + n_0$ is fixed by the gate voltage and $s_n = \pm[1 - (-1)^n]/2$. The latter correspond to a total z-component of the spin equal to 0 or $\pm 1/2$ for even or odd numbers of electrons in the dot. The chemical potential of the dot ground state is then

$$\mu_{\mathrm{GS}}(n, s_n) = \frac{1}{2}[E_\rho(2n + 1 - 2n_{\mathrm{g}}) + (-1)^n E_\sigma] . \tag{24}$$

The allowed parameter regions for the linear transport are given by the inequality $-eV/2 < \mu_{\mathrm{GS}}(n, s_n) < eV/2$ with $V \to 0$. The latter defines the resonance condition for the linear Coulomb blockade peaks at given gate voltages. Due to the spin, the separation of the peaks depends on the parity of n. At $T = 0$, the peak position of the $n \leftrightarrow n+1$ charge transition is given by

$$n_{\mathrm{g}}^{\mathrm{res}}(T = 0, n) = n + \frac{1}{2} + (-1)^n \frac{E_\sigma}{2E_\rho} . \tag{25}$$

The distance between two successive peaks is

$$\Delta n_{\mathrm{g}}^{\mathrm{res}}(n + 1, n) \equiv n_{\mathrm{g}}^{\mathrm{res}}(0, n + 1) - n_{\mathrm{g}}^{\mathrm{res}}(0, n) = 1 - (-1)^n \frac{E_\sigma}{E_\rho} . \tag{26}$$

The linear conductance can be calculated by solving the master equation for the current in the linear regime, in terms of the rates given in eqs. (19) and (20) and in the presence of the external gate voltage V_{g}.

For a given Coulomb peak, identified at $T = 0$ by $n_{\mathrm{g}}^{\mathrm{res}}(T = 0, n)$ in (25), we can extract an analytic expression for the conductance only in the limit $k_{\mathrm{B}}T \ll \varepsilon_\rho, \varepsilon_\sigma, E_\sigma$. Here, the temperature is low enough to avoid high spin polarization states and charge and spin density excitations. The dependence on the energy $\mu = E_\rho[n_{\mathrm{g}} - n_{\mathrm{g}}^{\mathrm{res}}(0, n)]$ and temperature is

$$G(\mu, T) = \frac{\beta e^2}{2\sqrt{2}} \frac{e^{-\beta\mu/2}}{\cosh[(\beta\mu + (-1)^n \ln 2)/2]} W_0^\rho(\varepsilon_\rho) W_0^\sigma(\varepsilon_\sigma)\, \gamma(\mu, T) . \tag{27}$$

At higher temperatures we obtain the conductance by solving numerically the master equation. These regimes are presented in Fig. 1. Here, the conductance of two successive peaks $(n, n+1)$ and $(n+1, n+2)$ with $n = 0$ is shown as a function of n_{g} at different temperatures.

Due to the degeneracy of the spin $s_n = \pm 1$ for the odd charge ground states, successive peak positions move in opposite directions when increasing the temperature, as indicated by the arrows in Fig. 1. At sufficiently low temperatures,

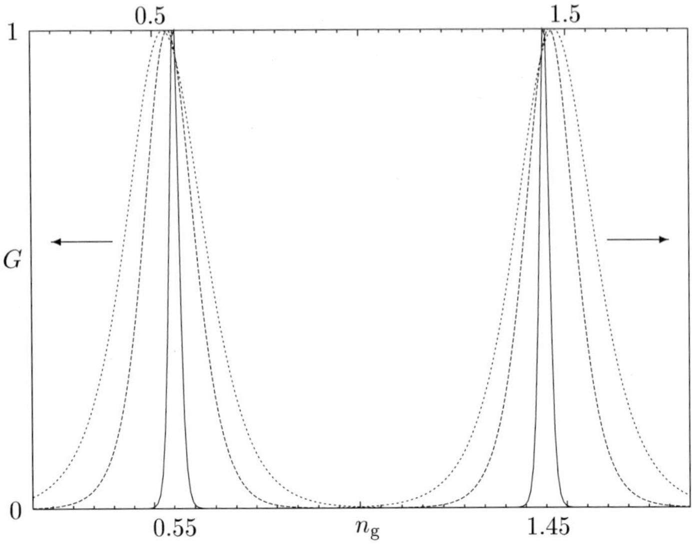

Fig. 1. The linear conductance G, normalized to its maximum value, as a function of the reference particle number n_g for two subsequent charge transitions $0 \leftrightarrow 1$ (*left*) and $1 \leftrightarrow 2$ (*right*) at different temperatures $k_B T/E_\sigma = 0.11$ (*solid curve*); 0.58 (*dashed curve*); 0.87 (*dotted curve*). Parameters: $g_\rho = 0.4$, $g_\sigma = 1$, $E_\rho/E_\sigma = 9$, $\varepsilon_\rho/E_\sigma = 5$, $\varepsilon_\sigma/E_\sigma = 2$, $E_\sigma/\omega_c = 10^{-6}$

$k_B T \ll E_\sigma$, one has

$$n_g^{res}(T, n) \approx n + \frac{1}{2} + (-1)^n \left[\frac{E_\sigma}{E_\rho} - \alpha \ln(2) \frac{k_B T}{E_\rho} \right] , \tag{28}$$

with a slope $\alpha(g_{eff})$ that depends on the interaction strength as shown in the inset (b) of Fig. 2. This generalizes the non-interacting case ($g_{eff} = 1$) where $\alpha(1) = 1/2$ gives the usual entropic factor [53]. For extremely strong interactions, $g_{eff} \to 0$, one obtains $\alpha(0) = 1$. At higher temperatures the peak positions approach the "spinless" value $n_g^{res}(n) = n + 1/2$ (cf. Fig. 2).

Also the maximum of the conductance $G^{max}(T)$ depends on T with two asymptotic behaviors

$$G^{max}(T) \propto T^{-2+1/g_{eff}} \qquad k_B T \ll E_\sigma, \varepsilon_\sigma \tag{29}$$

$$G^{max}(T) \propto T^{-2+1/g_{eff}+1/2g_\sigma} \qquad k_B T \gg E_\sigma, \varepsilon_\sigma \, . \tag{30}$$

At low temperatures $k_B T \ll E_\sigma, \varepsilon_\sigma$, the spin modes in the dot are not excited and the continuum of excitations is due to the modes in the leads only (cf. (27)). This implies a power law driven by the effective interaction g_{eff} in (21). At larger temperatures, $k_B T \gg E_\sigma, \varepsilon_\sigma$, the spin modes in the dot are also a continuum and contribute to the exponent of the temperature power law with the additional term $1/2g_\sigma$. The latter derives by the continuum limit ($\varepsilon_\sigma \to 0$) of the δ-functions present in the spectral density $J_\sigma(\omega)$ in (15). The inset (a) of Fig. 2 shows the details of these two different regimes.

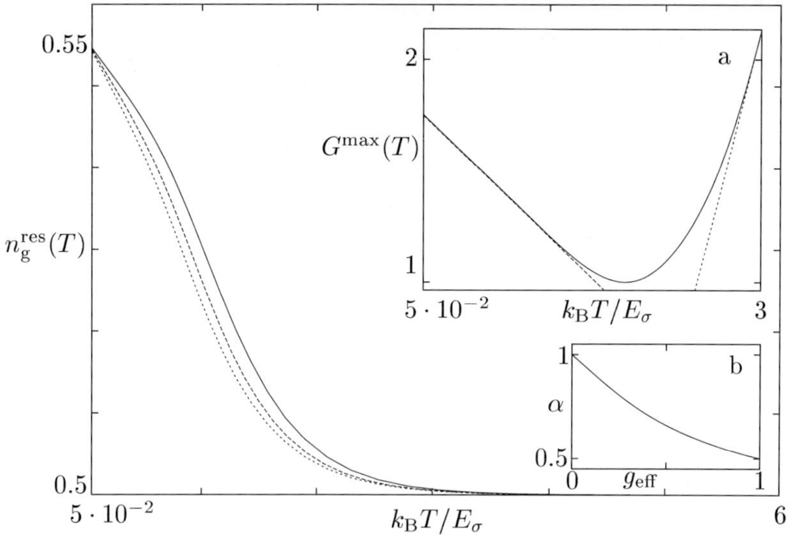

Fig. 2. Temperature dependence of the linear conductance peak position $n_g^{res}(T, n)$ for $n = 0$, charge transition $0 \leftrightarrow 1$, with zero exchange interaction $g_\sigma = 1$, and different charge interactions: $g_\rho = 1$ (*solid*), $g_\rho = 0.6$ (*dashed*), $g_\rho = 0.4$ (*dotted*). The other parameters are as in Fig. 1. Inset. **a** Double logarithmic plot of the temperature behavior of the maximum conductance normalized to its minimum value with arbitrary units and $g_\rho = 0.4$. The *dashed, dotted lines* correspond to the asymptotic power laws discussed in eqs. (29) and (30) respectively. **b** Slope α of the linear low temperature behavior of $n_g^{res}(T, n)$, (cf. (28)) as a function of the interaction g_{eff}

5.2 Non-linear Transport

When increasing the source drain voltage V the current-voltage characteristic shows several structures related to the possible higher transitions. They are defined by the inequality $-eV/2 < \mu(n, s_n) < eV/2$, where

$$\mu(n, s_n) = \frac{E_\rho}{2}(2n + 1 - 2n_g) + \frac{E_\sigma}{2}(s_{n+1}^2 - s_n^2) + l\varepsilon_\rho + m\varepsilon_\sigma \,. \qquad (31)$$

Positive or negative integers l and m denote the differences of the numbers of charge and spin excitation quanta of the corresponding $n + 1$ and n excited states. To describe these spectra we restrict our attention to spin polarization states with spin component larger than $s = 0, \pm 1$ and to spin waves only. We then restrict the voltage, $eV < E_\rho \pm E_\sigma, \epsilon_\rho$, in the region of two fixed charge states, n and $n + 1$, without any charge density excitations ($l = 0$).

Each time a new transport channel of high spin polarization or spin waves enter the differential conductance shows sharp peaks. At $T = 0$, as a function of the source-drain voltage, the conductance drops according to the interaction-induced non-Fermi liquid power law $(V - V_c)^{1/g_{eff}-2}$, with V_c the threshold of the entrance of a given channel. As an example of the rich spin-related fine structure

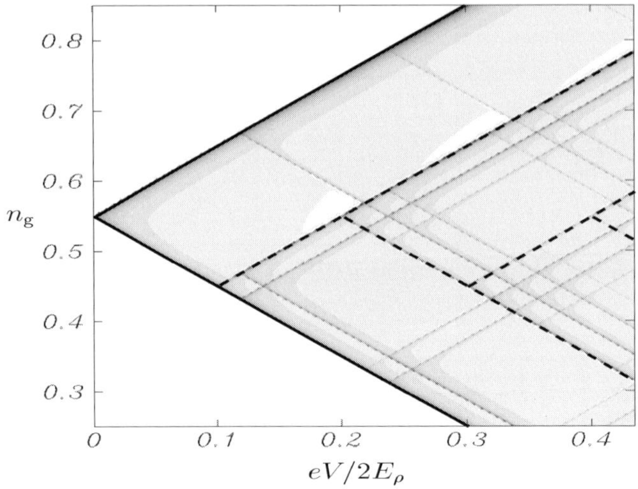

Fig. 3. Density plot of the differential conductance in the ($eV/2E_\rho$,n_g)-plane at $T = 0$. The units are arbitrary with a darker gray for higher values of the differential conducatance. Parameters: $g_\rho = 0.8$, $g_\sigma = 1.2$, $E_\rho = 10E_\sigma$, $\varepsilon_\rho = 16E_\sigma$, $\varepsilon_\sigma/E_\sigma = 2g_\sigma$, $E_\sigma/\omega_c = 10^{-6}$

in the transport spectrum a density plot of the differential conductance in the ($eV/2E_\rho$,n_g)-plane at $T = 0$ is shown in Fig. 3. Here we have considered the transitions $n \leftrightarrow n{+}1$ with n even. The white regions of the Coulomb blockade are well developed. With increasing bias voltage the complexity of the dot excitation spectrum rapidly increases and displays a considerable number of spin-related transitions. The two solid bold lines identify the $(n, s_n = 0) \leftrightarrow (n{+}1, s_{n+1} = \pm 1)$ ground state to ground state transitions. The other bold dashed lines correspond to higher spin value transitions $(n, s_n) \leftrightarrow (n + 1, s_{n+1})$ with $s_n = 2, 4, 6, \ldots$ and $s_{n+1} = 3, 5, 7, \ldots$.

In particular the transitions $(n + 1, s_{n+1}) \to (n, s_n)$ are represented by the lines

$$n_g = \frac{1}{2E_\rho} \left[E_\rho + E_\sigma + eV - 2E_\sigma s_n\right] , \tag{32}$$

while the transitions $(n, s_n) \to (n + 1, s_{n+1})$ are given by

$$n_g = \frac{1}{2E_\rho} \left[E_\rho + E_\sigma - eV - 2E_\sigma(s_{n+1} - 1)\right] . \tag{33}$$

For each of these high spin states one can generate also a sequence of spin density excitations $m\varepsilon_\sigma$ with $m = 0, \pm 1, \pm 2 \ldots$ (cf. (31)); they are represented by the weaker lines in Fig 3. With non-vanishing exchange interaction, $g_\sigma \neq 1$, accidental degeneracies due to spin addition and excitation are lifted; with

$\varepsilon_\sigma = 2g_\sigma E_\sigma$ one can then discriminate in the transport spectrum between spin addition energies and spin density waves.

6 Conclusion

We presented a theory for spin transport through a 1D quantum dot formed by a double barrier potential in the presence of Luttinger interactions between the electrons. Addition-energy terms, originating from the dissipative charge and spin bulk modes of the Luttinger liquid, have been found for both the charge and the spin. They are identified with the charge and spin addition energies in the case of the charge and spin excitations respectively.

We have shown that the spin dynamic affects the linear regime, leading to a parity effect in the Coulomb blockade peak spacing. Due to the degeneracy of the spin $s_n = \pm 1$ for the odd charge ground states we demonstrate that the positions of successive peaks move in opposite directions when increasing the temperature. In the nonlinear transport, charge/spin separation leads to conductance peaks related to spin excitations. Our results suggest that the electron spin cannot be neglected when quantitatively deducing the interaction parameter from experimental data, especially in the presence of backscattering impurities. The above scenario also predicts that the energetically lowest collective excitations in the quasi-1D quantum island used in the experiment [28] are spin excitations, consistent with an early suggestion based on a semi-phenomenological model in which exact diagonalization and rate equations have been combined [4].

Acknowledgments

This work has been supported by the European Union via the TMR and RTN programmes (FMRX-CT98-0180, HPRN-CT2000-00144), by the Italian MURST via PRIN02.

References

1. G.A. Prinz: Physics Today **282**, 58 (1995); Science **282**, 1660 (1998)
2. J.M. Kikkawa, D.D. Awschalom: Phys. Rev. Lett. **80**, 4313 (1998); D.D. Awschalom, J.M. Kikkawa: Phys. Today **52**, 33 (1999)
3. S. Tarucha, D.G. Austing, T. Honda, R.J. van der Hage, L.P. Kouwenhoven: Phys. Rev. Lett. **77**, 3613 (1996)
4. D. Weinmann, W. Häusler, B. Kramer: Phys. Rev. Lett. **74**, 984 (1995)
5. S. Tarucha, T. Honda, T. Saku: Solid State Commun. **94**, 413 (1995)
6. A. Yacoby, H.L. Stormer, N.S. Wingreen, L.N. Pfeiffer, K.W. Baldwin, K.W. West: Phys. Rev. Lett. **77**, 4612 (1996)
7. A. Yacoby, H.L. Stormer, K.W. Baldwin, L.N. Pfeiffer, K.W. West: Solid State Commun. **101**, 77 (1997)
8. M. Bockrath, D.H. Cobden, J. Lu, A.G. Rinzler, R.E. Smalley, L. Balents, P.L. McEuen: Nature **397**, 598 (1999)

9. J. Park, A.N. Pasupathy, J.I. Goldsmith, C. Chang, Y. Yaish, J.R. Petta, M. Rinkoski, J.P. Sethna, H.D. Abruña, P.L. McEuen, D.C. Ralph: Nature **417**, 722 (2002)
10. W. Liang, M.P. Shores, M. Bockrath, J.R. Long, H. Park: Nature **417**, 725 (2002)
11. A. Imamoglu, D.D. Awschalom, G. Burkard, D.P. DiVincenzo, D. Loss, M. Sherwin, A. Small: Phys. Rev. Lett. **83**, 4204 (1999)
12. P. Recher, E.V. Sukhorukov, D. Loss: Phys. Rev. Lett. **85**, 1962 (2000)
13. H.A. Engel, D. Loss: Phys. Rev. Lett. **86**, 4648 (2001)
14. A. Brathaas, Y. Nazarov, G.E.W. Bauer: Phys. Rev. Lett. **84**, 2481 (2000)
15. Q. Si: Phys. Rev. Lett. **81**, 3191 (1998)
16. L. Balents, R. Egger: Phys. Rev. Lett. **85**, 3464 (2000)
17. S. Tomonaga: Prog. Theor. Phys **5**, 544 (1950)
18. J.M. Luttinger: J. Math. Phys **4**, 1154 (1963)
19. F.D.M. Haldane: J. Phys. C **14**, 2585 (1981)
20. C.L. Kane, M.P.A. Fisher: Phys. Rev. Lett. **68**, 1220 (1992); Phys. Rev. B **46**, 15233 (1992)
21. A. Kawabata: J. Phys. Soc. Japan **65**, 30 (1995)
22. Y. Oreg, A. Finkel'stein: Phys. Rev. B **54**, R14265 (1996)
23. G. Cuniberti, M. Sassetti, B. Kramer: Phys. Rev. B **57**, 1515 (1998)
24. D.L. Maslov, M. Stone: Phys. Rev. B **52**, R5539 (1995)
25. V.V. Ponomarenko: Phys. Rev. B **52**, R8666 (1995)
26. I. Safi, H.J. Schulz: Phys. Rev. B **52**, R17040 (1995)
27. D.L. Maslov: Phys. Rev. B **52**, R14386 (1995)
28. O.M. Auslaender, A. Yacoby, R. de Picciotto, K.W. Baldwin, L.N. Pfeiffer, K.W. West: Phys. Rev. Lett **84**, 1764 (2000)
29. D.V. Averin, K.K. Likharev. In: *Mesoscopic Phenomena in Solids*, ed. by B.L.Altshuler, P.A. Lee, R.A. Webb (Elsevier, Amsterdam 1991) pp. 1–10
30. U. Meirav, M.A. Kastner, S.J. Wind: Phys. Rev. Lett. **65**, 771 (1990)
31. L.P. Kouwenhoven, C.M. Marcus, P.L. McEuen, S. Tarucha, R.M. Westervelt, N.S. Wingreen. In *Mesoscopic Electron Transport*, NATO Advanced Studies Institute, Ser. E **345**, ed. by L.L. Sohn, G. Schön, L.P. Kouwenhoven (Kluwer Acad. Publ., Dordrecht 1997) p. 105
32. A. Furusaki: Phys. Rev. B **57**, 7141 (1998)
33. A. Furusaki, N. Nagaosa: Phys. Rev. B **47**, 3827 (1993)
34. T. Kleimann, M. Sassetti, B. Kramer, A. Yacoby: Phys. Rev. B **62**, 8144 (2000)
35. A.T. Johnson, L.P. Kouwenhoven, W. de Jong, N.C. van der Vaart, C.J.P.M. Harmans, C.T. Foxon: Phys. Rev. Lett. **69**, 1592 (1992)
36. J. Weis, R.J. Haug, K. von Klitzing, K. Ploog: Phys. Rev. Lett. **71**, 4019 (1993)
37. J. Voit: Rep. Prog. Phys. **57**, 977 (1995)
38. M. Fabrizio, A.O. Gogolin, S. Scheidl: Phys. Rev. Lett. **72**, 2235 (1995)
39. N. Nagaosa, A. Furusaki: J. Phys. Soc. Japan. **63**, 413 (1995)
40. G. Cuniberti, M. Sassetti, B. Kramer: Europhys. Lett. **37**, 421 (1997)
41. M. Sassetti, B. Kramer: Phys. Rev. B **55**, 11213 (1997)
42. M. Sassetti, F. Napoli, U. Weiss: Phys. Rev. B **52** 11213 (1995)
43. R. Egger, H. Grabert: Phys. Rev. B **55**, 9929 (1997)
44. H. Maurey, T. Giamarchi: Europhys. Lett. **38**, 681 (1997)
45. A. Braggio, M. Sassetti, B. Kramer: Phys. Rev. Lett. **87**, 146802 (2001)
46. H.J. Schulz: Phys. Rev. Lett. **71**, 1864 (1993)
47. M. Sassetti, B. Kramer: Phys. Rev. Lett. **80**, 1485 (1998); Eur. Phys. J. B **4**, 357 (1998)

48. M. Sassetti, B. Kramer: Phys. Rev. B **54**, R5203 (1996)
49. T. Kleimann, F. Cavaliere, M. Sassetti, B. Kramer: Phys. Rev. B **66**, 165311 (2002)
50. U. Weiss: *Quantum Dissipative Systems,* 2nd edn. (World Scientific, Singapore 1999)
51. T. Brandes, B. Kramer: Phys. Rev. Lett **83**, 3021 (1999)
52. A.V. Khaetskii, Yu.V. Nazarov: Phys. Rev. B **61**, 12639 (2000)
53. G. Usaj, H.U. Baranger: Phys. Rev. B **64**, 201319R (2001)

Andreev-Lifshitz Supersolid for a Few Electrons on Small Periodic Square Lattices

Jean-Louis Pichard[1,2] and Zoltán Ádám Németh[1,3]

[1] CEA/DSM, Service de Physique de l'Etat Condensé,
Centre d'Etudes de Saclay, 91191 Gif-sur-Yvette Cedex, France
[2] Laboratoire de Physique Théorique et Modélisation,
Universtité de Cergy-Pontoise, 95031, Cergy-Pontoise Cedex, France
[3] Eötvös University, Departement of Physics of Complex Systems,
H-1117 Budapest, Pázmány Péter sétány 1/A, Hungary

1 Introduction

To describe how one goes from independent particle towards collective motion in two dimensions is a difficult, fascinating and fundamental question which has been investigated since almost half a century. The parameter governing this crossover is the dimensionless ratio r_s introduced by E. Wigner. In two dimensions, assuming a uniform density n_s of carriers interacting via a long range e^2/r Coulomb repulsion, $a_0 = 1/\sqrt{n_s}$ is the distance between the carriers. The Bohr radius $a_B = \hbar^2/me^2$ is the size of the hydrogen atom in its ground state, and characterizes the length where one should expect very strong quantum effects. For carriers of effective mass m^* in a medium of dielectric constant ϵ, a_B becomes an effective Bohr radius $a_B^* = \hbar^2\epsilon/m^*e^2$. The factor r_s is the ratio of those two characteristic scales:

$$r_s = \frac{a_0}{a_B^*} = \frac{E_{Coul}}{E_F}, \tag{1}$$

or the ratio of two characteristic energies, the first being the classical Coulomb energy:

$$E_{Coul} \sim \frac{e^2}{\epsilon a_0}, \tag{2}$$

while the second is the quantum kinetic energy:

$$E_F \sim \frac{\hbar^2}{2m^* a_0^2}, \tag{3}$$

i.e. the Fermi energy of the non interacting system.

When $r_s \gg 1$, a_0 is very large compared to a_B^*, the quantum effects are negligible and one has to solve a classical problem of electrostatics. In this limit, the carriers form a hexagonal lattice to minimize E_{Coul}, and one gets a Wigner crystal [1,2]. In the other limit, when $a_0 < a_B^*$, the quantum effects dominate, and the Coulomb interaction becomes negligible. This is mainly due to the Pauli principle, which yields a large value for $E_F \sim a_0^{-2}$, while $E_{Coul} \sim a_0^{-1}$. The ground state is a Fermi liquid, and the low energy dynamics can be described by weakly interacting Landau quasi-particles, with renormalized parameters. Therefore, if one increases the carrier density, one ontains a very fascinating crossover from the classical limit towards the quantum limit. The Wigner crystal has been observed for electrons on Helium surfaces [3], where r_s is very large. The 2d Fermi liquid has been achieved in doped semiconductor heterostuctures, where r_s is usually small (typically ≈ 1). However, one can create two dimensional electron (hole) gases in ultra clean heterostructures and, varying by a gate the density n_s, one can reach[4] a dilute limit where the Fermi energy is relatively small (typically $E_F \sim 1K$) and r_s relatively large (typically $r_s = 10-40$). Therefore, the study of a fundamental quantum-classical crossover becomes experimentally possible. This is time to return to basic questions, and to re-visit different scenarios which have been proposed long ago, combining the possibilities offered by today's computers and measurements using ultra-clean devices.

2 Overview of Quantum Monte Carlo Studies

From quantum Monte Carlo studies, it is usually assumed [5,6] that a single liquid-solid transition takes place at $r_s \approx 37$ in the continuous $2d$ thermodynamic limit. For instance, this single transition is given by fixed node calculations [5,7]. The existence of an intermediate polarized liquid phase separating the unpolarized liquid and the Wigner solid is still debated [8]. However, the quantum Monte Carlo calculations have the well-known "sign problem". This leads to fixed node approximations which are made to avoid the negative weights that would be generated otherwise by antisymmetric states, and gives only an upper bound to the ground state (GS) energy. In [5,7] for instance, two nodal structures are considered, given by two Slater-Jastrow wave functions adapted to describe the weak coupling Fermi liquid (nodal structure of a Slater determinant of plane waves) or the strong coupling Wigner solid (nodal structure of a Slater determinant of localized site orbitals). One cannot exclude that the existence of a single transition separating the Fermi liquid from the Wigner solid is a consequence of the assumed nodal structures.

3 Lattice Model

For revisiting the Fermi-Wigner crossover, one uses a two dimensional model describing N particles on $L \times L$ square lattice with periodic boundary conditions (BCs), i.e. with a torus topology. The most general Hamiltonian $\mathcal{H}$ of the lattice model one considers reads,

$$\mathcal{H} = -t \sum_{\langle ii' \rangle,\sigma} c^\dagger_{i',\sigma} c_{i,\sigma} + \sum_{i,\sigma} v_i n_{i,\sigma} + \frac{U}{2} \sum_{\substack{i,i' \\ i \neq i'}} \frac{n_{i,\sigma} n_{i',\sigma'}}{|i - i'|} + 2U \sum_i n_{i,\uparrow} n_{i,\downarrow}, \tag{4}$$

where the operators $c_{i,\sigma}$ ($c^\dagger_{i,\sigma}$) destroy (create) an electron of spin σ at the site i and $n_{i,\sigma} = c^\dagger_{i,\sigma} c_{i,\sigma}$. $\mathcal{H}$ consists of: (i) a hopping term $-t$ that couples nearest-neighbor sites, and accounts for the quantum kinetic energy, (ii) the pairwise electron-electron interaction, which itself consists of a $2U$ Hubbard repulsion when two electrons are at the same site i with opposite spins and a $U/|i - i'|$ spin independent Coulomb repulsion when they are separated by a distance $|i-i'|$ (smallest distance between the sites i and i' on a square lattice with periodic BCs), and eventually (iii) on site random potentials v_i which are uniformly distributed inside the interval $[-W/2, W/2]$. For a lattice spacing a, $\hbar^2/(2m^* a^2) \to t$, $e^2/(\epsilon a) \to U$, such that the factor r_s becomes in our model:

$$r_s = \frac{1}{\sqrt{\pi n_s} a^*_B} = \frac{U}{2t\sqrt{\pi n_e}}, \tag{5}$$

for a filling factor $n_e = N/L^2$.

We denote $\mathcal{S}$ and $\mathcal{S}_z$ the total spin and its component along an arbitrary direction z. Since $[\mathcal{S}^2, \mathcal{H}] = [\mathcal{S}_z, \mathcal{H}] = 0$, $\mathcal{H}$ can be written in a block-diagonal form, with $N+1$ blocks where $S_z = -N/2, \dots, N/2$ respectively. When $B = 0$, there is no preferential direction and the ground state energy E_0 does not depend on S_z. For a ground state of total spin S, $\mathcal{H}$ has $2S+1$ blocks with the same lowest eigenenergy $E_0(S^2)$ since $E_0(S^2) = E_0(S^2, S_z)$ where $S_z = -S, -S+1, ..., S-1, S$. Therefore, the number N_b of blocks of different S_z and of same lowest energy gives the GS magnetization $S = (N_b - 1)/2$.

If N and L are small enough, the ground state and the first excitations can be exactly calculated using the Lanczos algorithm. Let us focus on the case $N = 4$ and $L = 6$. When $B = 0$, we have the symmetry $\pm S_z$, and one has only to diagonalize the three sub-blocks with $S_z \geq 0$. $\mathcal{H}(S_z = 2)$ corresponds to fully polarized electrons (spinless fermions) where the orbital part of the wave-functions is totally anti-symmetric.

From exact diagonalization, one can study different quantities.

- The local persistent currents $\boldsymbol{J}(i)$ created at a site i by an Aharonov Bohm flux ϕ which is enclosed along the longitudinal l-direction. The flux ϕ can be included by taking appropriate longitudinal BCs (antiperiodic BCs corresponding to $\phi = \pi$ in our convention). The BCs along the transverse t-direction remain periodic. The currents $\boldsymbol{J}(i)$ are defined by their longitudinal and transverse components ($\boldsymbol{J}(i) = (J_{i,l}, J_{i,t})$), or by their angles $\theta_i = \arctan(J_{i,t}/J_{i,l})$ and their absolute values $J_i = |\boldsymbol{J_i}|$. We denote $\langle I_l \rangle = \left\langle \sum_{i_y=1}^{L} J_{i_y,l}(i_x = 1) \right\rangle$ the disorder average of the total longitudinal current enclosing a flux $\phi = \pi/2$.

- From the function $C(r) = N^{-1} \sum_i \rho_i \rho_{i-r}$, where $\rho_i = \langle \Psi | n_i | \Psi \rangle$ is the electronic density of the GS state $|\Psi\rangle$ at the site i, one defines a crystallization parameter γ by

$$\gamma = \max_r C(r) - \min_r C(r). \tag{6}$$

  Note that $\gamma = 1$ when the N particles are localized on N lattice sites and form a rigid solid and 0 when they are extended on the L^2 sites and form a homogeneous liquid.
- The distribution $P(s)$ of the first energy excitation

$$s = \frac{E_1 - E_0}{\langle E_1 - E_0 \rangle}, \tag{7}$$

  where the brackets denote disorder average, E_0 the GS energy and E_1 the energy of the first excited state.
- The magnetization S and the Zeeman energy necessary to polarize a non magnetized cluster. A parallel magnetic field B does not induce orbital effects, but defines the z-direction and removes the S_z degeneracy by the Zeeman energy $-g\mu BS_z$. The GS energy and its magnetization are given by the minimum of $E_0(S^2, S_z, B = 0) - g\mu BS_z$. For a $S = 0$ ground state without field, the value B^* for which $E_0(S_z) - g\mu B^* S_z = E_0(S_z = 0)$ defines the field necessary to polarize the system to $S \geq S_z$. When $N = 4$, the total $Q_2 = E_0(S_z = 2) - E_0(S_z = 0)$ and partial $Q_1 = E_0(S_z = 1) - E_0(S_z = 0)$ polarization energies give the Zeeman energies necessary to yield $S = 2$ and $S = 1$ respectively for a cluster with $S = 0$.

4 Overview of Exact Studies of Small Disordered Systems

We first consider disordered clusters and we summarize a few results given in [9–11]. Let us begin by considering $N = 4$ polarized electrons (spinless fermions) described by $\mathcal{H}(S_z = 2)$ for a size $L = 6$ and a disorder strength $W = 5$. The first observation [9] of an intermediate regime separating the weak and the strong coupling limits was given by the structure of the persistent currents, which exhibits three regimes, as shown in Fig. 1 (See also [12]). For weak coupling, the local currents flow randomly inside the cluster, due to elastic scattering on the site potentials. For intermediate coupling, the pattern of the persistent currents becomes oriented along the shortest direction enclosing ϕ. For large coupling, the oriented currents vanish.

To locate the first characteristic threshold r_s^F separating the two first regimes, one can consider the average total longitudinal persistent current $\langle I_l \rangle$ of the GS at $\phi = \pi/2$, and compare the exact quantity with the behavior given by the self consistent mean field Hartree-Fock (HF) approximation. For $N = 4$, $L = 6$, $W = 5$ and $S_z = 2$ (spinless fermions) one can see in Fig. 2 that below $r_s^F \approx 3$, the mean field approximation reproduces the exact $\langle I_l \rangle$, but strongly underestimates $\langle I_l \rangle$ above r_s^F. This sharp breakdown of the HF approximation means that strong correlation effects occur above r_s^F.

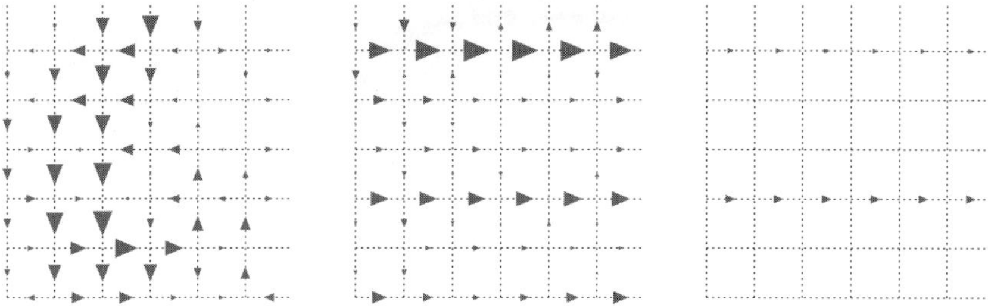

Fig. 1. Map of the local persistent currents in a given sample with $W = 5$ for small (*left*), intermediate (*middle*) and large (*right*) values of r_s

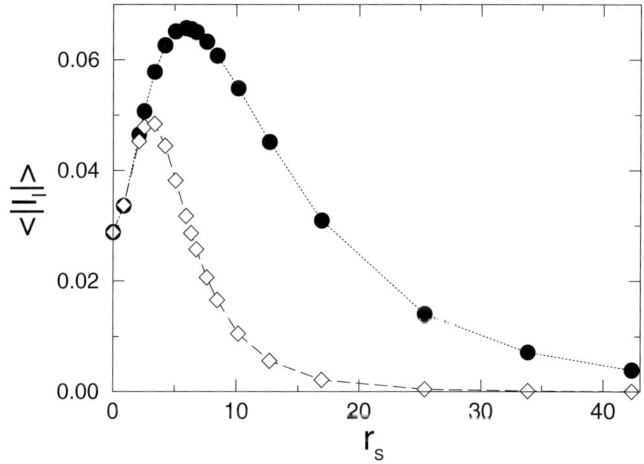

Fig. 2. Ensemble averaged longitudinal GS current $\langle I_l \rangle$ as a function of r_s for $N = 4$, $L = 6$ and $W = 5$. Exact values (*filled symbols*) and HF values (*empty symbols*)

One can also study the distribution $P(s)$ of the first energy excitation. Without interaction, the first excitation is a one particle excitation, and for moderate disorder ($W = 5$), the one particle dynamics being diffusive, one gets the well known Wigner surmise: $P_W(s) = (\pi s/2)\exp(-\pi s^2/4)$. It was noticed in [13] that the Wigner distribution holds quite accurately up to r_s^F before showing a weaker level repulsion above r_s^F. This signals also the onset of correlation effects.

For locating the second characteristic threshold r_s^W, one can study [10] the average amplitude $\langle J_i \rangle$ of the local currents: as shown in Fig. 3, $\langle J_i \rangle$ are essentially independent of r_s up to a second threshold $r_s^W \approx 10$ which exceeds r_s^F. Moreover, comparing in Fig. 3 the GS average crystallization parameter $\langle \gamma \rangle$ and $\langle J_i \rangle$, one can see that the suppression of the persistent currents coincides with the formation of a solid Wigner molecule inside the disordered clusters.

One concludes that between r_s^F and r_s^W, one has a correlated regime which does not correspond to the formation of a rigid Wigner molecule. This new

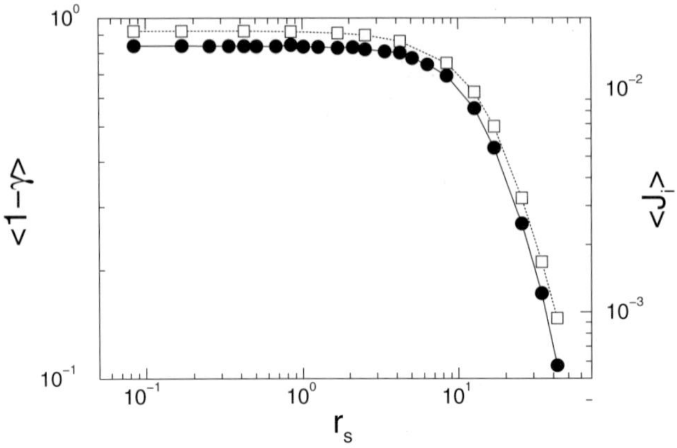

Fig. 3. Averages of the GS crystallization parameter $\langle 1-\gamma \rangle$ (*left scale, empty symbols*) and GS local current amplitude $\langle J_i \rangle$ (*right scale, filled symbols*) as a function of r_s

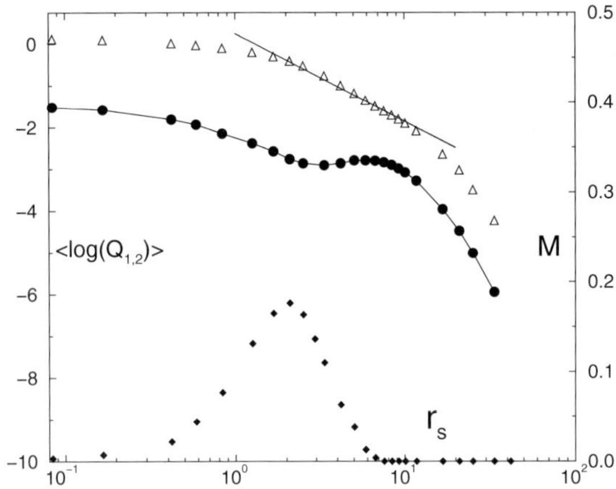

Fig. 4. As a function of r_s, fraction M of the statistical ensemble with $S = 1$ at $B = 0$ (*filled diamond, right scale*), partial $\langle \log Q_1 \rangle$ (*filled circle, left scale*) and total $\langle \log Q_2 \rangle$ (*empty triangle, left scale*) energies required to polarize $S = 0$ clusters to $S = 1$ and $S = 2$ respectively. The *straight line* corresponds to $0.25 - 2 \log r_s$

regime can also be seen if one studies the magnetization S. When $r_s = 0$ and $N = 4$, the two one body states of lowest energy are doubly occupied and $S = 0$ ($S = 1/2$ if N is odd). When r_s is large, the GS consists of 4 electrons forming an antiferromagnetic square Wigner molecule of side $L/2$, located on four sites j with the lowest substrate energy $\sum_{j=1}^{4} v_j$.

One can also include the spin degrees of freedom and study a statistical ensemble of systems described by the Hamiltonian (4) for $L = 6$ and $W = 5$. In Fig. 4, the fraction M of the statistical ensemble where $S = 1$ at $B = 0$ is given as a function of r_s. One can see the mesoscopic Stoner instability taking place at $r_s \approx 0.35$. The Stoner mechanism should eventually give fully polarized electrons. This is not the case, the increase of M breaks down when $r_s = r_s^{FS} \approx 2.2$, a value where the Stoner mechanism and hence the HF approximation break down. In the same clusters, we have seen that the HF approximation fails to describe the persistent currents of the fully polarized sub-block of $\mathcal{H}$ (polarized electrons) when $r_s > r_s^{FP} \approx 3$. r_s^{FP} takes a smaller value r_s^{FS} when the spin degrees of freedom are included. Above r_s^{FS}, M regularly decreases to reach a zero value for $r_s^{WS} \approx 9$ where an antiferromagnetic square molecule is formed. In the intermediate regime, there is a competition between the Stoner ferromagnetism and the Wigner antiferromagnetism. Since the $S = 0$ clusters are characterized by log-normal distributions, the ensemble averages $\langle \log Q_1 \rangle$ and $\langle \log Q_2 \rangle$ (without taking into account the $S = 1$ spontaneously magnetized clusters) define the typical fields B necessary to yield $S = 1$ or $S = 2$ in a $S = 0$ cluster. Figure 4 provides two magnetic signatures confirming the existence of a novel intermediate regime between the Fermi glass ($r_s < r_s^{FS}$) and the Wigner glass ($r_s > r_s^{WS}$): $\langle \log Q_1 \rangle$ becomes roughly independent of r_s, while $Q_2 \propto r_s^{-2}$. This last behavior is very interesting, since recent experiments [14] have shown that the parallel magnetic field necessary to yield full polarization varies linearly with n_s, which is compatible with our behavior $\propto r_s^{-2}$ The question is then to explain what kind of new dynamics characterizes the intermediate values of r_s. This can be related to an old conjecture proposed at the end of the sixties.

5 Andreev-Lifshitz Supersolid

In 1969, it was conjectured by Andreev and Lifshitz [15] that at zero temperature, delocalized defects may exist in a quantum solid, as a result of which the number of sites of an ideal crystal lattice may not coincide with the total number of particles. Originally, this conjecture was proposed for three dimensional quantum solids made of atoms (He^3, He^4, ...) which do not interact via Coulomb repulsion. If one considers the quantum melting of the solid from the dilute limit (large r_s), the nature of the relevant defects is not an easy question. One can imagine a particle being put into an interstitial site of the Wigner lattice, creating a vacancy-interstitial pair at a certain electrostatic cost δU. Classically, this defect remains localized. But quantum tunneling may lead to delocalization and to the appearance of a band of defects of finite width B_d which increases when r_s decreases. When B_d exceeds δU, one can imagine two possibilities: either the total quantum melting of the crystal (simple solid-liquid transition), or a partial melting leading to the persistence of a floppy crystal with delocalized defects. If a delocalized defect appears in the quantum crystal, the crystal remains perfectly periodic, but the number of crystal lattice sites becomes smaller than the total number of particles. This is the supersolid scenario proposed by Andreev-Lifshitz for He physics.

If one considers charge crystallization from the other limit, where the density n_s is large (small r_s), one can argue that the interaction will create correlated pairs of particles near the Fermi surface, but will not reorganize the one particle states well below the Fermi surface. Such a possibility has been proposed by

Bouchaud et al [16] for liquid He^3. In this picture, the system is thought as made of unpaired fermions with a *reduced* Fermi energy, co-existing with a strongly paired, nearly solid assembly. To the concept of a crystal with a reduced number of crystal lattice sites, as discussed by Andreev and Lifshitz from the solid limit, corresponds the concept of unpaired fermions with a reduced Fermi energy, as discussed by Bouchaud et al from the liquid limit.

6 A Supersolid Molecule Made of 4 Polarized Electrons

One can ask if the intermediate regime which we have seen might not be a supersolid glass. The supersolid being first a possible property of the clean limit, one considers [13] the lattice model without impurities.

To understand the nature of the intermediate GS, we study the distribution of the different inter-particle spacings for $W = 0$, $N = 4$ and $L = 6$. For all the states $c_i^\dagger c_j^\dagger c_k^\dagger c_l^\dagger |0\rangle$, one defines the 6 spacings $d_{ijkl}(1) \leq d_{ijkl}(2) \leq \ldots \leq d_{ijkl}(6)$ ordered by increasing values. The nth moment $d^n(p)$ of the pth GS inter-particle spacing at r_s is given by:

$$\langle d^n(p)\rangle = \sum_{ijkl=1}^{N_H} d^n_{ijkl}(p)|\langle\Psi_0(r_s)|c_i^\dagger c_j^\dagger c_k^\dagger c_l^\dagger |0\rangle|^2. \tag{8}$$

We show in Fig. 5 how the 6 mean GS inter-particle spacings $\langle d_p\rangle$ vary as a function of r_s. Without disorder, the total momentum K is conserved, and we have followed the state of minimum energy having $K = 0$. This state is the ground state above $r_s \approx 10$ for $L = 6$ where one has a level crossing between GSs of different total momenta [13]. When $r_s \to \infty$, the 3×3 Wigner molecule gives $d_1 = d_2 = d_3 = d_4 = 3$ and $d_5 = d_6 = 3\sqrt{2}$. As r_s decays, one can see that one of the largest spacings out of two and two of the smallest spacings out of four remain close their asymptotic values, in contrast to the others. This shows us that when r_s decreases, one particle out of four begins to delocalize, yielding a floppy solid made of three particles only, forming a 'triangle' of low Coulomb energy.

The behaviors of the relative fluctuations

$$u_d(p) = \sqrt{\frac{\langle d^2(p)\rangle}{\langle d(p)\rangle^2} - 1} \tag{9}$$

of the 6 inter-particle spacing $d(p)$ are given in Fig. 5. When $r_s \to \infty$, the fluctuations of the square molecule can be calculated [13] using a t/U lattice expansion. At first order, one can move only a single particle, which modifies three inter-particle spacings out of six. The fluctuations of the three remaining spacings are obtained by moving two particles, which requires to go to the second order. This explains the three r_s^{-1} decays and the three r_s^{-2} decays characterizing the correlated lattice regime. The behaviors in the intermediate regime below $r_s^W \approx 27$ are remarkable:

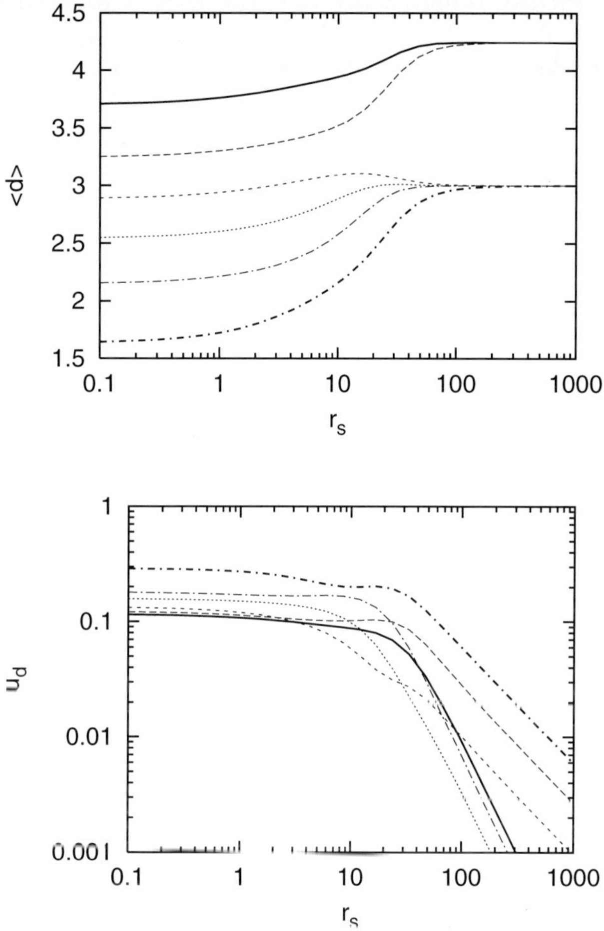

Fig. 5. *Above:* The 6 mean inter-particle spacings $\langle d(p) \rangle$ of the GS with $K = 0$ as a function of r_s. *Below:* Corresponding relative fluctuations $u_d(p)$ as a function of r_s. Note that three spacings have an almost r_s independent fluctuations below $r_s^W \approx 27$

- The relative fluctuations of three spacings out of six decay as r_s increases, as one can expect if a floppy 3 particle triangular molecule becomes more rigid.
- The three others spacings (notably the smallest) have relative fluctuations which are nearly independent of r_s, as one can expect if the 4th particle remains delocalized.

The behaviors of the different inter-particle spacings are consistent with the Andreev-Lifshitz conjecture for the considered mesoscopic lattice model. Notably, the number of crystal lattice sites is indeed smaller than the total number of particles.

7 A Supersolid Ansatz for 3 Polarized Electrons

Let us now introduce [17] a simple analytical ansatz to describe $N = 3$ polarized electrons which we have studied when $L = 6$ and $W = 0$. In that case, one can show that the GS momentum is $K = (2\pi/6, 2\pi/6)$. This ansatz is based on the concept of partially melted triangular molecules (PMTMs). A x-oriented PMTM (x-PMTM) is a rigid two particle Wigner molecule (2PWM) with $d_{min} = L/2$ combined with a third particle free to move with a wave vector k_{3x} parallel to the 2PWM at a distance $L/2$, as sketched in Fig. 6. The x-PMTM wave function of momentum K reads:

$$|\Psi_x(K, k_{3x})\rangle = \frac{1}{6\sqrt{2}} \sum_j e^{i(K_x - k_{3x})j_x + K_y j_y} c^\dagger_{j+a} c^\dagger_{j+b} c^\dagger_{k_{3x}, j_y + c_y} |0\rangle , \qquad (10)$$

where $a = (0, 0), b = (3, 0), c = (0, 3)$, and

$$c^\dagger_{k_{3x}, j_y + c_y} = \frac{1}{\sqrt{6}} \sum_{j_{x'}} e^{i k_{3x} j_{x'}} c^\dagger_{j_{x'}, j_y + c_y} . \qquad (11)$$

$(6/2\pi)(K_x - k_{3x})$ must be odd, which leads to $k_{3x} = 0, \pm 2\pi/3$ for $K = (2\pi/6, 2\pi/6)$. The y-oriented PMTM wave function $|\Psi_y(K, k_{3y})\rangle$ is defined in a similar way. The final ansatz for a PMTM of momentum K is a normalized combination of the x and y-PMTMs, which reads:

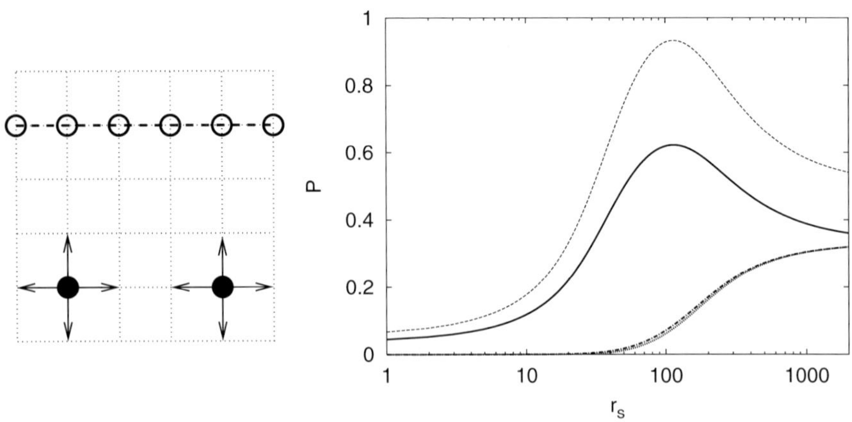

Fig. 6. *Left:* Scheme of a x-oriented partially melted triangular molecule (PMTM) for $L = 6$ and $N = 3$ polarized electrons. *Right:* GS projections $P_x(K, k_x = 0)$ (*solid line*) $P_x(K, k_x = 2\pi/3)$ (*dashed-dotted line*), $P_x(K, k_x = -2\pi/3)$ (*dotted line*) and $P(K, k = (0, 0)$ (upper thin dashed line) over x-PMTMs of momentum k_x and over the PMTM of momentum $k = (0, 0)$ respectively. Total momentum $K = (2\pi/6, 2\pi/6)$

$$|\Psi(K,k_3)\rangle = \frac{|\Psi_x(K,k_{3x})\rangle - |\Psi_y(K,k_{3y})\rangle}{\sqrt{2-2\langle\Psi_x(K,k_{3x})\mid\Psi_y(K,k_{3y})\rangle}}$$
$$= \sqrt{\frac{3}{8}}\left(|\Psi_x(K,k_{3x})\rangle - |\Psi_y(K,k_{3y})\rangle\right). \qquad (12)$$

and the constraint $k_{3x} = k_{3y}$ makes it invariant under $x-y$ permutation.

In Fig. 6, the three values $P_x(K,k_x) = |\langle\Psi_0(K)\mid\Psi_x(K,k_{3x})\rangle|^2$ taken by the projections of the exact GS $|\Psi_0(K)\rangle$ over the x-PMTMs of wave vectors (K,k_{3x}) are given as a function of r_s, together with the GS projection

$$P(K,k_3) = |\langle\Psi_0(K)\mid\Psi(K,k_3)\rangle|^2 \qquad (13)$$

over the PMTM of momenta $(K,k_3 = (0,0))$. One can see that $P_{K,k_3=(0,0)} \approx 93\%$ at $r_s \approx 100$ and that when $r_s \leq 50$, only the PMTMs with $k_{3x} = 0$ ($k_{3y} = 0$) contribute. This corresponds to a situation where the third particle is fully delocalized. Following the three projections over the x-PMTMs of different wave vector k_x, one can see how the third particle gets progressively localized in the x-direction as r_s increases, the simple triangular molecule corresponding to $P_x(K,k_{3x}) = 1/3$ for the three possible k_{3x}. For $r_s \geq 50$, a good GS description would require to take into account the three k_{3x} (k_{3y}) momenta. But for $r_s \approx 50$, the discrepancy between the exact GS and our ansatz mainly comes from the assumed rigid character of the 2PWM rather than from the assumed total delocalization of the third particle. This can be easily fixed [17] using a t/U expansion for the localized part of the PMTM, assuming one or two nearest neighbor hops for the particles making the 2PWM, and keeping the third particle in its delocalized plane wave state.

The improvements coming from this partial U/t expansion of the PMTM ansatz are given in Fig. 7, where one can see the behaviors of the bare ansatz, of the ansatz corrected to first order and second order of the partial U/t expansion. In the upper figures, the GS projections and the relative errors $\Delta E(p)/E$ are shown, where E denotes the exact GS energy, $\Delta E(p) = E_A(p) - E$, $E_A(p)$ being the ansatz energy at the pth order of the partial U/t expansion. Not only the GS description is improved, but lower values of r_s can be reached by the partial U/t expansions. In the lower figures, the three GS interparticle spacings d_{min}, d_{int} and d_{max} are given, and compared to the corresponding values of the partial t/U expansions of the ansatz. As underlined by the arrows, both the averages and the variances of those spacings are now accurately recovered for $r_s \approx 40$.

8 Conclusion

One cannot exclude that the observed supersolid regime is favored by the chosen geometry, because of the number of particles and the underlying square lattice, but not favored at all in the continuous limit, which has not a square symmetry but a spontaneously broken hexagonal symmetry. An investigation of this issue using larger sizes has begun in [18] and is still in progress. However, one can point out two things in conclusion:

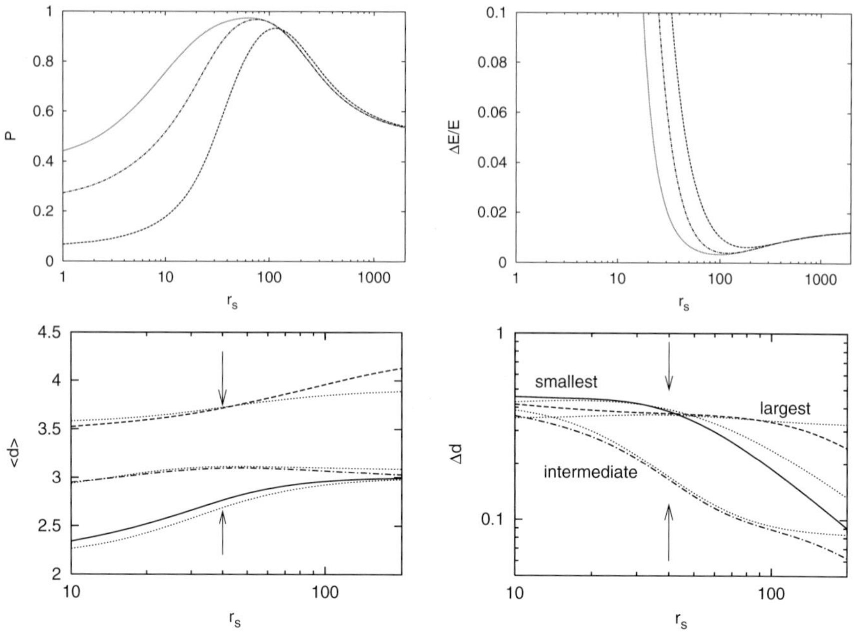

Fig. 7. *Above left:* GS projections over $k = (0,0)$ PMTM ansatz; *Above right:* Relative errors $\Delta E/E$; The *thin lines* gives the bare ansatz behaviors (*dashed line*), corrected to first order (dashed-dotted line) and second order (*dotted line*) of the partial t/U expansion; *Below left:* Average inter-particle spacings $\langle d \rangle$; *Below right:* Fluctuations Δd of the inter-particle spacings d as a function of r_s. Exact behaviors (*thick lines*), bare ansatz (*thin dashed line*) and partial second order t/U expansion (*dotted line*)

- A Monte Carlo study [19] of a few electrons in an harmonic trap has shown that mesoscopic Wigner crystallization proceeds in two stages: (i) via radial ordering of electrons on shells and (ii) freezing of the intershell rotation. This crystallization in two steps, with a particular intermediate behavior, could be attributed to the non uniform density characterizing the harmonic trap. We have shown that mesoscopic Wigner crystallization takes also place in two stages when the particles are confined on a periodic lattice with a uniform density.
- If this intermediate regime is not a pure mesoscopic lattice effect valid only in small systems, but the mesoscopic trace of the Andreev-Lifshitz supersolid, where unpaired fermions with reduced Fermi energy co-exist with a floppy solid, this might explain a few recent experimental studies [20–22] of the metallic $2d$ hole gas in GaAs, where a not well identified metallic phase (for a review see [4]) seems to coexist with a more usual Fermi liquid phase, responsible of usual weak localization behaviors. This two-phase coexistence scenario can be simply explained by a supersolid phase, without having to necessarily take into account the disorder effects [23].

Acknowledgments

We thank Giuliano Benenti, Georgios Katomeris, Moisés Martínez, Franck Selva and Xavier Waintal for their collaboration in a series of works which we have used in this short review. We thank Houman Falakshahi for stimulating discussions. Z.Á. Németh acknowledges the financial support provided through the European Community's Human Potential Programme under contract HPRN-CT-2000-00144 and the Hungarian Science Foundation OTKA TO34832.

References

1. E. Wigner, Phys. Rev. **46**, 1002 (1934) and Trans. Faraday Soc. **34**, 678 (1938)
2. W.J. Carr, Phys. Rev. **122**, 1437 (1961)
3. E.Y. Andrei, *Two-Dimensional Electron Systems on Helium other Cryogenic Substrates*, Kluwer Academic Publishers (1997)
4. E. Abrahams, S.V. Kravchenko and M.P. Sarachik, Rev. Mod. Phys. **73**, 251 (2001) and refs therein
5. B. Tanatar and D.M. Ceperley, Phys. Rev. B **39**, 5005 (1989)
6. M. Imada and M. Takahashi, J. Phys. Soc. Jpn. **53**, 3770 (1984)
7. Ladir Cândido, Philip Philipps and D.M. Ceperley, Phys. Rev. Lett. **86**, 492 (2001)
8. C. Attaccalite, S. Moroni, P. Gori-Giorgi and G.B. Bachelet, Phys. Rev. Lett. **88**, 256601 (2002)
9. G. Benenti, X. Waintal and J.-L. Pichard, Phys. Rev. Lett. **83**, 1826 (1999)
10. G. Benenti, X. Waintal and J.-L. Pichard, Europhys. lett. **51**, 89 (2000)
11. F. Selva and J.-L. Pichard, Europhys. Lett. **55**, 518 (2001)
12. R. Berkovits and Y. Avishai, *Phys. Rev.* B **57**, R15076 (1998)
13. G. Katomeris, F. Selva and J.-L. Pichard, preprint cond-mat/0206404
14. A.A. Shashkin, S.V. Kravchenko, T.M. Klapwijk, Phys. Rev. Lett. **87**, 266402 (2001)
15. A.F. Andreev and I.M. Lifshitz, Sov. Phys. JETP **29**, 1107 (1969)
16. J.P. Bouchaud and C. Lhuillier, Europhys. Lett. **3**, 481 (1987) and Europhys. Lett. **3**, 1273 (1987); J.P. Bouchaud, A. Georges and C. Lhuillier, J. Phys. France **49**, 553 (1988)
17. Z.Á. Németh and J.-L. Pichard, Europhys. Lett. **58**, 744 (2002)
18. M. Martínez and J.-L. Pichard, Eur. Phys. J. B **30**, 93 (2002)
19. A.V. Filinov, M. Bonitz and Yu. E. Lozovik, Phys. Rev. Lett. **86**, 3851 (2001)
20. Xuan P.A. Gao et al., Phys. Rev. Lett. **88**, 106803 (2002)
21. Xuan P.A. Gao et al., Phys. Rev. Lett. **89**, 016801 (2002)
22. S. Ilani et al., Science, bf 292, 1324 (2002)
23. Junren Shi and X.C. Xie, Phys. Rev. Lett. **88**, 86401 (2002)

Spin Blockades in the Transport through Quantum Dots

Dietmar Weinmann

Institut de Physique et Chimie des Matériaux de Strasbourg, UMR 7504 (CNRS-ULP), 23 rue du Loess, BP 43, 67034 Strasbourg Cedex 2, France

1 Introduction

The low-temperature electronic transport properties of sub-micrometer sized quantum dots, coupled to leads via tunnel barriers [1], are among the most promising areas of present-day research in mesoscopic physics [2,3]. In particular, the interest of numerous research groups is focused on spin effects in quantum dots. This effort is motivated by the recent development of a novel kind of electronics which makes use of the spin degree of freedom of the electrons (spintronics), and proposals of using spins in quantum dots for quantum computing [4,5].

Moreover, the electronic properties of quantum dots contain information about the many-electron system inside the dot, providing a marvellous laboratory to study systematically the many-body effects in confined structures. For this reason, quantum dots are sometimes called "artificial atoms" [6]. The subject is therefore of large fundamental interest as well as extremely promising in the view of possible applications.

Spin effects in the electronic transport through quantum dots have become a hot subject in recent years. This was anticipated by Bernhard Kramer when he opened the research line in his group very early, initiating among others the work on spin blockades.

2 Transport through a Quantum Dot

The transport properties of a quantum dot connected to leads (as sketched in Fig. 1) strongly depend on the voltage applied to the gate. This gate voltage V_{G} influences the electrostatic potential in the dot and thereby the energies of the confined electrons.

The most prominent interaction effect is the Coulomb blockade [7] which suppresses the linear conductance through the dot by fixing the electron number in the dot to the energetically most favourable value. Significant linear conductance appears only at specific values of V_{G} where different electron numbers in the dot can co-exist. This leads to the emergence of regular conductance peaks as a function of the gate voltage and allows for the manipulation of the number of electrons in the dot one by one. Such a device is often called single electron transistor [8].

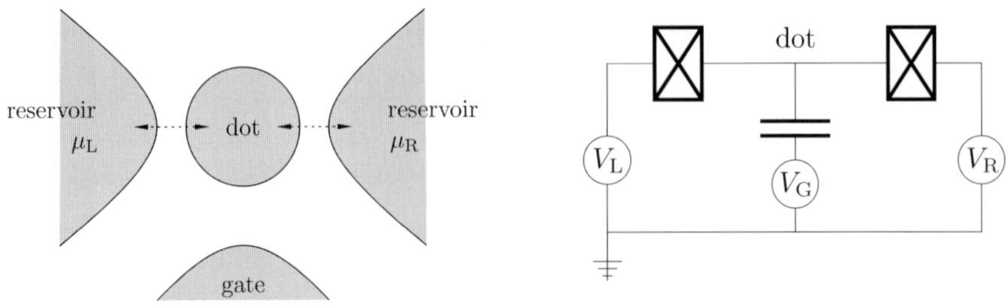

Fig. 1. *Left:* Sketch of the experimental situation we have in mind. A quantum dot is weakly connected to electron reservoirs via tunnelling barriers and capacitively influenced by a gate. *Right:* Circuit modelling the situation of a transport experiment through a quantum dot. The tunnel junctions between the leads and the dot are thought of as highly resistive capacitors. The gate is coupled capacitively to the dot

2.1 Coulomb Blockade and Conductance Oscillations

In a simplified view, the phenomenon of Coulomb blockade can be understood starting from the many-body states $|\psi_i\rangle$ of the isolated dot. They possess different electron numbers N_i and eigenenergies E_i. The energies are influenced by the electrostatic potential in the dot which is a function of the voltages applied to the leads and the gate. From a simple capacitor model (see Fig. 1), one gets the potential

$$\phi = \frac{C_G V_G + C_L V_L + C_R V_R}{C_\Sigma} , \tag{1}$$

where $C_\Sigma = C_G + C_L + C_R$ is the total capacitance of the dot, and contributions $E_i^\phi = -eN_i\phi$ to the many-body energies.

The current through the dot can be thought of as electrons entering from one lead, causing a transition from an N-electron state to an $N+1$-electron state of the dot, and electrons tunnelling from the dot to the other lead, accompanied by transitions of the dot towards states having lower particle number.

Energy Considerations. Except for very short time scales (irrelevant at low enough current), tunnelling processes into and out of the dot conserve the total energy of the system. For an electron at energy ϵ_1, which is tunnelling from the left reservoir into the dot, causing a transition from an N-electron state i of the dot to an $N+1$-electron state j, the condition of energy conservation reads

$$\epsilon_1 + E_i = E_j \qquad \text{with} \qquad N_j = N_i + 1 . \tag{2}$$

In the limit of low temperatures, the energy of electrons in the left reservoirs cannot exceed the electrochemical potential μ_L, and the processes are possible only if $\mu_L > \epsilon_1 = E_j - E_i$. In the opposite case of an electron tunnelling out of the dot, one gets $E_j - E_i = \epsilon_2 > \mu_R$. Combining these two conditions, which are illustrated in Fig. 2, one finds that transitions between states of the isolated dot

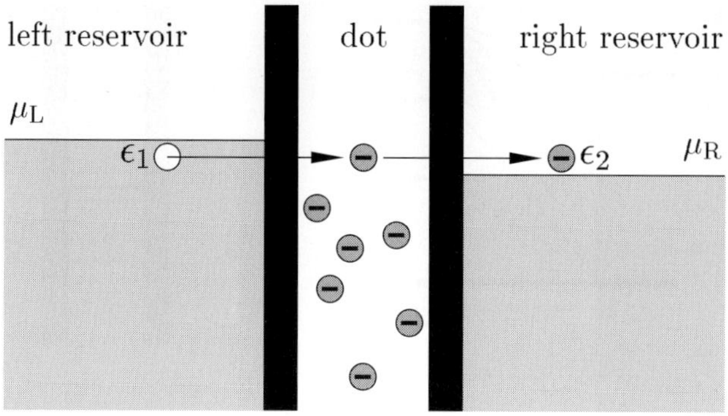

Fig. 2. Schematic view of the tunnelling through a quantum dot

allow for entering and leaving of electrons (and thereby are able to contribute to transport) if

$$\mu_L > E_j - E_i > \mu_R \,. \tag{3}$$

However, if the "chemical potential" of the dot $\mu_{dot}(N) = E_0(N+1) - E_0(N)$, the difference between the ground state energies of adjacent electron numbers, is outside the energy interval defined by the two chemical potentials of the leads, μ_L and μ_R, the dot is blocked in one of its ground states.[1] The dot is in the N-electron ground state if $\mu_{dot}(N) > \mu_L, \mu_R$ and all transitions to the ground state or an excited state of the $N+1$-particle spectrum require energies which are higher than the available ones.

In the opposite case, when $\mu_{dot}(N) < \mu_L, \mu_R$, the dot is blocked in its $N+1$-particle ground state and no transition to an N-particle state is possible.

Conductance Oscillations. In the limit of low transport (source-drain) voltage $eV_{sd} = \mu_L - \mu_R \ll \delta$, with δ being the spacing of the many-body levels in the dot, notable current is observed only when $\mu_{dot}(N)$ is aligned with the chemical potentials of the leads.

If the ground state energy of N particles is dominated by interaction effects, it can be roughly estimated as $E_0(N) \approx (E_C/2)N(N-1) - eN\phi$, with the charging energy E_C describing the interaction energy needed to add an electron to the dot. Then, its chemical potential is $\mu_{dot}(N) = E_0(N+1) - E_0(N) \approx E_C N - e\phi$ and, setting the chemical potentials in the leads to zero, one gets conductance peaks at the values $eV_G = (C_\Sigma/C_G)E_C N$ of the gate voltage. This represents regular conductance peaks separated by the Coulomb blockade regions whose width is increased by the interaction strength, as observed in experiments (see Fig. 4c).

[1] For simplicity, we discuss the case where the chemical potential of the dot $\mu_{dot}(N)$ is closer to μ_L and μ_R than the μ_{dot} corresponding to other particle numbers.

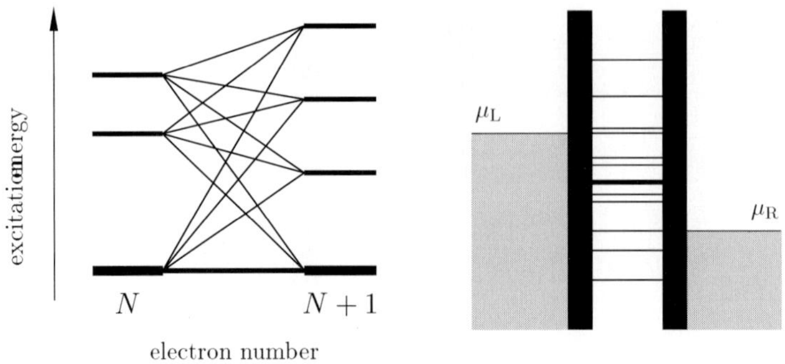

Fig. 3. *Left:* Sketch of N- and $N+1$-electron dot spectra, and the transitions between their states. *Right:* The energy differences corresponding to the transitions. The ground state to ground state transition is represented by the thick line. In this example, several transitions have energy differences between the two chemical potentials and can contribute to low-temperature transport

Thus, a transport experiment through a quantum dot in the linear regime of low voltage provides informations on the ground state energies of the dot.

2.2 Transport Spectroscopy

The excited levels of the quantum dot become involved in the transport processes once the energy necessary for their excitation is available. Possible sources are temperature, finite frequency or transport voltage exceeding the excitation energy δ. In the sequel, we will focus on the non-linear regime of finite transport voltage. In this situation, the current through the dot depends on the "open" transitions (satisfying the energy condition (3)) between dot states having different electron numbers N and $N+1$. However, in order to be able to contribute significantly to the current, the initial state of a transition needs to be populated with a non-negligible probability. This is possible only if it is reachable from one of the ground states via an uninterrupted row of open transitions.

When the transport voltage increases, more transitions contribute to the current through the dot. At temperatures lower than δ, the resulting change in the current as a function of the voltage is sufficiently sharp for the detection of the opening of a given transition as a peak in the differential conductance $\partial I/\partial V_{\mathrm{sd}}$. These peak positions can be obtained from (3), yielding

$$\mu_{\mathrm{L}} = E_j - E_i \qquad \text{and} \qquad \mu_{\mathrm{R}} = E_j - E_i\,, \tag{4}$$

corresponding to the transition energy crossing the chemical potential in the left and in the right lead, respectively. From the dependence of the electrostatic potential on the gate- and transport voltages, one finds that these conditions lead to linear structures in the V_{G}–V_{sd} plane. The slopes of the lines are determined by the various capacitances and the offset in V_{G} is a signature of the energy difference $E_j - E_i$.

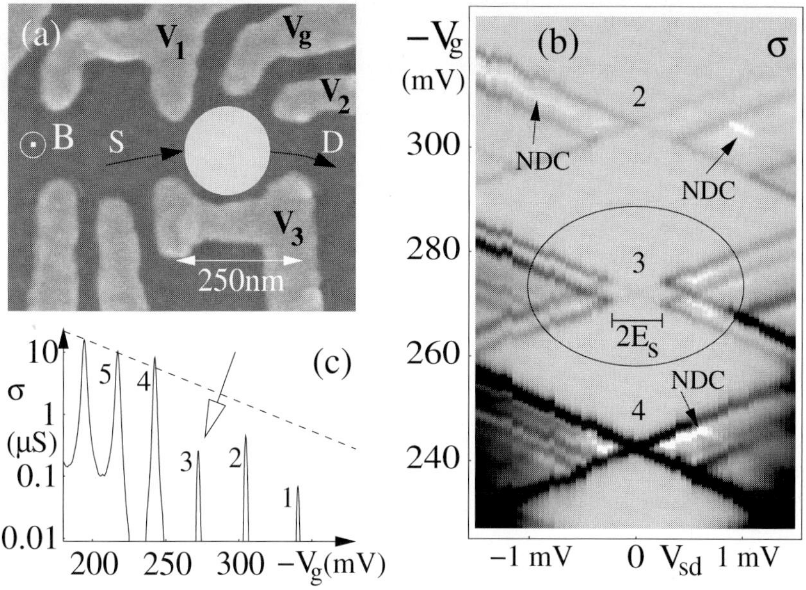

Fig. 4. **a** Top view of the quantum dot examined in [9]. The dot is confined electrostatically by metallic gates, and located approximately in the grey area. **b** The differential conductance through this dot, measured as a function of the gate- and transport (source-drain) voltages V_{sd}, which are plotted on the vertical and horizontal axis, respectively. The light grey diamonds are the Coulomb blockade zones where the conductance is suppressed. In the single electron tunnelling regime outside the blockade regions, linear structures parallel to the borders of the Coulomb-blockade region appear. Dark and bright regions correspond to high and low differential conductance, respectively. **c** The linear conductance as a function of the gate-voltage, plotted on a logarithmic scale

Measuring the current as a function of transport- and gate voltage, one can therefore obtain informations not only about the ground state energies of the electrons confined inside the quantum dot, but also about the excited many-body levels [10–12]. A recent experimental result for the transport properties of a small quantum dot at an electron temperature of $T_{\mathrm{el}} \approx 95$ mK, defined in a GaAs/GaAlAs heterostructure and containing about $N \sim 50$ electrons [9] is shown in Fig. 4.

3 Model and Method

3.1 Tunnelling Hamiltonian

For the theoretical investigation of electronic correlation effects in the transport properties through a quantum dot, one needs a model description of interacting confined electrons, connected to the continuum of the leads. We use the so-called tunnelling Hamiltonian

$$H = H_{\mathrm{L}} + H_{\mathrm{L}}^{\mathrm{T}} + H_{\mathrm{D}} + H_{\mathrm{R}}^{\mathrm{T}} + H_{\mathrm{R}} \,, \tag{5}$$

which is based on an independent description of the dot and the leads, and takes into account the tunnelling by small coupling terms.

The dot part H_{D} describes interacting electrons in the isolated quantum dot. It can be written in the form

$$H_{\mathrm{D}} = \sum_{m,\sigma} \varepsilon_m d_{m,\sigma}^{+} d_{m,\sigma} + \sum_{\substack{m_1,m_2,m_3,m_4\\ \sigma_1,\sigma_2}} V_{m_3,m_4}^{m_1,m_2} \, d_{m_1,\sigma_1}^{+} d_{m_2,\sigma_2}^{+} d_{m_3,\sigma_2} d_{m_4,\sigma_1} \,. \tag{6}$$

The operators $d_{m,\sigma}^{+}$ create an electron with spin σ in the one-particle dot state m, and the electron-electron interaction is contained in the two-body interaction matrix elements $V_{m_3,m_4}^{m_1,m_2}$. The latter could be specified, as well as the one-electron energies ε_m, in order to treat concrete models of quantum dots. However, our aim is not to discuss the spectra of specific models for the dot, but to provide level scenarios and the characteristic transport properties that can arise from certain spectra.

Electrons in the leads left (L) and right (R) of the dot are assumed to be non-correlated and described by

$$H_{\mathrm{L/R}} = \sum_{k,\sigma} \varepsilon_k^{\mathrm{L/R}} c_{\mathrm{L/R},k,\sigma}^{+} c_{\mathrm{L/R},k,\sigma} \,, \tag{7}$$

where the operators $c_{\mathrm{L/R},k,\sigma}^{+}$ create (quasi-)electrons with energy $\varepsilon_k^{\mathrm{L/R}}$ and spin σ. To neglect the electron-electron correlations may be justified by the typically large dimensionality of the leads and the high electron density there. These aspects favour screening and reduce the importance of the interactions.

The weak connection between the leads and the dot is accounted for phenomenologically by the tunnelling terms

$$H_{\mathrm{L/R}}^{\mathrm{T}} = \sum_{k,m,\sigma} \left(T_{k,m}^{\mathrm{L/R}} c_{\mathrm{L/R},k,\sigma}^{+} d_{m,\sigma} + h.c. \right) \,, \tag{8}$$

where $T_{k,m}^{\mathrm{L/R}}$ is the spin-independent tunnelling amplitude between a state k in the left/right lead and a one-particle dot state m.

Such a model Hamiltonian is consistent only in the limit of very weak tunnelling [13]. Working in this regime, we treat $H_{\mathrm{L/R}}^{\mathrm{T}}$ as a perturbation of the isolated dot, and the solution of the many-body Hamiltonian H_{D} is the 0th order of this procedure. The eigenstates $|\psi_i\rangle$ of H_{D} are characterised by the energy E_i, the electron number N_i, the total spin S_i, and the magnetic quantum number M_i.

In 2nd order in the tunnelling, one obtains [14] the transition rates between the states of the isolated dot. They read

$$\begin{aligned}
\Gamma_{j,i}^{\mathrm{L/R},+} &= \left| \langle S_i, M_i, \tfrac{1}{2}, \pm\tfrac{1}{2} \mid S_j, M_j \rangle_{\mathrm{CG}} \right|^2 \times \frac{t^{\mathrm{L/R}}}{2} f_{\mathrm{L/R}}(E_j - E_i) \delta_{N_j,N_i+1}, \\
\Gamma_{i,j}^{\mathrm{L/R},-} &= \left| \langle S_i, M_i, \tfrac{1}{2}, \pm\tfrac{1}{2} \mid S_j, M_j \rangle_{\mathrm{CG}} \right|^2 \times \frac{t^{\mathrm{L/R}}}{2} [1 - f_{\mathrm{L/R}}(E_j - E_i)] \delta_{N_i,N_j-1},
\end{aligned} \tag{9}$$

for processes where an electron tunnels from the left/right lead in (+) and out (-) of the dot. The transmittance of the tunnelling barriers is

$$t^{\mathrm{L/R}} = \frac{2\pi}{\hbar} \sum_k \left| T^{\mathrm{L/R}}_{k,m} \right|^2 \delta(\varepsilon^{\mathrm{L/R}}_k - E) \approx \frac{2\pi}{\hbar} \left| T^{\mathrm{L/R}} \right|^2 \rho^{L/R}(E), \tag{10}$$

where, for simplicity, the energy dependence of the coupling matrix elements $T^{\mathrm{L/R}}$ and their dependence on the quantum numbers k and m is neglected. The Clebsch-Gordan coefficients $\langle S_i, M_i, \frac{1}{2}, \pm\frac{1}{2} \mid S_j, M_j \rangle_{\mathrm{CG}}$ account for the combination of the spin of the initial dot state with the spin 1/2 of the tunnelling electron to the spin of the final state. The tunnelling rates vanish except for the transitions where the selection rules

$$N_j = N_i \pm 1\,, \qquad S_j = S_i \pm 1/2\,, \quad \text{and} \quad M_j = M_i \pm 1/2, \tag{11}$$

are fulfilled. Within this idealised approach, the transition rates conserve the total spin of the system and transition rates between many-body states differing in total spin and/or magnetic quantum number by more than 1/2 vanish. In realistic situations, this is approximately true if the spin is conserved on a long enough time scale. Since the spin coherence time is typically much longer than the phase coherence time, the assumption of spin conservation in the tunnelling processes is a good starting point. In Sect. 5, we nevertheless take into account weak coupling terms which do not conserve the total spin of the system, and show that they qualitatively change its non-linear transport properties.

3.2 Rate Equation Approach

Assuming sequential tunnelling (no interference of subsequent tunnelling processes), the current through the dot can be calculated from a rate equation approach [15–17]. To work in the regime of sequential tunnelling is justified when the typical time between tunnelling processes e/I is much longer than the phase coherence time τ_ϕ of the electrons in the quantum dot or when finite temperature leads to sufficient smearing of the initial states. However, we also assume that the discrete nature of the electronic levels in the dot survives this smearing, and that the spin of the states remains well defined, requiring that τ_ϕ is much larger than the Heisenberg time $\hbar/\delta$. This is achieved at low temperatures and very low currents, consistent with the usage of the tunnelling Hamiltonian and the perturbative treatment of $H^{\mathrm{T}}_{\mathrm{L/R}}$. In this regime, the rate equation

$$\frac{\mathrm{d}}{\mathrm{d}t} P_i = \sum_{j\,(j \neq i)} \left(\Gamma_{i,j} P_j - \Gamma_{j,i} P_i \right) \tag{12}$$

describes the incoherent time-evolution of the occupation probabilities P_i of the dot states $|\psi_i\rangle$, taking into account the rates $\Gamma_{i,j} = \sum_{\mathrm{X=L,R}} \Gamma^{\mathrm{X},\pm}_{i,j}$ for transitions $j \to i$. In the dc limit, the occupation probabilities satisfy the stationary rate equation $\mathrm{d}P_i/\mathrm{d}t = 0$. They are used to calculate the dc current through the dot

$$I \equiv I^{\mathrm{R}} = e \sum_{i,j\,(j \neq i)} P_j \left(\Gamma^{\mathrm{R},-}_{i,j} - \Gamma^{\mathrm{R},+}_{i,j} \right) . \tag{13}$$

4 Spin Blockades

Interestingly, the suppression of some transition rates due to the spin selection rules equation (11) can cause qualitatively new features in the transport through quantum dots. Striking effects — the so-called spin blockades [18–20] — appear when a high spin state has low energy, close to the ground state.

4.1 Spin Blockade I: Negative Differential Conductance

The first kind of spin blockade arises when a high spin state is close to the ground state [18,20]. This is the case for the example of Coulomb-interacting electrons in a one-dimensional box of length L. In the low-density limit $L \gg a_{\mathrm{B}}$ corresponding to strong electronic correlations (a_{B} is the Bohr radius), fully polarised states occur at very low excitation energy [21].

The spin blockade strikes when the voltage allows for a population of the high spin state. Since the only way to leave this state is via a transition in which the spin is lowered, there are less transitions starting from this state, and the current is reduced. Thus, an increase of the voltage can result in a decrease ("spin blockade") of the current [18,20]. Such a scenario is consistent with the experimental observation of negative differential conductances (see, for example, [9,11,12]).

4.2 Spin Blockade II: Suppression of Linear Conductance Peaks

The most striking spin effect is the suppression of a linear conductance peak. This occurs when a transition between ground states of the dot corresponding to adjacent electron numbers N and $N+1$ is forbidden by the spin selection rule (11) [19,20]. This happens when the difference between the spins of the involved ground states exceeds 1/2. The corresponding conductance peak disappears because the transition between the ground states is the only one which contributes to the conductance at low voltage.

Spectra leading to this spin blockade have been obtained for models of Coulomb-interacting electrons in a two-dimensional harmonic potential [22], and in a square-shaped quantum dot [23], in the regime of low density (strong correlations). In spite of considerable efforts [24–26], it is at present not clear under which general conditions electronic interactions induce large ground state spins, and in particular large differences in the ground state spins of adjacent electron numbers.

A strong suppression of some of the linear conductance peaks has recently been detected in the low-temperature ($T \approx 200$ mK) transport through a very small silicon dot (diameter about 15 nm) containing $N = 0 \ldots 30$ electrons [27]. The suppression was observed around $N = 6$ and has been related to a particularly strong magnetic-field dependence of the peak position. This indicates that the suppressed transitions correspond indeed to large spin differences, larger than 1/2, and provides strong evidence for the spin blockade mechanism being at the origin of the suppression. However, the same experiment shows high peaks at larger electron number in the dot, where the magnetic-field dependence is strong as well. While a weak violation of the spin selection rules equation (11) (for example due to spin-orbit coupling or higher order tunnelling processes) is always present, it is not yet understood why the spin blockade appears to be completely lifted in this case.

5 Spin Blockade with Weakly Violated Spin Conservation

In the experiment [9], shown in Fig. 4, peak #3 is suppressed by more than an order of magnitude, as compared to the exponential dependence on the gate voltage of the surrounding peaks (see Fig. 4c). In this case, a complete measurement of the non-linear transport properties is available (Fig. 4b).

The measured behaviour of the differential conductance in the non-linear regime can not be understood within the standard spin blockade scenario, using simple level configurations. The result of such simulations always looks like the upper left picture of Fig. 5. Besides the fact that the linear conductance peak is completely suppressed due to the spin blockade, as discussed in the previous section, the finite voltage behaviour is qualitatively different from the experimentally observed pattern encircled in Fig. 4b.

At zero gate- and low transport voltage, the energy of the transition between the N-electron ground state ($S = 2$) and the excited $N+1$-electron state ($S = 5/2$) is situated at the energy δ above the chemical potentials in the leads, while the transition between the excited N-electron state ($S = 3$) and the $N+1$-electron ground state ($S = 7/2$) is situated δ below the chemical potentials in the leads. Therefore, electrons could leave the dot via the former and enter the dot via the latter transition, both of them ending in one of the ground states.

In this situation, no significant current is possible at low temperature $k_{\mathrm{B}}T \ll \delta$, since the dot remains blocked in one of its ground states, and the excited states are not populated. If the voltage is large enough to open one of the transitions in both directions, while the other one remains unidirectional as in the low-voltage limit, there is still one ground state in which the dot can get trapped and from which the system cannot escape.

A significant current can appear only if the voltage V_{sd} is large enough to open both transitions leaving the ground states at N and $N+1$ shown in Fig. 5 (top right). This explains the absence of conductance in Fig. 5 (top left) for voltages $|eV_{\mathrm{sd}}| < 2\delta$.

However, if one takes into account a weak violation of the spin selection rules, the calculated behaviour changes qualitatively [9], as shown in Fig. 5 (bottom left), and becomes similar to the experimental result (Fig. 4).

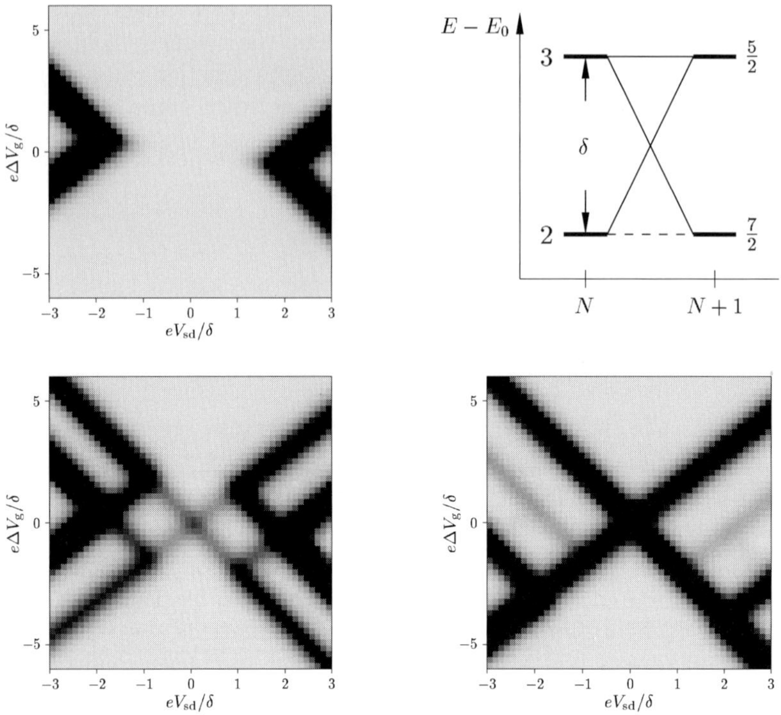

Fig. 5. *Top left:* The differential conductance calculated in the case of full spin conservation, for the level scenario shown in the upper right picture. *Bottom left:* The same situation, but now including small transition rates (14a,14b) between the ground states, violating the spin selection rules. *Bottom right:* With violated spin conservation, at finite magnetic field $g\mu_B B/\delta = 1$

In order to model processes which lead to a violation of the spin selection rules, we have introduced small transition rates

$$\bar{\Gamma}_{j,i}^{L/R,+} = \frac{t_{SF}^{L/R}}{2} f_{L/R}(E_j - E_i)\delta_{N_j,N_i+1}, \tag{14a}$$

$$\bar{\Gamma}_{i,j}^{L/R,-} = \frac{t_{SF}^{L/R}}{2} [1 - f_{L/R}(E_j - E_i)]\delta_{N_i,N_j-1}, \tag{14b}$$

which are independent of the spin quantum numbers of the involved many-body states. A very small value $t_{SF}^{L/R} = t^{L/R}/100$ was chosen for the prefactor in the calculations leading to Fig. 5 (bottom), much smaller than the one for the allowed transitions $t^{L/R}$ and consistent with the amount of suppression of the linear conductance peak in the experiment.

These small transition rates violate the spin conservation of the system and allow for transitions between the two ground states. Then, the system can no longer be trapped in one of the ground states. This has different effects depen-

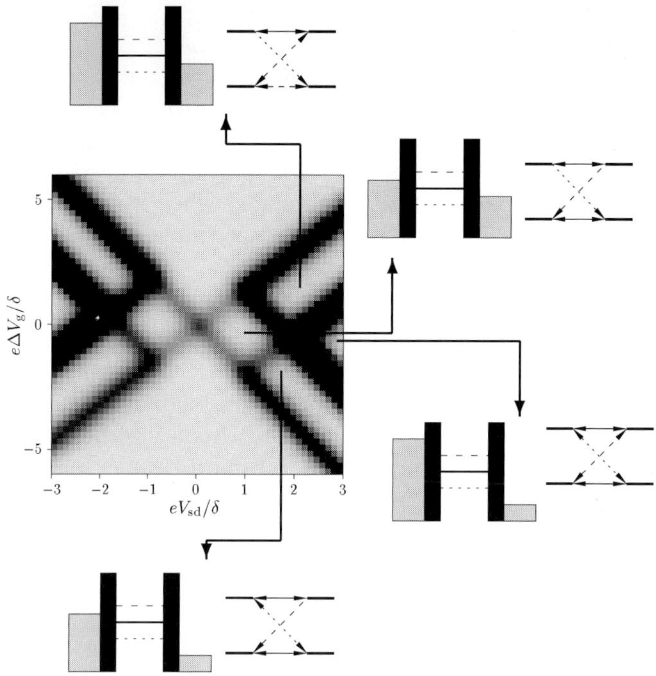

Fig. 6. The mechanism which is responsible for the characteristic pattern of the differential conductance at zero magnetic field, in the presence of weakly violated spin blockade (Fig. 5 (*bottom left*)). The energies of the different transitions as compared to the chemical potentials and the direction in which the transitions are open are depicted for various voltage regions

ding on whether one or both of the transitions involving excited states are open in one or both directions. A small conductance peak is recovered at low voltage. Most important, the dot is no longer trapped in one of its ground states while a transition involving the other ground state is open and could contribute to transport. This leads to pronounced structures at voltages $|eV_{\rm sd}|$ between δ and 2δ, and the characteristic pattern of Fig. 5 (bottom left). Given the striking similarity with the differential conductance measured in [9], such a weakly violated spin blockade scenario might very well be the mechanism at work in the experiment. The different situations arising from different voltage regimes, together with a schematic view of the levels and transitions, are shown in Fig. 6.

When one takes into account a finite magnetic field through the Zeeman energy of the many-body states, the lowest component of the N-electron state with $S = 3$ becomes energetically lower than the lowest component of the state having $S = 2$ when $|g\mu_{\rm B}B| > \delta$. Then, the transition between the ground states is allowed and the spin blockade disappears, as can be seen in Fig. 5 (bottom right). This is consistent with the experimental findings [9].

6 Conclusions

In conclusion, we have shown important spin effects in the transport properties of quantum dots. Striking spin blockades appear if strong electronic correlations lead to dot states with high spin at low energies. If the transition between ground states of adjacent electron numbers is forbidden by the spin selection rules, the finite-voltage behaviour is qualitatively modified when a small transition rate violating the spin selection rules is taken into account.

Acknowledgements

I am grateful to R.H. Blick and A.K. Hüttel for a close collaboration in the analysis of their experimental data. I thank R.A. Jalabert for enlightening discussions. Financial support from the European Union via the RTN program is gratefully acknowledged. Moreover, I thank Bernhard Kramer for always friendly advice and continued support, and wish him many more years of great fun in physics.

References

1. L.P. Kouwenhoven, C.M. Marcus, P.L. McEuen, S. Tarucha, R.M. Westervelt, N.S. Wingreen: in *Mesoscopic electron transport*, ed. by L.L. Sohn, L.P. Kouwenhoven, and G. Schön (Kluwer Series **E345**, 1997), p. 105
2. T. Dittrich, P. Hänggi, G.-L. Ingold, B. Kramer, G. Schön, W. Zwerger: *Quantum Transport and Dissipation*, Wiley-VCH (Weinheim 1998)
3. Y. Imry: *Introduction to Mesoscopic Physics*, Oxford University Press (New York 1997)
4. D. Loss, D.P. DiVicenzo: Phys. Rev. A **57**, 120 (1998)
5. L.M.K. Vandersypen, R. Hanson, L.H. Willems van Beveren, J.M. Elzerman, J.S. Greidanus, S. De Franceschi, L.P. Kouwenhoven: to appear in *Quantum Computing and Quantum Bits in Mesoscopic Systems*, Kluwer Academic Plenum Publishers (2002)
6. S. Tarucha, D.G. Austing, Y. Tokura, W.G. van der Wiel, L.P. Kouwenhoven: Phys. Rev. Lett. **84**, 11 (2000)
7. H. Grabert, M. Devoret: *Single Charge Tunneling*, NATO ASI Series **B294**, Plenum Press (New York 1992).
8. M.A. Kastner: Rev. Mod. Phys. **64**, 849 (1992)
9. A.K. Hüttel, H. Qin, A.W. Holleitner, R.H. Blick, K. Neumaier, D. Weinmann, K. Eberl, J.P. Kotthaus: cond-mat/0109104
10. J. Weis, R.J. Haug, K. v. Klitzing, K. Ploog: Phys. Rev. B **46**, 12837 (1992)
11. A.T. Johnson, L.P. Kouwenhoven, W. de Jong, N.C van der Vaart, C.J.P.M. Harmans: Phys. Rev. Lett. **69**, 1592 (1992)
12. J. Weis, R.J. Haug, K. v. Klitzing, K. Ploog: Phys. Rev. Lett. **71**, 4019 (1993)
13. R. E. Prange: Phys. Rev. **131**, 1083 (1963).
14. D. Weinmann: *Quantum Transport in Nanostructures*, (Ph. D. thesis, Universität Hamburg), PTB-Bericht PTB-PG-4, ISBN 3-89429-542-2 (1994)
15. D.V. Averin, A.N. Korotkov: Journ. of Low Temp. Phys. **80**, 173 (1990).

16. C.W.J. Beenakker: Phys. Rev. B **44**, 1646 (1991)
17. D.V. Averin, A.N. Korotkov, K.K. Likharev: Phys. Rev. B **44**, 6199 (1991)
18. D. Weinmann, W. Häusler, W. Pfaff, B. Kramer, U. Weiss: Europhys. Lett. **26**, 467 (1994)
19. D. Weinmann, W. Häusler, B. Kramer: Phys. Rev. Lett. **74**, 984 (1995)
20. D. Weinmann, W. Häusler, B. Kramer: Ann. Phys. (Leipzig) **5**, 652 (1996)
21. W. Häusler, B. Kramer: Phys. Rev. B **47**, 16353 (1993)
22. R. Egger, W. Häusler, C.H. Mak, H. Grabert: Phys. Rev. Lett. **82**, 3320 (1999)
23. W. Häusler: Festkörperprobleme/Adv. Solid State Phys. **34**, 171 (Vieweg, Braunschweig/Wiesbaden 1994)
24. Ph. Jacquod, A.D. Stone: Phys. Rev. Lett. **84**, 3938 (2000)
25. I.L. Kurland, R. Berkovits, B.L. Altshuler: Phys. Rev. Lett. **86**, 3380 (2001)
26. Y Oreg, P.W. Brouwer, X. Waintal, B.I. Halperin: arXiv:cond-mat/0109541, to appear in *Nano-Physics and Bio-Electronics*, ed. by T. Chakraborty, F. Peeters, U. Sivan (Elsevier)
27. L.P. Rokhinson, L.J. Guo, S.Y. Chou, D.C. Tsui: Phys. Rev. B **63**, 035321 (2001)

Conductance Quasi-quantization of Quantum Point Contacts: Why Tight Binding Models Are Insufficient

Stefan Kirchner[1], Johann Kroha[2], Peter Wölfle[1], and Elke Scheer[3]

[1] Institut für Theorie der Kondensierten Materie, Universität Karlsruhe, 76128 Karlsruhe, Germany
[2] Physikalisches Institut der Universität Bonn, 53115 Bonn, Germany
[3] Fachbereich Physik, Universität Konstanz, 78457 Konstanz, Germany

1 Introduction

Recent advances in miniaturization techniques have proliferated the theoretical and experimental interest in controlling transport through contacts with minimal possible size. From a purely classical point of view, where the problem resembles the one of calculating the passage of a dilute gas through a narrow hole, one obtains for the resistance R of such a point contact [1]

$$R = \frac{p_F}{e^2 D^2 n_e}, \tag{1}$$

where p_F is the Fermi momentum, n_e is the electron density and D is the diameter of the contact. This resistance is commonly referred to as the Sharvin resistance. For small contacts quantum effects will become important, since for small D only a limited number of eigenmodes of the electronic system under consideration can fit into the contact, and consequently the quantization of the transverse momentum limits the transport through the system. In the case of non-interacting electrons each transmission channel or eigenmode will carry one quantum of conductance $G_0 = 2e^2/h$, corresponding to a resistance of 12.9kΩ. The factor 2 in this formula is due to spin degeneracy. In general, the probability τ_i of channel i can be any number between zero and one characterizing the conductance of each channel in units of the quantum of conductance. The total transmission probability $\mathcal{T}$ is the sum of all single channel transmission coefficients,

$$\mathcal{T} = \sum_{i=1}^{M} \tau_i, \quad \text{with} \quad \tau_i \in [0,1], \tag{2}$$

and can be any number between zero and M, where M is the number of channels present. The total conductance is therefore given by

$$G = 2\frac{e^2}{h}\mathcal{T}. \tag{3}$$

In the linear response regime the conductance $G = \frac{dI}{dV}|_{V=0}$ is related to the local density of states (LDoS) of the interacting region provided the coupling to left and right lead is symmetric [2]. Hence, it seems sufficient to determine the number of channels M and the independent transmissions τ_i. In the following we will briefly review the experimental findings and then discuss which problems may arise in a simple tight-binding modelling of the junction. A more complete treatment of these problems leads to an interacting model. In the case of strong interactions we are able to connect the experimental findings to an approximate sum rule.

2 Experimental Observations

Experiments on a large ensemble of metallic contacts have demonstrated the *statistical* tendency of atomic-size contacts to have preferred values of conductance G [3]. The experimental evidence stems from conductance histograms calculated from repeated recordings of breaking curves from contacts fabricated by different experimental methods such as scanning tunnelling microscopes (STM) [4], dangling wires [5] or mechanically controllable breakjunctions (MCB) [6].

In the case of monovalent metals the preferred values are often close to integer multiples of G_0 [4,6,5]. The natural explanation of this finding is a set of transport channels that are either fully open ($\tau_i = 1$) or completely closed ($\tau_i = 0$), i.e. that there is an underlying "transmission quantization". This interpretation is supported by measurements of the shot noise [7], the thermo-power [8] or the conductance fluctuation amplitude [9] of gold few-atom contacts fabricated with the mechanically controllable break-junction technique. For all three properties a minimum is expected and observed when fully open transport channels are present. Here, it is thought that each individual transport channel is made up of the *single* valence orbital of a monovalent atom and that a contact with conductance $M\ G_0$ is comprised of M such atoms. Interestingly also some of the multivalent metals do show histograms with a pronounced peak structure with spacings of the order of G_0 [10], see Fig. 1.

For the case of aluminum where the first histogram peak is located around $0.8G_0$, it has been shown, that contacts with this conductance do transmit more than one channel, mostly three channels [11,12]. In this experiment the channel ensemble has been determined by analyzing the nonlinear current-voltage characteristics of superconducting atomic contacts. In contrast to the observations for monovalent metal contacts, each of these channels has a transmission well below one. It has been argued that these findings might be either due to strong disorder in the contact region induced by the particular sample fabrication method that involves thin evaporated films [13] or by the influence of the determination procedure relying on superconductivity. However, additional evidence for not completely open channels is again found in the shot-noise signal of aluminum contacts in the normal state fabricated with the MCB method from bulk aluminum [7]. The fact that more than one channel contributes to the conductance of a single-atom contact is naturally explained by a quantum che-

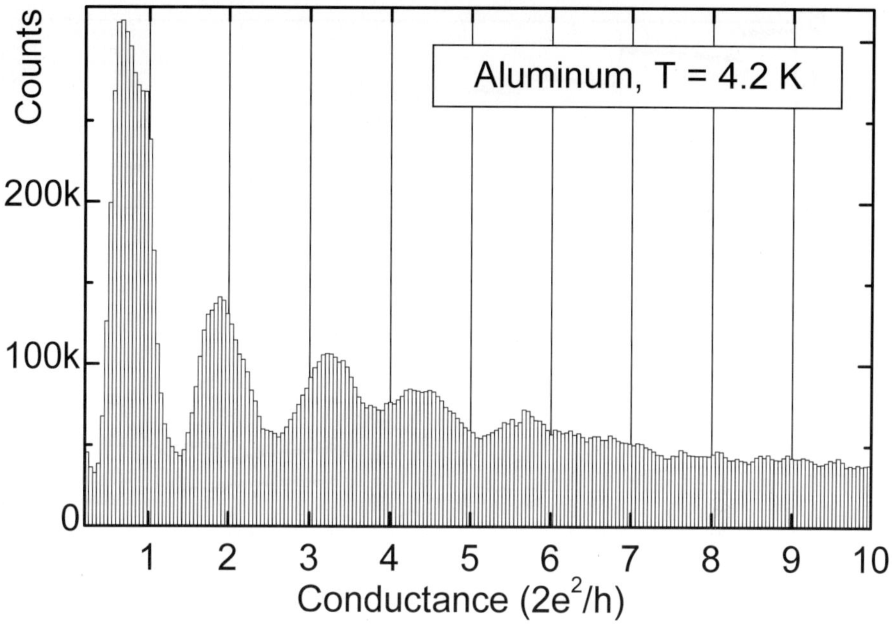

Fig. 1. Conductance histogram constructed from over 30000 individual opening curves for two different samples made of aluminum fabricated by the MCB technique. Each curve was recorded at 4.2 K while stretching contact to break. From [10]

mical model that calculates the transport channel in a tight-binding formalism starting from the valence orbitals of the metal [14,15]. This description implies that the conductance properties of atomic-size contacts are dominated by atomic arrangements. E.g. jumps in the conductance when stretching a contact would be a consequence of a rearrangements of the atoms as suggested by the experiment from Rubio *et al.* [16] who showed that the jumps in the *conductance* appear simultaneously with a jump in the strain force. A possible explanation of the preferred conductance values could thus be the existence of preferred atomic arrangements of the contact. This would imply the appearance of preferred transmission coefficients. However, this interpretation still lacks a complete explanation why the transmissions of the individual channels of a single Al atom add up to a total conductance value close to 1. The latter is evidenced in Fig. 2. The bottom panel of Fig. 2 shows in detail the evolution of $\{\tau_i\}$ when a contact is opened. The upper panel of Fig. 2 shows the evolution of the total transmission $\mathcal{T}$ as obtained from the sum of all individual transmissions. There are several remarkable features in this evolution. First, the abrupt changes in $\mathcal{T}$ correspond generally to a *complete rearrangement* of the transmission set. Second, even during the more continuous evolution on the tilted plateaus the variations of $\mathcal{T}$ arise from changes in *several* of the individual channels. Interestingly the variations of the total conductance are smaller than the variations of the individual τ_i, since some of the τ_i increase while others decrease. Similar results are observed when

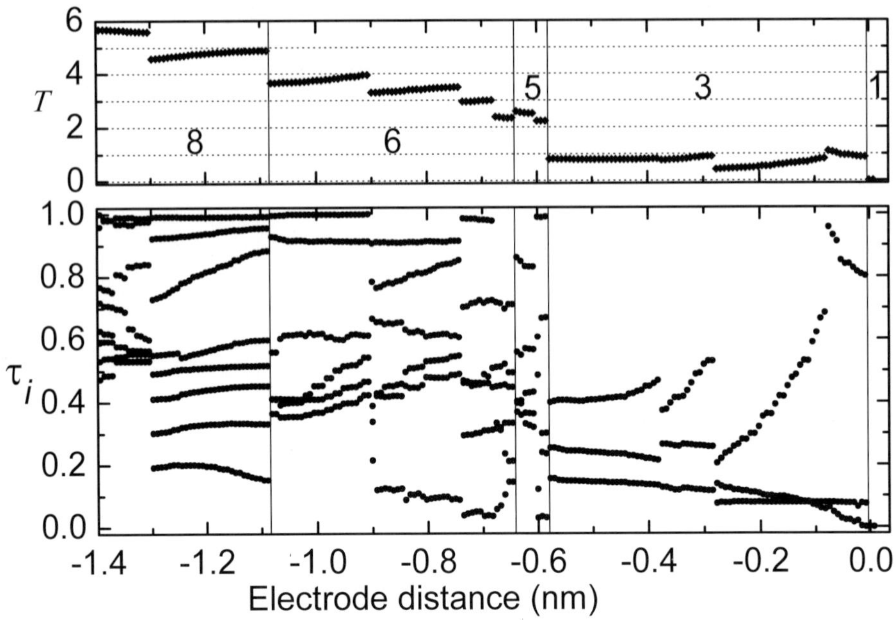

Fig. 2. *Top panel:* total transmission $\mathcal{T}=\sum \tau_i$ as a function of time while opening an aluminum sample at 0.5 pm/s and below 100 mK. *Bottom panel:* evolution of individual transmission coefficients τ_i. The vertical lines correspond to conductance jumps with change of the number of channels. The x-axis scale indicates the approximate variation of the distance between anchors. The origin of the distance axis has been set to the point where the contact breaks and enters the tunnel regime

closing the contacts as shown in Fig. 3. Even within a plateau rearrangements of the $\{\tau_i\}$ occur while the total conductance remains almost unchanged (see e.g. the plateau with $M = 3$ in Fig. 2 or $M = 5$ in Fig. 2. Thus, there seems to be a tendency for the contacts to adopt such contacts that have a preferred value of the total conductance, i.e. the sum of all transmission, regardless of the transmission ensemble itself. The *continuous* evolution of the transmission without abrupt rearrangements can again be explained by a tight binding model which describes the evolution of the LDoS and consequently of the $\{\tau_i\}$ [17]. However, the mechanism giving rise to channel rearrangements without change of G remains unclear.

In what follows we describe the non-self-consistent tight binding calculation, stress the special role of the constriction and present a possible explanation for the observed behavior.

3 Tight-Binding Modelling for Break-Junctions

It is a hallmark of Fermi liquid theory that for low-lying excitations the electron-electron interaction leads only to a renormalization expressed in terms of Fermi

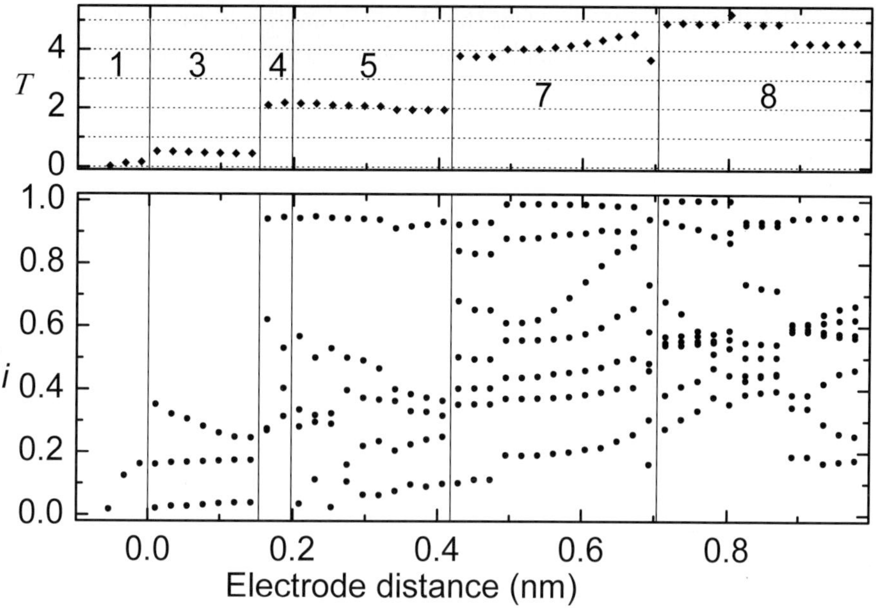

Fig. 3. Same as Fig. 2 when closing the contact at a speed of 1.1 pm/s

liquid parameters. In calculating bulk properties of so-called 'simple' metals - in contrast to e.g. narrow band materials- the electron-electron interaction can safely be neglected altogether. In those cases the band structure and, therefore, the LDoS is easily obtained from a band structure calculation using, e.g., a non-self-consistent tight binding model of the material under consideration [18,19]. In general the tight binding method aims at replacing the exact many-body Hamiltonian H by a parametrized Hamiltonian matrix in a basis of well localized functions ('atomic orbitals'). The omission of the Coulomb interaction among the electrons leads to a Hamiltonian matrix that does not depend on the distribution of electrons. Consider a usual tight-binding Hamiltonian H_{TB}:

$$H_{TB} = \sum_{\substack{i,\alpha \\ \sigma}} \epsilon_{i\alpha} c^\dagger_{i\alpha\sigma} c_{i\alpha\sigma} + \sum_{<i,j>} \sum_{\substack{\alpha,\beta \\ \sigma}} t_{i\alpha,j\beta} c^\dagger_{i\alpha\sigma} c_{j\beta\sigma}, \tag{4}$$

where i runs over all the atoms, α, β are band indices and $\sum_{<i,j>}$ is a restricted sum over nearest neighbors only. In the following N will be the total number of atoms and n will denote the number of bands.

The hopping element $t_{i\alpha,j\beta}$ of (4) is obviously equal to $\langle \phi_{i\alpha} | H_{TB} | \phi_{j\beta} \rangle$, where $|\phi_{i\alpha}\rangle = c^\dagger_{i\alpha} |0\rangle$. The Slater-Koster two-center approximation approximates this integral in the case of $i \neq j$ by [20,21]

$$t_{i\alpha,j\beta}(\boldsymbol{b}_i - \boldsymbol{b}_j) = \int d^3r \, \tilde{\psi}^*_{i\alpha}(\boldsymbol{r} - \boldsymbol{b}_i) \tilde{H} \tilde{\psi}_{j\beta}(\boldsymbol{r} - \boldsymbol{b}_j), \tag{5}$$

where $\tilde{H}$ is the two-center part of the Hamiltonian H consisting of the kinetic energy operator and the (considered spherically symmetric) part of the single-particle potential on atom i and j located at position $\boldsymbol{b}_i$ and $\boldsymbol{b}_j$. The $\tilde{\psi}_{j\beta}$ are Löwdin orbitals. These hopping elements depend on the magnitude - usually the lattice constant - and the orientation of $\boldsymbol{u}$, determined by the lattice type. In the case where the two atoms are identical, the $t_{i\alpha,j\beta}(\boldsymbol{u})$ can be parameterized by ten independent 'Slater-Koster parameters' usually denoted by $h_{\alpha\beta\gamma}$ where α and β specify the orbital angular momenta (s, p, d), and $\gamma = \sigma, \pi, \delta$ specifies the angular momentum component relative to the vector $\boldsymbol{u}$ connecting atom i and atom j. The relation between the $t_{i\alpha,j\beta}$ of (4) and the Slater Koster parameters is relatively simple for s- and p-like orbitals and fixed distance or lattice constant.

A frequently used method to go beyond this simple tight-binding approach and to include some aspects of the electron-electron interaction for bulk properties is via the local charge neutrality condition (LCNC). The tight-binding Hamiltonian, (4) might give rise to spurious charge transfer resulting in a local charge different from the ionic charge. In a good metal, this net charge is usually screened on a scale -the screening length- smaller than the lattice constant, thereby restoring a uniform charge density. This is modelled by the LCNC by enforcing an electron density that equals on each site the ionic charge. The tight-binding method augmented with the condition of local charge neutrality, which enforces the same occupation on each site has been employed by A. Yeyati [14] and J. C. Cuevas [19,15,17] to obtain the evolution of the channel transmissions in atomic break-junctions when stretching the contacts.

4 A Multi-level Impurity Model

The opening of the quantum point contact just before rupture might be modelled by varying the overlap of the wavefunctions of the atom or atoms in the constriction with its neighbors. According to Harrison [22] a lattice constant dependence can easily be built into the $t_{i\alpha,j\beta}$ to obtain the band structure at different lattice spacings by scaling the Slater-Koster parameters as d^{-2} for the (s, s), (s, p), and (p, p) parameters, where d is the ratio between new and old lattice spacing. Other parameters show a more complex scaling. The (s, d) elements for example scale as $d^{-7/2}$. Harrison's scaling argument might be oversimplified in many instances. In any case the overlap of the wavefunctions and hence the $t_{i\alpha,j\beta}$ have to vanish exponentially for large enough d.

We will now rewrite the tight-binding Hamiltonian in order to stress the central role of the atom(s) in the constriction which will be referred to as the central atom or impurity.

After redistributing the indices i in such a way that $i = 0$ corresponds to the central atom, H_{TB} ofn (4) can be rewritten as:

$$
\begin{aligned}
H_{TB} &\equiv H_B + H_C \\
H_B &= \sum_{\substack{i\neq 0\\ \alpha,\sigma}} \epsilon_{i\alpha} c^\dagger_{i\alpha\sigma} c_{i\alpha\sigma} + \sum_{\substack{<i,j>\\ i\neq 0, j\neq 0}} \sum_{\substack{\alpha,\beta\\ \sigma}} t_{i\alpha,j\beta} c^\dagger_{i\alpha\sigma} c_{j\beta\sigma} \\
H_C &= \sum_{m,\sigma} \epsilon_m c^\dagger_{0m\sigma} c_{0m\sigma} \\
&+ \sum_{\substack{i=\mathtt{n.N.}\\ \beta}} \sum_{m,\sigma} \left(t_{0m,i\beta} c^\dagger_{0m\sigma} c_{i\beta\sigma} + t_{i\beta,0m} c^\dagger_{i\beta\sigma} c_{0m\sigma} \right).
\end{aligned} \tag{6}
$$

$\sum_{i=\mathtt{n.N.}}$ denotes a sum over the nearest neighbors of the central atom. In order to have H_{TB} hermitian, the hopping elements have to satisfy $t_{i\alpha,j\beta} = t^*_{j\beta,i\alpha}$. This then implies that H_B is an hermitian operator and hence can be diagonalized by a unitary transformation:

$$
\tilde{\mathbf{U}} H_{TB} \tilde{\mathbf{U}}^{-1} = \tilde{\mathbf{U}} H_B \tilde{\mathbf{U}}^{-1} + \tilde{\mathbf{U}} H_C \tilde{\mathbf{U}}^{-1}, \tag{7}
$$

where $\tilde{U} H_B \tilde{U}^{-1}$ is diagonal in the new basis. With the following transformation of the annihilation and creation operators from the old to the new representation

$$
c_{k\sigma} = \sum_{i\alpha} U_k^{i\alpha} c_{i\alpha,\sigma} \tag{8}
$$

the hybridization term between the central atom and the leads assumes the following form (for simplicity: $t_{0m,i\beta} \to t_{m,i\beta}$ and $c^\dagger_{0m\sigma} \to d^\dagger_{m\sigma}$):

$$
\begin{aligned}
\sum_{\substack{i=\mathtt{n.N.}\\ \beta,m,\sigma}} t_{m,i\beta} d^\dagger_{m\sigma} c_{i\beta\sigma} &= \sum_{\substack{k,\sigma\\ m}} \sum_{\substack{i=\mathtt{n.N.}\\ \beta}} t_{m,i\beta} d^\dagger_{m\sigma} (U^k_{i\beta})^* c_{k\sigma} \\
&\equiv \sum_{\substack{k,\sigma\\ m}} V_k^m d^\dagger_{m\sigma} c_{k\sigma},
\end{aligned} \tag{9}
$$

where V_k^m is given by

$$
V_k^m = \sum_{\substack{i=\mathtt{n.N.}\\ \beta}} t_{m,i\beta} (U^k_{i\beta})^*. \tag{10}
$$

In this basis the tight-binding Hamiltonian equation (4) can therefore be written as:

$$
\begin{aligned}
H_{TB} &= \sum_{k,\sigma} \epsilon_k c^\dagger_{k\sigma} c_{k\sigma} + \sum_{m,\sigma} \epsilon_m d^\dagger_{m\sigma} d_{m\sigma} \\
&+ \sum_{k,\sigma} \sum_m (V_k^m d^\dagger_{m\sigma} c_{k\sigma} + \text{h.c.}).
\end{aligned} \tag{11}
$$

This is nothing but a multi-level impurity model without Coulomb repulsion on the impurity ($U = 0$). In general $t_{i\alpha,j\beta} \neq 0$ for $\alpha \neq \beta$, which is –in parts– a consequence of s-p hybridization. Therefore the resulting model (even for only one local level) is different from the usual $SU(n) \times SU(2)$ Anderson model although several bands are involved.

5 Electronic Density and Static Screening in the Constriction

In the following we will argue that the LCNC breaks down in the constriction and therefore a tight-binding approach together with a local charge neutrality constraint treats the Coulomb interaction in an oversimplified way.

In order to understand how a local charge neutrality constraint can modify the transmission through the constriction we will employ a semiclassical argument to see that the geometric constriction will alter the screening length. In a metallic system a local impurity potential Φ shifts the energy levels locally by an amount Φ. In order to guarantee a spatially uniform Fermi energy E_F throughout the system electrons will have to be redistributed such that the potential generated by the density change $\delta n(\boldsymbol{r})$ according to Poisson's law cancels the impurity potential. Therefore the impurity potential is related to the local electron density via

$$\nabla^2 \Phi(\boldsymbol{r}) = \delta n(\boldsymbol{r})\, e/\epsilon_o \approx D(\boldsymbol{r}, E_F)\, \Phi(\boldsymbol{r})\, e^2/\epsilon_o, \tag{12}$$

where $D(\boldsymbol{r}, E_F)$ is the local density of states and $\Phi \ll E_F$ has been assumed. In spherical coordinates and with open boundary conditions we obtain the Thomas Fermi screening length for the bulk:

$$\lambda_T = \sqrt{\epsilon_o/e^2 D(E_F)}. \tag{13}$$

For aluminum the bulk screening length is roughly $\lambda_T \sim 0.5$ Å. The lattice constant is approximately $a \sim 4$ Å and therefore $\lambda_T \ll a$ for the bulk.

In order to obtain the corresponding screening length in a quantum point contact where the finite boundary conditions will change the screening properties of the electrons, not only (12) in the presence of the new boundary conditions has to be solved. In addition, the Schrödinger equation must be solved to take the effect on the local electron density into account. In standard perturbation theory this would correspond to solving the random phase approximation (RPA) for the chosen geometry.

In the following we consider a simple toy model to simulate the effect of the finite geometry. To this end we model the elongation of the quantum wire in oblate spheroidal coordinates (ζ, η, ϕ). The coordinate surfaces of this system are confocal ellipses and hyperbolas rotated around the minor axis. We will use this set of coordinates since a suitable approximation of the surface of the sample is obtained by having $\eta = \pm\eta_0$ with $0 < \phi < \pi$ and $1 < \zeta < \infty$. The elongation of the quantum wire is then modelled by a decrease in η_0. The minimal possible η_0 is assumed while the wire breaks and should be below a lattice constant. Fig. 4 shows this surface for various η_0. For details of this model and oblate spheroidal coordinates see [23,24]. J. Torres et al have generalized the Landauer-Büttiker formula to a wire with similar geometry [25]. Kassubek *et al.* have used a free-electron model for two- and three-dimensional wires [26]. Imposing hard wall boundary conditions and neglecting Coulomb interaction among the electrons

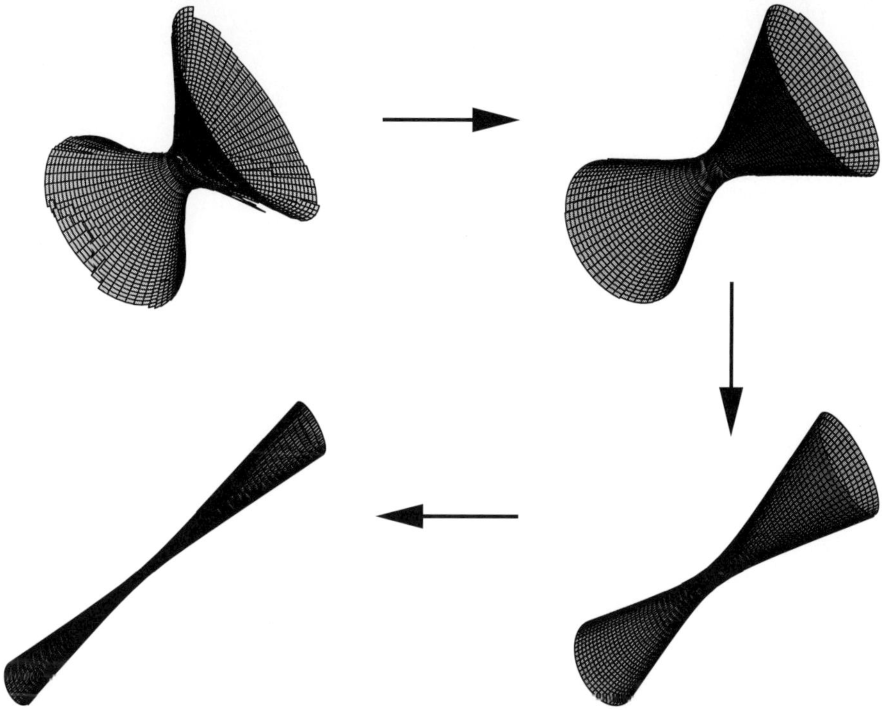

Fig. 4. Modelling the break-junction experiment: While opening the contact the central region of the quantum wire gets thinner until it breaks. In this simple model of the experiment we describe the wire by a surface of constant η_0 in oblate spheroidal coordinates (ζ, η, ϕ) and the opening of the contact by a decrease in η_0. For details, see [23,24]

we solve for the one-particle density. As expected, the increase in kinetic energy leads to a depletion of the density in the constricted region. Figure 5 shows our result for the electron density where we assumed the first 200 eigenstates to be occupied. In the numerical evaluation we chose a cut-off ζ_0 large enough such that the density in the constriction ($\zeta \approx 1$) did not depend on it.

Although we neglected the Coulomb interaction in our toy model, which will try to balance any density fluctuations, it is clear that the competition between kinetic energy and interaction cannot restore a uniform density. LCNC on the contrary enforces a uniform density by assuming that $\lambda_T \ll a$ everywhere. Forcing the system to a constant screening length even in the constriction will of course modify the LDoS at the Fermi energy, $D(E_F)$. This is analogous to fixing the screening length according to $\lambda_T \ll a$, see (13). Consequently the current and hence the conductance will be modified accordingly. In order to prevent this, the Coulomb interaction in the constriction has to be explicitly taken into account in the tight binding Hamiltonian, (4) without resorting to the

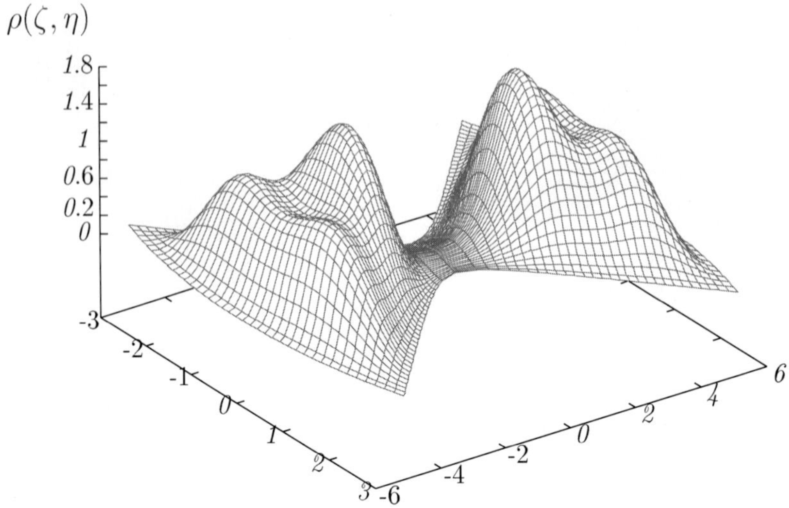

Fig. 5. One-particle density for an opening angle of the point contact corresponding to $\eta_0 = 0.5$. The density does not depend on ϕ due to the symmetry of the chosen hard wall potential. The density in the constriction is depleted in the central region

LCNC. The Hamiltonian we will use to describe the break-junction experiments therefore assumes the form:

$$H = H_{TB} + \frac{1}{2} \sum_{(m,\sigma)\neq(m',\sigma')} U_{m,m'} \hat{n}_{m\sigma} \hat{n}_{m'\sigma'}, \tag{14}$$

where we introduced intra- and interlevel Coulomb matrix elements $U_{m,m'}$ in the central region.

In the last section we will derive an approximate sum rule for the total transmission probability $\mathcal{T}$ of the Hamiltonian of (14) with 2 orbitals on the central atom.

6 Conductance in the Strongly Correlated Regime

The current through the quantum point contact described by (14) in the case of symmetric coupling to left and right lead can be related to the LDoS according to [2]

$$I = \frac{e}{h} \sum_{\sigma} \int d\omega \, [f(\omega) - f(\omega + \frac{eV}{\hbar})] \, \mathrm{Im}\,\mathrm{tr}\{\Gamma \cdot \mathcal{G}_\sigma(\omega)\}.$$

Here, $\mathcal{G}_\sigma$ is the local Green function and the lead-to-orbital coupling matrix Γ is given by [2]:

$$\Gamma_{nm} = 2\pi \sum_k \rho_k(\epsilon) V_k^n V_k^{m*}, \tag{15}$$

where $\rho_k(\epsilon)$ is the density of states of the leads. The conductance follows immediately:

$$G = \left.\frac{dI}{dV}\right|_{V=0} = \mathrm{Im}\{\Gamma \cdot \mathcal{G}_\sigma(0)\}. \tag{16}$$

The lattice constant dependence of the elements of Γ is obtained from (10) and Harrison's scaling law as

$$\Gamma_{nm} = 2\pi\rho(0) \sum_{\substack{i=\mathtt{n.N.} \\ \alpha}} t_{m,i\alpha} t_{i\alpha,n} \sim d^{-4}, \tag{17}$$

which is a rather strong lattice constant dependence. Since the ground state of H in (14) is a spin singlet [27], the quantum dot acts for low enough temperatures as a pure potential scatterer for electrons traversing the system, and the following Fermi liquid relations hold [28,24],

$$\Sigma''_m(\omega) = \frac{(\hbar\omega)^2 + (\pi k_B T)^2}{k_B T_K} \qquad \omega, T < T_K \tag{18}$$

$$\int_{-\infty}^{0} d\omega \; \mathrm{tr}\Big\{\frac{\partial \Sigma(\omega)}{\partial \omega} \cdot \mathcal{G}_{d\sigma}(\omega)\Big\} = 0, \tag{19}$$

where T_K is a dynamically low energy scale of the system, analytically given in [29]. The averaged electron number in the dot per spin, $n_{d,\sigma}$, can now be evaluated using the general relation $\frac{d}{d\omega}\ln(\mathcal{G}_d^{-1}) = (1 - \frac{d\Sigma}{d\omega}) \cdot \mathcal{G}_d$ and the Luttinger theorem (19),

$$n_{d\sigma} = \mathrm{Im} \int_{-\infty}^{0} \frac{d\omega}{\pi} \mathrm{tr}\; \mathcal{G}_{d\sigma}(\omega) = \frac{\mathrm{Im}}{\pi} \Big[\mathrm{tr}\{\ln \mathcal{G}_{d\sigma}(\omega)^{-1}\}\Big]_{-\infty}^{0}.$$

It may be re-expressed, using $\mathrm{tr}\ln \mathcal{G}_{d\sigma}{}^{-1} = \ln\det \mathcal{G}_{d\sigma}{}^{-1}$, as

$$n_{d\sigma} = \frac{1}{\pi}\mathrm{arccot}\left[\frac{\mathrm{Re}\det \mathcal{G}_{d\sigma}(0)^{-1}}{\mathrm{Im}\det \mathcal{G}_{d\sigma}(0)^{-1}}\right]. \tag{20}$$

The scattering T-matrix of the device, $\Gamma \cdot \mathcal{G}_{d\sigma}$, which for symmetric coupling to left/right lead completely determines the conductance G. Using the Fermi liquid property equation (18) together with the Dyson equation relating $\mathcal{G}_{d\sigma}$ and $\Sigma_\sigma(\omega)$ and (20), we obtain at the Fermi energy ($\omega = 0$, $T \ll T_K$) for $n = 2$,

$$\mathrm{Im}\;\mathrm{tr}\;(\Gamma \cdot \mathcal{G}_\sigma(0)) = \sin^2(\pi n_{d\sigma}) + \sin(2\pi n_{d\sigma}) \frac{\mathrm{Re}[\det(i\Gamma - \Sigma'(0))]}{\Gamma_{11}(\varepsilon_{d,2} + \Sigma'_{22}(0)) + \Gamma_{22}(\varepsilon_{d,1} + \Sigma'_{11}(0))}. \tag{21}$$

This is an exact result, valid for arbitrary microscopic parameters Γ_{nm}, ε_m and $U_{m,m'}$. It is the generalization of the well-known unitarity rule of the single-level Anderson impurity problem to the case of several impurity levels [30]. Having at least one of the local levels significantly below the Fermi level ($\varepsilon_{m=1} < 0$, $|\varepsilon_{m=1}|/\Gamma_{nm} > 1$) and the Coulomb repulsion large enough ($U_{m,m'}/\Gamma_{nn'} \gg 1$) to enforce $n_{d\sigma} \approx 1/2$ we obtain a conductance close to the conductance unit. This resembles the observed behavior. The condition $U_{m,m'}/\Gamma_{nn'} \gg 1$ can be met through the decrease in the hopping amplitude while opening the contact. This situation is analogous to the situation in narrow band materials where the small band width leads to a strongly correlated state. Provided the wire does not break and the temperature is well below the low energy scale T_K, which depends exponentially on the entries of the coupling matrix, it seems that we will always reach this regime while elongating the wire according to (17).

Acknowledgments

It is a pleasure to acknowledge stimulating discussions with F. Evers, H. v. Löhneysen, A. Mildenberger, and J. Paaske. Finally, we have enjoyed fruitful interaction with D. Averin, J.C. Cuevas and A. Levy Yeyati, and we thank them for providing us with their respective computer codes. The experiments have been performed in collaboration with C. Urbina at the CEA Saclay, France. We thank J.M. van Ruitenbeek for the allowance to reproduce his data. This work was partially supported by the Center for Functional Nanostructures (CFN) and Sonderforschungsbereiche SFB513 and SFB195 of the Deutsche Forschungsgemeinschaft.

References

1. Y.V. Sharvin. Sov. Phys. JETP **21**, 655 (1965)
2. Y. Meir, N. Wingreen. Phys. Rev. Lett. **68**, 2512 (1992)
3. J.M. van Ruitenbeek. In *Mesoscopic Electron Transport*, eds. L.L. Sohn, L.P. Kouwenhoven, G. Schön (Kluwer Academic, Dordrecht, 1997), pp. 549–579; and references therein
4. L. Olesen, E. Laegsgaard, I. Stensgaard, F. Besenbacher, J. Schiøtz, P. Stoltze, K. Jacobsen, J. Nørskov. Phys. Rev. Lett. **72**, 2251 (1994)
5. J.L. Costa-Krämer, N.P. García-Mochales, P.A. Serena. Surface Science **342**, L1144 (1995)
6. J.M. Krans, J.M. van Ruitenbeek, V.V. Fisun, I.K. Yanson, L.J. de Jongh. Nature **375**, 767 (1995)
7. H. van den Brom, J.M. van Ruitenbeek. Phys. Rev. Lett. **82**, 1530 (1999)
8. B. Ludoph, M.H. Devoret, D. Esteve, C. Urbina, J.M. van Ruitenbeek. Phys. Rev. Lett. **82**, 1530 (1999)
9. B. Ludoph, J.M. van Ruitenbeek. Phys. Rev. B **59**, 12290 (1999)
10. A. Yanson, J.M. van Ruitenbeek. Phys. Rev. Lett. **79**, 2157 (1997)
11. E. Scheer, P. Joyez, D. Esteve, C. Urbina, M. Devoret. Phys. Rev. Lett. **78**, 3535 (1997)
12. E. Scheer, N. Agraït, J. Cuevas, A. Yeyati, B. Ludoph, A. Martin-Rodero, G. Bollinger, J.M. van Ruitenbeek, C. Urbina. Nature **394**, 154 (1998)
13. J.M. van Ruitenbeek, A. Alvarez, I. Piñeyro, C. Grahmann, P. Joyez, M.H. Devoret, D. Esteve, C. Urbina. Rev. Sci. Inst. **67**, 108 (1996); and references therein

14. A.L. Yeyati, A. Martin-Rodero, F. Flores. Phys. Rev. B. **56**, 10369 (1997)
15. J. Cuevas, A. Yeyati, A. Martin-Rodero. Phys. Rev. Lett. **80**, 1066 (1997)
16. G. Rubio, N. Agraït, S. Vieira. Phys. Rev. Lett. **76**, 2302 (1996)
17. J. Cuevas, A. Yeyati, A. Martin-Rodero, G. Bollinger, C. Untiedt, N. Agraït. Phys. Rev. Lett. **81**, 2990 (1998)
18. C. Goringe, D. Bowler, E. Hernández. Rep. Prog. Phys. **60**, 1447 (1997)
19. J. Cuevas, A. Martin-Rodero, A. Yeyati. Phys. Rev. B. **54**, 7366 (1996)
20. P. Slater, G. Koster. Phys. Rev. **94**, 1498 (1954)
21. D. Papaconstantopoulos. *Handbook of the Band Structure of Elemental Solids* (Plenum Press, New York, 1986)
22. W. Harrison. *Electronic Structure and the Properties of Solids* (W.H. Freeman and Co., 1980)
23. M. Abramowitz, A. Stegun, eds. *Handbook of Mathematical Functions* (Dover Publications, Inc., New York, 1965)
24. S. Kirchner. Shaker, Aachen (2002), ISBN 3-8322-0183-1.
25. J. Torres, J. Sáenz. Phys. Rev. Lett. **77**, 2245 (1996)
26. F. Kassubek, C.A. Stafford, H. Grabert. Phys. Rev. B **59**, 7560 (1999)
27. A. C. Hewson. *The Kondo Problem to Heavy Fermions* (Cambridge University Press, Cambridge, 1993)
28. A. Yoshimori, A. Zawadowski. J. Phys. C **15**, 5241 (1982)
29. J. Kroha, S. Kirchner, G. Sellier, P. Wölfle, D. Ehm, F. Reinert, S. Hüfner, C. Geibel. to appear in Physica B (2003)
30. S. Kirchner, J. Kroha, E. Scheer. In *Kondo Effect and Dephasing in Low-Dimensional Metallic Systems*, eds. V. Chandrasekhar, C.V. Haesendonck, A. Zawadowski (Kluwer Academic Publishers, Dordrecht, 2001)

Druck:	Strauss Offsetdruck, Mörlenbach
Verarbeitung:	Schäffer, Grünstadt